T0210649

Methods of Contemporary Gauge Theory

This book introduces the quantum theory of gauge fields. Emphasis is placed on four nonperturbative methods: path integrals, lattice gauge theories, the $1/N$-expansion, and reduced matrix models, all of which have important contemporary applications. Written as a textbook, it assumes a knowledge of quantum mechanics and elements of perturbation theory, while many relevant concepts are pedagogically introduced at a basic level in the first half of the book. The second half comprehensively covers large-N Yang–Mills theory. The book uses a modern approach to gauge theories based on path-dependent phase factors known as the Wilson loops, and contains problems with detailed solutions to aid understanding.

Suitable for advanced graduate courses in quantum field theory, the book will also be of interest to researchers in high-energy theory and condensed matter physics as a survey of recent developments in gauge theory. This title, first published in 2002, has been reissued as an Open Access publication on Cambridge Core.

YURI MAKEENKO received his PhD in theoretical and mathematical physics from the Institute of Theoretical and Experimental Physics (ITEP), Moscow. He has been a staff member at the ITEP since 1983, and since 1993 has been a visiting professor at the Niels Bohr Institute, Copenhagen. He has given lecture courses at universities throughout Europe and has published numerous papers and review articles in journals.

Professor Makeenko has been working on nonperturbative quantum field theory since 1975. His research has covered four-dimensional conformal field theories and the $1/N$-expansion of QCD, where he derived in 1979 an equation presently known as the Makeenko–Migdal equation or the "loop equation", in collaboration with A.A. Migdal. In the early 1980s he explored lattice gauge theories, including the Monte Carlo simulations, and went on to work on matrix models and string theory in the 1990s. His recent work has focused on matrix theory and noncommutative gauge theories.

CAMBRIDGE MONOGRAPHS ON
MATHEMATICAL PHYSICS
General editors: P. V. Landshoff, D. R. Nelson, S. Weinberg

† Issued as a paperback

Methods of
Contemporary Gauge Theory

YURI MAKEENKO

Institute of Theoretical and Experimental Physics, Moscow
The Niels Bohr Institute, Copenhagen

CAMBRIDGE
UNIVERSITY PRESS

Shaftesbury Road, Cambridge CB2 8EA, United Kingdom

One Liberty Plaza, 20th Floor, New York, NY 10006, USA

477 Williamstown Road, Port Melbourne, VIC 3207, Australia

314–321, 3rd Floor, Plot 3, Splendor Forum, Jasola District Centre, New Delhi – 110025, India

103 Penang Road, #05-06/07, Visioncrest Commercial, Singapore 238467

Cambridge University Press is part of Cambridge University Press & Assessment, a department of the University of Cambridge.

We share the University's mission to contribute to society through the pursuit of education, learning and research at the highest international levels of excellence.

www.cambridge.org
Information on this title: www.cambridge.org/9781009402057

DOI: 10.1017/9781009402095

First published 2002
Reissued as OA 2023

A catalogue record for this publication is available from the British Library.

ISBN 978-1-009-40205-7 Hardback
ISBN 978-1-009-40210-1 Paperback

Contents

Preface

He reached a much higher plane of creativity when he blacked out everything but *a*, *an* and *the*. That erected more dynamic intralinear tensions.

J. HELLER, *Catch-22*

This textbook is based on lecture courses originally given at:

(1) Autonoma University of Madrid, Winter Semester of 1993;
(2) Leipzig University, Winter Semester of 1995;
(3) Moscow Physical and Technical Institute, Spring Semester of 1995;

and then repeated with some modifications at several Universities, Schools of Physics, etc.

My intention in these courses was to introduce graduate students to selected nonperturbative methods of contemporary gauge theory. The term "nonperturbative" means literally "beyond the scope of perturbation theory". Therefore, it is assumed that the reader is familiar with quantum mechanics as well as with the standard methods of perturbative expansion in quantum field theory and, in particular, with the theory of renormalization.

Another purpose was to make the courses useful for more experienced researchers (including those working in condensed-matter theory), as a survey of ideas, terminology and methods, which have been developed in Gauge Theory since the beginning of the 1970s. For this reason, these notes do not go into great detail, and so some subjects are only touched upon briefly. Correspondingly, the subjects which are usually covered by modern courses in string theory, such as two-dimensional conformal field theories, are not examined. It is assumed that such a course will follow this one.

The main body of the book deals with lattice gauge theories, large-N methods, and reduced models. These three parts are preceded by Part 1, which is devoted to the method of path integrals. The path-integral approach is loosely used in quantum field theory and statistical mechanics. In Part 1, I shall pay most attention to aspects of the path integrals, which are then used in the next three parts.

xi

At the beginning of each part, I try to stay as close to the original papers, where the methods were first proposed, as possible. The list of these papers includes:

1. FEYNMAN R.P. 'An operator calculus having applications in quantum electrodynamics'. *Phys. Rev.* **84** (1951) 108.
2. WILSON K.G. 'Confinement of quarks'. *Phys. Rev.* **D10** (1974) 2445.
3. 'T HOOFT G. 'A planar diagram theory for strong interactions'. *Nucl. Phys.* **B72** (1974) 461.
4. EGUCHI T. AND KAWAI H. 'Reduction of dynamical degrees of freedom in the large-N gauge theory'. *Phys. Rev. Lett.* **48** (1982) 1063.

The lectures were followed by seminars where some more involved problems were solved on a blackboard. They are inserted in the text as problems, which may be omitted at first reading. Some more information is also added as remarks after the main text. Both of them contain some relevant references.

The references, which are collected at the end of each part of the book, are usually given only to either a first paper (or papers) in a series or those containing a pedagogic presentation of the material. With the modern electronic database at `http://www.slac.stanford.edu/spires/hep` (SLAC), a list of subsequent papers can, in most cases, be retrieved by downloading citations of the first paper.

The selection of the material for this book is, as usual, personal and dictated by the author's research activity in the area of quantum field theory over the last almost 30 years. In fact, many important developments, in particular, in supersymmetric gauge theories are not included.

I would like to thank my students for their attention, patience, and questions. I am grateful to Martin Gürtler for his help in preparing the lecture notes. I am also indebted to my colleagues – too numerous to be listed personally – for their invaluable comments, suggestions, and encouragement.

Y.M.

Part 1
Path Integrals

The path integral is a method of quantization which is equivalent to the operator formalism. It recovers the operator formalism in quantum mechanics and perturbation theory in quantum field theory (QFT).

The approach based on path integrals has several advantages over the operator formalism. It provides a useful tool for nonperturbative studies including:

(1) instantons,

(2) analogy with statistical mechanics,

(3) numerical methods.

A standard way of deriving the path integral is from the operator formalism:

We shall proceed in the opposite direction, following the original paper by Feynman [Fey51].

1

Operator calculus

The operator calculus developed by Feynman [Fey51] makes it possible to represent functions of (noncommuting) operators as path integrals, with the integrand being the path-ordered exponential of operators, the order of which is controlled by a parameter that varies along the trajectory. This procedure is termed *Feynman disentangling*. It is also applicable to functions of matrices (say, γ-matrices which are associated with a spinor particle). When applied to the evolution operator, this procedure results in the standard path-integral representation of quantum mechanics.

In this chapter we first demonstrate the general technique using the simplest example, a free propagator in Euclidean space, and then consider the path-integral representation of quantum mechanics, as well as propagators in an external electromagnetic field.

1.1 Free propagator

Let us first consider the simplest propagator of a free scalar field which is given in the operator formalism by the vacuum expectation value of the T-product*

$$G(x - y) \;=\; \langle 0 | \boldsymbol{T}\boldsymbol{\varphi}(x)\,\boldsymbol{\varphi}(y) | 0 \rangle \tag{1.1}$$

with φ being the field-operator.

The T-product (1.1) obeys the equation

$$\left(-\partial^2 - m^2\right) G(x - y) \;=\; \mathrm{i}\,\delta^{(d)}(x - y)\,, \tag{1.2}$$

where $d = 4$ is the dimension of space-time, however the formulas are applicable at any value of d. In the operator formalism, Eq. (1.2) is a

---

* The ordered products of operators were introduced by Dyson [Dys49]. This paper and other classical papers on quantum electrodynamics are collected in the book edited by Schwinger [Sch58].

consequence of the free equations

$$\left.\begin{array}{rcl}(-\partial^2 - m^2)\,\varphi(x)|\,0\rangle & = & 0,\\ \langle\,0\,|\,(-\partial^2 - m^2)\,\varphi(x) & = & 0\end{array}\right\} \tag{1.3}$$

and canonical equal-time commutators

$$\left.\begin{array}{rcl}[\varphi(t,\vec{x})\,,\dot{\varphi}(t,\vec{y})] & = & \mathrm{i}\,\delta^{(d-1)}(\vec{x}-\vec{y})\,,\\ [\varphi(t,\vec{x})\,,\varphi(t,\vec{y})] & = & 0.\end{array}\right\} \tag{1.4}$$

The delta-function $\delta^{(1)}(x_0 - y_0)$ emerges when $(\partial/\partial x_0)^2$ is applied to the operator of the T-product in (1.1).

Problem 1.1 Derive Eq. (1.2) in the operator formalism.

Solution Let us apply the operator on the left-hand side (LHS) of Eq. (1.2) to the T-product which is defined by

$$T\varphi(x)\,\varphi(y) \;=\; \theta(x_0 - y_0)\,\varphi(x)\,\varphi(y) + \theta(y_0 - x_0)\,\varphi(y)\,\varphi(x) \tag{1.5}$$

with

$$\theta(x_0 - y_0) \;=\; \left\{\begin{array}{l} 1 \text{ for } x_0 \geq y_0\\ 0 \text{ for } x_0 < y_0\,.\end{array}\right. \tag{1.6}$$

Equation (1.3) implies a nonvanishing result to emerge only when $(\partial/\partial x_0)^2$ is applied to the operator of the T-product. One obtains

$$\begin{array}{rcl}(-\partial^2 - m^2)\,\langle\,0|\,T\varphi(x)\,\varphi(y)\,|0\rangle & = & -\dfrac{\partial}{\partial x_0}\,\langle\,0|\,T\dot{\varphi}(x)\,\varphi(y)\,|0\rangle\\[2mm] & = & \delta^{(1)}(x_0 - y_0)\,\langle\,0|\,[\varphi(y)\,,\dot{\varphi}(x)]\,|0\rangle\\[2mm] & = & \mathrm{i}\,\delta^{(d)}(x - y)\,,\end{array} \tag{1.7}$$

where the canonical commutation relations (1.4) are used.

The explicit solution to Eq. (1.2) for the free propagator is well-known and is most simply given by the Fourier transform:

$$G(x - y) \;=\; \int \frac{\mathrm{d}^d p}{(2\pi)^d}\,\mathrm{e}^{\mathrm{i}p(x-y)}\,\frac{\mathrm{i}}{p^2 - m^2 + \mathrm{i}\varepsilon}. \tag{1.8}$$

An extra $\mathrm{i}\varepsilon$ (with $\varepsilon \to +0$) in the denominator is due to the T-product in the definition (1.1) and unambiguously determines the integral over p_0. The propagator (1.8) is known as the Feynman propagator that respects causality.

Problem 1.2 Perform the Fourier transformation of the free momentum-space propagator in the energy p_0:

$$G_\omega(t - t') \;=\; \int\limits_{-\infty}^{+\infty} \frac{\mathrm{d}p_0}{2\pi}\,\mathrm{e}^{\mathrm{i}p_0(t-t')}\,\frac{\mathrm{i}}{p_0^2 - \omega^2 + \mathrm{i}\varepsilon},\qquad \omega \;=\; \sqrt{\vec{p}^2 + m^2}. \tag{1.9}$$

Solution The poles of the momentum-space propagator are at

$$p_0 = \pm\omega \mp i\varepsilon. \tag{1.10}$$

For $t > t'$ ($t < t'$), the contour of integration can be closed in the upper (lower) half-plane which gives

$$\begin{aligned} G_\omega(t - t') &= \theta(t - t')\frac{e^{-i\omega(t-t')}}{2\omega} + \theta(t' - t)\frac{e^{i\omega(t-t')}}{2\omega} \\ &= \frac{e^{-i\omega|t-t'|}}{2\omega}. \end{aligned} \tag{1.11}$$

The Green function (1.11) obeys the equation

$$\left(-\frac{\partial^2}{\partial t^2} - \omega^2\right) G_\omega(t - t') = i\delta^{(1)}(t - t') \tag{1.12}$$

and therefore coincides with the causal Green function for a harmonic oscillator with frequency ω.

Remark on operator notations

In mathematical language, the Green function $G(x - y)$ is termed the *resolvent* of the operator on the LHS of Eq. (1.2), and is often denoted as the matrix element of the inverse operator

$$G(x - y) = \left\langle y \left| \frac{i}{-\partial^2 - m^2} \right| x \right\rangle. \tag{1.13}$$

The operators act in an infinite-dimensional Hilbert space, the elements of which in Dirac's notation [Dir58] are the *bra* and *ket* vectors $\langle g|$ and $|f\rangle$, respectively. The coordinate representation emerges when these vectors are chosen to be the eigenstates of the position operator x_μ:

$$x_\mu|x\rangle = x_\mu|x\rangle. \tag{1.14}$$

These basis vectors obey the completeness condition

$$\int d^d x \, |x\rangle\langle x| = 1, \tag{1.15}$$

while the wave functions, associated with $\langle g|$ and $|f\rangle$, are given by

$$\langle g \,|\, x \rangle = g(x), \qquad \langle x \,|\, f \rangle = f(x). \tag{1.16}$$

These wave functions appear in the expansions

$$|f\rangle = \int d^d x \, f(x)|x\rangle, \qquad \langle g| = \int d^d y \, g(y)\langle y|. \tag{1.17}$$

The action of a linear operator O on the bra and ket vectors in Hilbert space is determined by its matrix element $\langle y\,|O|\,x\rangle$, which is also known as the *kernel* of the operator O and is denoted by

$$\langle y\,|O|\,x\rangle \;=\; O(y,x)\,. \tag{1.18}$$

Using the expansion (1.17), one obtains

$$\langle g\,|O|\,f\rangle \;=\; \int \mathrm{d}^d x \int \mathrm{d}^d y\, g(y)\, O(y,x)\, f(x)\,. \tag{1.19}$$

Since the kernel of the unit operator is the delta-function,

$$\langle y\,|1|\,x\rangle \;=\; \langle y|x\rangle \;=\; \delta^{(d)}(x-y)\,, \tag{1.20}$$

the formula

$$\langle y\,|O|\,x\rangle \;=\; O\,\delta^{(d)}(x-y) \tag{1.21}$$

can also be written down as a direct consequence of Eq. (1.20), where the operator O on the right-hand side (RHS) acts on the variable x.

Therefore, when the operator acts on a function $f(x)$, the result is expressed via the kernel by the standard formula

$$O f(y) \;\equiv\; \langle y\,|O|\,f\rangle \;=\; \int \mathrm{d}^d x\, O(y,x)\, f(x)\,. \tag{1.22}$$

Equation (1.21) is obviously reproduced when f is substituted by a delta-function, while Eq. (1.19) takes the form

$$\langle g\,|O|\,f\rangle \;=\; \int \mathrm{d}^d x\, g(x)\, O f(x)\,. \tag{1.23}$$

If space-time is approximated by a discrete set of points, then the operator O is approximated by a matrix with elements $\langle y\,|O|\,x\rangle$.

1.2 Euclidean formulation

Equation (1.8) can be obtained alternatively by inverting the operator on the LHS of Eq. (1.2). Before doing that, it is convenient to make an analytic continuation in the time-variable t, and to pass to the Euclidean formulation of quantum field theory (QFT) where one substitutes

$$t \;=\; -\mathrm{i}\, x_4\,. \tag{1.24}$$

The four-momentum operator in Minkowski space reads as

$$p_{\mathrm{M}}^{\mu} \;=\; \mathrm{i}\, \partial_{\mathrm{M}}^{\mu} \;\equiv\; \left(\mathrm{i}\frac{\partial}{\partial t}, -\mathrm{i}\frac{\partial}{\partial \vec{x}}\right) \quad \boxed{\text{Minkowski space}}\,, \tag{1.25}$$

while its Euclidean counterpart is given by

$$\boldsymbol{p}_E^\mu = -i\,\partial_E^\mu \equiv \left(-i\frac{\partial}{\partial\vec{x}}, -i\frac{\partial}{\partial x_4}\right) \qquad \boxed{\text{Euclidean space}}\,. \qquad (1.26)$$

These two formulas together with Eq. (1.24) yield

$$E \equiv p_0 = -i\,p_4 \qquad (1.27)$$

for the relation between energy and the fourth component of the Euclidean four-momentum.

The passage to Euclidean space results in changing the Minkowski signature of the metric $g_{\mu\nu}$ to the Euclidean one:*

$$(+---) \longrightarrow (++++)$$

$$\boxed{\text{Minkowski signature}} \longrightarrow \boxed{\text{Euclidean signature}}\,. \qquad (1.28)$$

As such, one finds

$$p_M^2 = p_0^2 - \vec{p}^{\,2} \longrightarrow -p_E^2 = -\vec{p}^{\,2} - p_4^2\,. \qquad (1.29)$$

The exponent in the Fourier transformation changes analogously:

$$-p_\mu x^\mu = -Et + \vec{p}\vec{x} \longrightarrow p_E^\mu x_E^\mu = \vec{p}\vec{x} + p_4 x_4\,. \qquad (1.30)$$

This reproduces the standard Fourier transformation in Euclidean space

$$\left.\begin{aligned} f(p) &= \int d^d x\, e^{-ipx} f(x)\,, \\[2mm] f(x) &= \int \frac{d^d p}{(2\pi)^d}\, e^{ipx} f(p)\,. \end{aligned}\right\} \qquad (1.31)$$

We shall use the same notation v^μ for a four-vector in Minkowski and Euclidean spaces:

$$\left.\begin{aligned} v_M^\mu &= (v_0, \vec{v}) \qquad \boxed{\text{Minkowski space}}\,, \\[2mm] v_E^\mu &= (\vec{v}, v_4) \qquad \boxed{\text{Euclidean space}}\,, \end{aligned}\right\} \qquad (1.32)$$

---

* An older generation will be familiar with the Euclidean notation which is used throughout the book by Akhiezer and Berestetskii [AB69]. In contrast, the two canonical books on quantum field theory by Bogoliubov and Shirkov [BS76] and by Bjorken and Drell [BD65] use the Minkowskian notation instigated by Feynman. The modern generation of textbooks on quantum field theory includes those by Brown [Bro92] and Weinberg [Wei98].

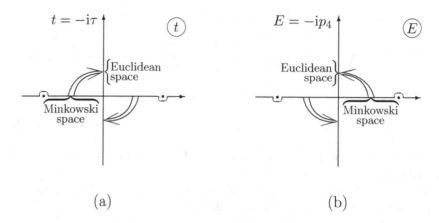

(a)   (b)

Fig. 1.1.   Direction of Wick's rotation from Minkowski to Euclidean space (indicated by the arrows) for (a) time and (b) energy. The dots represent singularities of a free propagator in (a) coordinate and (b) momentum spaces. The contours of integration in Minkowski space are associated with causal Green functions. They can obviously be deformed in the directions of the arrows.

with

$$v_0 \;=\; -iv_4 \,. \tag{1.33}$$

The only difference resides in the metric. We do not distinguish between upper and lower indices in Euclidean space.

Using Eqs. (1.24) and (1.26), we see that in Euclidean space Eq. (1.2) takes the form

$$\left(-\partial^2 + m^2\right) G(x - y) \;=\; \delta^{(d)}(x - y) \tag{1.34}$$

with a positive sign in front of m^2.

The passage to the Euclidean formulation is justified in perturbation theory where it is associated with the Wick rotation. The direction in which the rotation is performed is unambiguously prescribed by the $+i\varepsilon$ term in Eq. (1.8), and is depicted in Fig. 1.1. The variable $t = x_0$ rotates through $-\pi/2$, while $E = p_0$ rotates through $\pi/2$.

Figure 1.1a explains the sign in Eq. (1.24). Figure 1.1b and Eq. (1.27) implies that the integration over p_4 goes in the opposite direction, so that

$$\int_{-\infty}^{+\infty} \frac{dp_0}{2\pi} \cdots \;=\; i \int_{-\infty}^{+\infty} \frac{dp_4}{2\pi} \cdots \,. \tag{1.35}$$

Thus when passing into Euclidean variables, Eq. (1.8) becomes

$$G(x - y) \;=\; \int \frac{d^d p}{(2\pi)^d} \, e^{ip(y-x)} \frac{1}{p^2 + m^2} \,. \tag{1.36}$$

Note that the RHS of Eq. (1.36) is nothing but the Fourier transform of the free momentum-space Euclidean propagator, and there is no need to retain an $i\varepsilon$ in the denominator since the integration prescription is now unambiguous.

It is now clear why we keep the same notation for the coordinate-space Green functions: the Feynman propagator in Minkowski space and the Euclidean propagator. They are the same analytic function of the time-variable.

Problem 1.3 Repeat the calculation of Problem 1.2 in Euclidean space.

Solution According to Eq. (1.36) we need to calculate

$$
G_\omega(\tau - \tau') = \int\limits_{-\infty}^{+\infty} \frac{dp_4}{2\pi}\, e^{ip_4(\tau'-\tau)} \frac{1}{p_4^2 + \omega^2}. \tag{1.37}
$$

The integral on the RHS can be calculated for $\tau > \tau'$ ($\tau < \tau'$) by closing the contour in the lower (upper) half-plane, and taking the residues at $p_4 = -i\omega$ ($p_4 = i\omega$), respectively. This yields

$$
\begin{aligned}
G_\omega(\tau - \tau') &= \theta(\tau - \tau')\frac{e^{\omega(\tau'-\tau)}}{2\omega} + \theta(\tau' - \tau)\frac{e^{\omega(\tau-\tau')}}{2\omega} \\
&= \frac{e^{-\omega|\tau-\tau'|}}{2\omega}.
\end{aligned} \tag{1.38}
$$

The Euclidean Green function (1.38) can obviously be obtained from the Minkowskian one, Eq. (1.11), by the substitution

$$
\tau = it, \qquad \tau' = it' \tag{1.39}
$$

and vice versa. $G_\omega(\tau - \tau')$ obeys the equation

$$
\left(-\frac{\partial^2}{\partial\tau^2} + \omega^2\right) G_\omega(\tau - \tau') = \delta^{(1)}(\tau - \tau') \tag{1.40}
$$

and, therefore, is the Green function for a Euclidean harmonic oscillator with frequency ω.

As we shall see in a moment, the Euclidean formulation makes path integrals well-defined, and allows nonperturbative investigations analogous to statistical mechanics to be carried out. There are no reasons, however, why Minkowski and Euclidean formulations should always be equivalent nonperturbatively.

Remark on Euclidean γ-matrices

The γ-matrices in Minkowski space satisfy

$$
\{\gamma_M^\mu, \gamma_M^\nu\} = 2g^{\mu\nu}\,\mathbb{I}, \tag{1.41}
$$

where $\mathbb{I}$ denotes the unit matrix. Therefore, γ_0 is Hermitian while the Minkowskian spatial γ-matrices are anti-Hermitian.

Analogously, the Euclidean γ-matrices satisfy

$$\{\gamma_\mu, \gamma_\nu\} \;=\; 2\,\delta_{\mu\nu}\,\mathbb{I}, \tag{1.42}$$

so that all of them are Hermitian. We compose them from 2×2 matrices as

$$\gamma_4 \;=\; \gamma_0 \;=\; \begin{pmatrix} \mathbb{I} & 0 \\ 0 & -\mathbb{I} \end{pmatrix} \tag{1.43}$$

and

$$\vec{\gamma} \;=\; \begin{pmatrix} 0 & -i\vec{\sigma} \\ i\vec{\sigma} & 0 \end{pmatrix}, \tag{1.44}$$

where $\vec{\sigma}$ are the usual Pauli matrices. Note that the Euclidean spatial γ-matrices differ from the Minkowskian ones by a factor of i.

The free Dirac equation in Euclidean space reads as

$$\left(\hat{\partial} + m\right)\psi \;=\; 0\,, \qquad \hat{\partial} \;=\; \gamma_\mu \partial_\mu \tag{1.45}$$

or

$$(i\widehat{\boldsymbol{p}} + m)\,\psi \;=\; 0 \tag{1.46}$$

with $\boldsymbol{p}$ given by Eq. (1.26).

1.3 Path-ordering of operators

There are no problems in defining a function of an operator A, say via the Taylor series. For instance,

$$e^A \;=\; \sum_{n=0}^{\infty} \frac{1}{n!} A^n. \tag{1.47}$$

However, it is more complicated to define a function of several noncommuting operators (or matrices), e.g. A and B having

$$[A, B] \;\neq\; 0\,, \tag{1.48}$$

since the order of operators is now essential. In particular, one has

$$e^{A+B} \;\neq\; e^A\, e^B, \tag{1.49}$$

so that the law of addition of exponents fails. Certainly, the exponential on the LHS is a well-defined function of $A + B$, but since A and B

are intermixed in the Taylor expansion, this expansion is of little use in practice. We would like to have an expression where all Bs are written, say, to the right of all As. Generically, this is a problem of representing a symmetric ordering of operators via a normal ordering.

This can be achieved by the following formal trick [Fey51].

Let us write

$$
\begin{aligned}
\mathrm{e}^{A+B} &= \lim_{M\to\infty} \left[1 + \frac{1}{M}(A+B)\right]^M \\
&= \lim_{M\to\infty} \underbrace{\left[1 + \frac{1}{M}(A+B)\right] \cdots \left[1 + \frac{1}{M}(A+B)\right]}_{M \text{ times}}.
\end{aligned} \quad (1.50)
$$

The structure of the product on the RHS prompts us to introduce an index i running from 1 to M and replace $(A+B)$ in each multiplier by $(A_i + B_i)$. Therefore, one writes

$$
\begin{aligned}
\mathrm{e}^{A+B} &= \lim_{M\to\infty} \prod_{i=1}^{M} \left[1 + \frac{1}{M}(A_i + B_i)\right] \\
&= \lim_{M\to\infty} \left[1 + \frac{1}{M}(A_M + B_M)\right] \cdots \left[1 + \frac{1}{M}(A_1 + B_1)\right],
\end{aligned} \quad (1.51)
$$

where the index i controls the order of the operators which are all treated *differently*. The ordering is such that the larger i is, the later the operator with the index i acts. This order of operators is prescribed by quantum mechanics, where initial and final states are represented by ket and bra vectors, respectively.

Equation (1.51) can be rewritten as

$$
\mathrm{e}^{A+B} = \boldsymbol{P} \lim_{M\to\infty} \exp\left[\frac{1}{M} \sum_{i=1}^{M} (A_i + B_i)\right], \quad (1.52)
$$

where the symbol $\boldsymbol{P}$ denotes the ordering operation. There is no ambiguity on the RHS of Eq. (1.52) concerning ordering A_i and B_i with the same index i, since such terms are $\mathcal{O}(M^{-2})$ and are negligible as $M \to \infty$.

To describe the continuum limit as $M \to \infty$, one introduces the continuum variable $\sigma = i/M$ which belongs to the interval $[0, 1]$. The continuum limit of Eq. (1.52) reads as

$$
\mathrm{e}^{A+B} = \boldsymbol{P} \exp\left\{\int_0^1 \mathrm{d}\sigma \,[A(\sigma) + B(\sigma)]\right\}, \quad (1.53)
$$

where $A(i/M) = A_i$ and $B(i/M) = B_i$, while the operator $A(\sigma) + B(\sigma)$ acts at order σ.

Equation (1.53) is, in fact, obvious since it only involves the operator $A + B$, which commutes with itself. For commuting operators there is no need for ordering so that $A(\sigma) + B(\sigma)$ does not depend on σ in this case. The integral in the exponent on the RHS of Eq. (1.53) can then be performed, and reproduces the LHS.

Equation (1.53) can however be manipulated as though $A(\sigma)$ and $B(\sigma)$ were just functions rather than operators since the order would be specified automatically by the path-ordering operation. This is analogous to the well-known fact that operators can be written in an arbitrary order under the T-product. Therefore, we can rewrite Eq. (1.53) as

$$
e^{A+B} \;=\; \boldsymbol{P}\, e^{\int_0^1 d\sigma' A(\sigma')}\, e^{\int_0^1 d\sigma B(\sigma)} . \tag{1.54}
$$

This is the operator analog of the law of addition of exponents.

Problem 1.4 Calculate explicitly the first term of the expansion of $\exp{(A+B)}$ in B.

Solution Expanding the RHS of Eq. (1.54) in B, one finds

$$
e^{A+B} \;=\; e^A + \int_0^1 d\sigma\, e^{\int_\sigma^1 d\sigma' A(\sigma')} B(\sigma)\, e^{\int_0^\sigma d\sigma' A(\sigma')} + \cdots . \tag{1.55}
$$

There is no need for a path-ordering sign in this formula, since the order of the operators A and B is written explicitly. There is also no ambiguity in defining the exponentials of the operator A as already explained.

Since the order is explicit, one drops the formal dependence of A and B on the ordering parameter which gives

$$
e^{A+B} \;=\; e^A + \int_0^1 d\sigma\, e^{(1-\sigma)A} B\, e^{\sigma A} + \cdots . \tag{1.56}
$$

Formulas (1.55) and (1.56) are known from time-dependent perturbation theory in quantum mechanics.

Problem 1.5 Using Eq. (1.56), derive

$$
\frac{1}{A+B} \;=\; \frac{1}{A} - \frac{1}{A} B \frac{1}{A} + \cdots \tag{1.57}
$$

for small B.

Solution Exponentiating and using Eq. (1.56), we obtain

$$
\frac{1}{A+B} = \int_0^\infty d\tau\, e^{-\tau(A+B)}
$$

$$
= \int_0^\infty d\tau \left[e^{-\tau A} - \tau \int_0^1 d\sigma\, e^{\tau(\sigma-1)A} B\, e^{-\tau\sigma A} \right] + \cdots . \tag{1.58}
$$

Introducing the new variables

$$
\tau_1 = \tau(1-\sigma), \qquad \tau_2 = \tau\sigma, \tag{1.59}
$$

we rewrite the RHS of Eq. (1.58) as

$$
\frac{1}{A} - \int_0^\infty d\tau_1\, e^{-\tau_1 A} B \int_0^\infty d\tau_2\, e^{-\tau_2 A} + \cdots = \frac{1}{A} - \frac{1}{A} B \frac{1}{A} + \cdots \tag{1.60}
$$

which proves Eq. (1.57).

1.4 Feynman disentangling

The operator on the LHS of Eq. (1.34) can be inverted as follows:

$$
G(x - y) = \frac{1}{-\partial^2 + m^2} \delta^{(d)}(x - y)
$$

$$
= \frac{1}{2} \int_0^\infty d\tau\, e^{\frac{1}{2}\tau(\partial^2 - m^2)} \delta^{(d)}(x - y)
$$

$$
= \frac{1}{2} \int_0^\infty d\tau\, e^{-\frac{1}{2}m^2\tau}\, \boldsymbol{P}\, e^{\frac{1}{2}\int_0^\tau dt\, \partial^2(t)} \delta^{(d)}(x - y), \tag{1.61}
$$

where we have formally labeled the derivatives using an ordering parameter $t \in [0, \tau]$, which is an analog of σ from the previous section. This is
the general procedure upon which the Feynman disentangling is built.

Since the operators ∂_μ and ∂_ν commute in the free case, we could manage without introducing the t-dependence, however the operators do not
commute in general. The simple example of the nonrelativistic Hamiltonian and the propagator in an external electromagnetic field are considered later in this chapter. Other cases where the disentangling is needed
are related to inverting an operator which is also a matrix in some symmetry space.

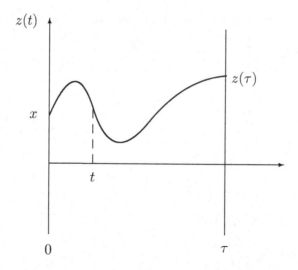

Fig. 1.2.   Trajectory $z_\mu(t)$. The operator $\partial_\mu(t)$ acts at the order t.

Continuing with the disentangling, the RHS of Eq. (1.61) can be rewritten as

$$G(x - y) = \frac{1}{2} \int\limits_0^\infty d\tau \, e^{-\frac{1}{2}m^2\tau} \int\limits_{z_\mu(0)=x_\mu} \mathcal{D}z_\mu(t) \, e^{-\frac{1}{2}\int_0^\tau dt \, \dot{z}_\mu^2(t)}$$

$$\times \boldsymbol{P} \, e^{\int_0^\tau dt \, \dot{z}_\mu(t)\partial_\mu(t)} \, \delta^{(d)}(x - y), \qquad (1.62)$$

where the integration runs over all trajectories $z_\mu(t)$ which begin at the point x, as depicted in Fig. 1.2.

Since the operator $\partial_\mu(t)$ acts at the order t, these operators are ordered along the trajectory $z_\mu(t)$ with $\boldsymbol{P}$, in Eq. (1.62), denoting the path-ordering operator. Note, that $\dot{z}_\nu(t)$ and $\partial_\mu(t)$ commute since

$$\partial_\mu(t)\dot{z}_\nu(t) = \frac{d}{dt}\delta_{\mu\nu} = 0 \qquad (1.63)$$

so that their order is not essential in Eq. (1.62). With these rules of manipulation, Eq. (1.62) can be proven by the "translation"

$$z_\mu(t) \rightarrow z'_\mu(t) = z_\mu(t) + \int\limits_0^t dt' \, \partial_\mu(t') \qquad (1.64)$$

of the integration variable $z_\mu(t)$ in the Gaussian integral.

The integral over the functions $z_\mu(t)$ in Eq. (1.62) is called a *path integral* or a *functional integral*.

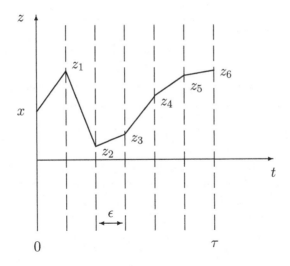

Fig. 1.3. Discretization of trajectory $z_\mu(t)$ (depicted for $M = 6$).

The continual path integral can be approximated by a finite one. To this end, let us choose M points $t_i = i\epsilon$, where ϵ is the discretization step, and $M = \tau/\epsilon$. We then connect the points

$$z_0 = x, \quad z_i = z(i\epsilon) \quad i = 1, 2, \ldots, M \tag{1.65}$$

by straight lines. Such a discretization of the trajectory $z_\mu(t)$ is depicted in Fig. 1.3. The measure in Eq. (1.62) can then be discretized by

$$\int \mathcal{D}z_\mu(t) \cdots = \prod_{i=1}^{M} \int \frac{\mathrm{d}^d z_i}{(2\pi\epsilon)^{d/2}} \cdots . \tag{1.66}$$

The explicit form of the operator ∂_μ in Eq. (1.34) was not essential in deriving Eq. (1.62). If ∂_μ in Eq. (1.34) is replaced by an arbitrary operator D_μ with noncommuting components, then Eq. (1.62) holds with $\partial_\mu(t)$ substituted by $D_\mu(t)$. The discretized path-ordered exponential of a general operator $D_\mu(t)$ is given by

$$\boldsymbol{P} e^{\int_0^\tau \mathrm{d}t\, \dot{z}^\mu(t) D_\mu(t)} = \lim_{\epsilon \to 0} \prod_{i=1}^{M} [1 + (z_i - z_{i-1})^\mu D_\mu(i\epsilon)] . \tag{1.67}$$

The order of multiplication here is the same as in Eq. (1.51).

The explicit form of the operator ∂_μ is essential when we calculate how it acts on the delta-function as prescribed by the RHS of Eq. (1.62). For the free case, when the t-dependence of $\partial_\mu(t)$ is not essential, one simply finds

$$\boldsymbol{P}\,\mathrm{e}^{\int_0^\tau \mathrm{d}t\,\dot{z}^\mu(t)\partial_\mu(t)} \;=\; \exp\left\{[z^\mu(\tau)-x^\mu]\frac{\partial}{\partial x^\mu}\right\}, \tag{1.68}$$

which is nothing but the shift operator. Applying the operator on the RHS of Eq. (1.68) to the delta-function, one obtains

$$\boldsymbol{P}\,\mathrm{e}^{\int_0^\tau \mathrm{d}t\,\dot{z}^\mu(t)\partial_\mu(t)}\,\delta^{(d)}(x-y) \;=\; \delta^{(d)}(z(\tau)-y). \tag{1.69}$$

Therefore, $z_\mu(\tau)$ has to coincide with y_μ owing to the delta-function, which disappears after the integration over $z_\mu(\tau)$ has been performed. Thus the final answer is

$$G(x-y) \;=\; \frac{1}{2}\int_0^\infty \mathrm{d}\tau\,\mathrm{e}^{-\frac{1}{2}\tau m^2}\int_{\substack{z_\mu(0)=x_\mu\\z_\mu(\tau)=y_\mu}} \mathcal{D}z_\mu(t)\,\mathrm{e}^{-\frac{1}{2}\int_0^\tau \mathrm{d}t\,\dot{z}_\mu^2(t)}. \tag{1.70}$$

This path integral goes over all trajectories $z_\mu(t)$ that connect the initial point x_μ and the final point y_μ.

Problem 1.6 Derive Eqs. (1.62) and (1.70) by introducing a path integral over velocity $v_\mu(t) = \dot{z}_\mu(t)$.

Solution The operator on the RHS of Eq. (1.61) can be disentangled using the following Gaussian path integral:

$$\boldsymbol{P}\,\mathrm{e}^{\frac{1}{2}\int_0^\tau \mathrm{d}t D_\mu^2(t)} \;=\; \int \mathcal{D}v_\mu(t)\,\mathrm{e}^{-\frac{1}{2}\int_0^\tau \mathrm{d}t\,v_\mu^2(t)}\,\boldsymbol{P}\,\mathrm{e}^{\int_0^\tau \mathrm{d}t\,v^\mu(t)D_\mu(t)}. \tag{1.71}$$

This formula holds for an arbitrary operator D_μ and can be proven formally by calculating the Gaussian integral after shifting $v_\mu(t)$.

Substituting $D_\mu(t) = \partial_\mu(t)$ and calculating the action of the path-ordered exponential on $\delta^{(d)}(x-y)$, we obtain

$$G(x-y) \;=\; \frac{1}{2}\int_0^\infty \mathrm{d}\tau\,\mathrm{e}^{-\frac{1}{2}\tau m^2}\int \mathcal{D}v_\mu(t)\,\mathrm{e}^{-\frac{1}{2}\int_0^\tau \mathrm{d}t\,v_\mu^2(t)}\,\delta^{(d)}\left(x+\int_0^\tau \mathrm{d}t\,v(t)-y\right). \tag{1.72}$$

The integration over $\mathcal{D}v_\mu(t)$ in this formula has no restrictions.

To derive Eq. (1.70) from Eq. (1.72), let us note that the discretized velocities read as

$$v_i^\mu \;=\; \frac{z_i^\mu - z_{i-1}^\mu}{\epsilon}. \tag{1.73}$$

Since

$$\int_0^\tau \mathrm{d}t\,v^2(t) \;\to\; \epsilon\sum_{i=1}^M v_i^2, \tag{1.74}$$

the measure

$$\int \mathcal{D}v_\mu(t) \cdots = \prod_{i=1}^{M} \int \frac{d^d v_i}{(2\pi/\epsilon)^{d/2}} \cdots \qquad (1.75)$$

obviously recovers Eq. (1.66) after calculating the Jacobian from the variables v_i to the variables z_i. Therefore, Eq. (1.72) reproduces Eq. (1.70) provided

$$z^\mu(t) = x^\mu + \int_0^t dt' v^\mu(t'). \qquad (1.76)$$

Remark on definition of the measure

The discretized trajectory in Fig. 1.3 can be written analytically as the expansion

$$z^\mu(t) = \sum_{i=1}^{M} z_i^\mu f_i(t) + x^\mu(1 - t/\epsilon)\, \theta(\epsilon - t), \qquad (1.77)$$

where the basis functions

$$f_i(t) = \begin{cases} 1 + (t/\epsilon - i) & \text{for } t \in [(i-1)\epsilon, i\epsilon], \\ 1 - (t/\epsilon - i) & \text{for } t \in [i\epsilon, (i+1)\epsilon], \\ 0 & \text{otherwise} \end{cases} \qquad (1.78)$$

are nonvanishing only for the ith and $(i+1)$th intervals. The measure (1.66) is defined, therefore, via the coefficients z_i as a multiple product of dz_i.

While the basis functions $f_i(t)$ are not orthogonal:

$$\frac{1}{\epsilon} \int_0^\tau dt\, f_i(t) f_j(t) = \frac{2}{3}\delta_{ij} + \frac{1}{6}\delta_{i(j+1)} + \frac{1}{6}\delta_{i(j-1)}, \qquad (1.79)$$

the orthogonal set appears in the expansion of the velocity

$$\dot{z}^\mu(t) = \sum_{i=1}^{M} (z_i^\mu - z_{i-1}^\mu)\phi_i(t), \qquad (1.80)$$

where

$$\phi_i(t) = \begin{cases} 1/\epsilon & \text{for } t \in [(i-1)\epsilon, i\epsilon], \\ 0 & \text{otherwise}. \end{cases} \qquad (1.81)$$

This shows why the discretized velocities from Problem 1.6 are natural variables.

One can choose, instead, another set of (orthogonal) basis functions
and expand

$$z^\mu(t) = \sum_{n=1}^{M} c_n^\mu \phi_n(t) \tag{1.82}$$

with some coefficients c_n^μ. Then the measure (1.66) takes the form

$$\mathcal{D}z_\mu(t) \cdots \propto \prod_{n=1}^{M} \mathrm{d}^d c_n \cdots \tag{1.83}$$

modulo a c-independent Jacobian. Mathematically, this implies that one
approximates the functional space by M-dimensional spaces.

1.5 Calculation of the Gaussian path integral

The Gaussian path integral (1.70) can be calculated easily using the following trick.* Let us substitute the variable $z_\mu(t)$ by a new variable $\xi_\mu(t)$,
which are related by the formula

$$z_\mu(t) = \frac{y_\mu - x_\mu}{\tau} t + \xi_\mu(t) + x_\mu. \tag{1.84}$$

The boundary conditions for the variable $\xi(t)$ are determined by Eq. (1.84)
to be

$$\xi_\mu(0) = \xi_\mu(\tau) = 0. \tag{1.85}$$

On substituting Eq. (1.84) into the exponent in Eq. (1.70), one finds

$$\int_0^\tau \mathrm{d}t\, \dot z^2(t) = \frac{(y-x)^2}{\tau} + 2\frac{(y-x)}{\tau} \left[\xi(\tau) - \xi(0)\right] + \int_0^\tau \mathrm{d}t\, \dot\xi^2(t). \tag{1.86}$$

The second term on the RHS vanishes owing to the boundary conditions
(1.85) so that the propagator becomes

$$G(x-y) = \frac{1}{2} \int_0^\infty \mathrm{d}\tau\, e^{-\frac{1}{2}\tau m^2}\, e^{-(y-x)^2/2\tau} \int\limits_{\xi_\mu(0)=\xi_\mu(\tau)=0} \mathcal{D}\xi_\mu\, e^{-\frac{1}{2}\int_0^\tau \mathrm{d}t\, \dot\xi_\mu^2(t)}. \tag{1.87}$$

The path integral over ξ on the RHS of Eq. (1.87) is a function solely
of τ:

$$\int\limits_{\xi_\mu(0)=\xi_\mu(\tau)=0} \mathcal{D}\xi_\mu\, e^{-\frac{1}{2}\int_0^\tau \mathrm{d}t\, \dot\xi_\mu^2(t)} = \mathcal{F}(\tau). \tag{1.88}$$

---

* See, for example, the book by Feynman [Fey72], Chapter 3.

This expression is to be compared with the proper-time representation of the Euclidean free propagator which reads as

$$G(x-y) = \int \frac{d^d p}{(2\pi)^d} e^{ip(x-y)} \frac{1}{2} \int_0^\infty d\tau \, e^{-\frac{\tau}{2}(p^2+m^2)}$$

$$= \frac{1}{2} \int_0^\infty d\tau \, e^{-\frac{1}{2}\tau m^2} e^{-(x-y)^2/2\tau} \frac{1}{(2\pi\tau)^{d/2}}. \tag{1.89}$$

These two expressions coincide provided that

$$\mathcal{F}(\tau) = \frac{1}{(2\pi\tau)^{d/2}}. \tag{1.90}$$

Problem 1.7 Calculate $\mathcal{F}(\tau)$ from the discretized path integral.

Solution The discretized version of the path integral in Eq. (1.70) is

$$\int_{\substack{z_\mu(0)=x_\mu \\ z_\mu(\tau)=y_\mu}} \mathcal{D}z_\mu(t) \, e^{-\frac{1}{2}\int_0^\tau dt \, \dot{z}_\mu^2(t)} = \frac{1}{(2\pi\epsilon)^{d/2}} \int \prod_{i=1}^{M-1} \frac{d^d z_i}{(2\pi\epsilon)^{d/2}} e^{-\frac{1}{2\epsilon}\sum_{i=1}^M (z_i-z_{i-1})^2}, \tag{1.91}$$

where $z_0 = x$ and $z_M = y$. The integral can be calculated using the well-known formula for the Gaussian integral

$$\int \frac{d^d z}{(2\pi)^{d/2}} \exp\left[-\frac{(x-z)^2}{2\tau_1} - \frac{(z-y)^2}{2\tau_2}\right] = \left(\frac{\tau_1\tau_2}{\tau_1+\tau_2}\right)^{d/2} \exp\left[-\frac{(x-y)^2}{2(\tau_1+\tau_2)}\right]. \tag{1.92}$$

After applying this formula $M-1$ times, one arrives at Eq. (1.90). Note that ϵ cancels in the final answer.

Problem 1.8 Which trajectories are essential in the path integral?

Solution It is seen from the discretization on the RHS of Eq. (1.91) that only trajectories with

$$|z_i - z_{i-1}| \sim \sqrt{\epsilon} \tag{1.93}$$

are essential as $\epsilon \to 0$. Such trajectories are typical *Brownian* trajectories. They are continuous as $\epsilon \to 0$ but not smooth ($|z_i - z_{i-1}| \sim \epsilon$ for smooth trajectories). In mathematical language, these functions are said to belong to the Lipshitz class $1/2$.

Remark on mathematical structure

The measure (1.66) for integration over functions is sometimes called the Lebesgue measure. It was introduced in mathematics by Wiener [Wie23] in connection with the problem of Brownian motion. With the Gaussian factor incorporated, it is also known as the Wiener measure while the proper path integral is known as the Wiener integral.* The measure (1.66) is defined on the space L_2 (i.e. the space of functions whose square is integrable, in the sense of the Lebesgue integral, $\int dt\, z^2(t) < \infty$). The integration on L_2 goes over trajectories $z_\mu(t)$, which are generically discontinuous. However, the extra weight factor $\exp\left[-\frac{1}{2}\int_0^\tau dt\, \dot{z}^2(t)\right]$ restricts the trajectories in the above path integrals to be continuous.

1.6 Transition amplitudes

As is well-known in quantum mechanics, $G(x - y)$ is the probability for a (scalar) particle to propagate from x to y. A convenient notation for a trajectory $z_\mu(t)$ that connects x_μ and y_μ is

$$\Gamma_{yx} \equiv \{z_\mu(t);\ 0 \le t \le \tau,\quad z_\mu(0) = x_\mu,\ z_\mu(\tau) = y_\mu\}. \tag{1.94}$$

Note that Γ_{yx} denotes a trajectory as a geometric object, while $z_\mu(t)$ is a function that describes a given trajectory in some parametrization t. This function (but not the geometric object itself) depends on the choice of parametrization and changes under the *reparametrization* transformation

$$t \ \rightarrow \ \sigma(t),\qquad \frac{d\sigma}{dt} \ge 0, \tag{1.95}$$

with σ being a new parameter.

A convenient parametrization is via the proper length of Γ_{yx} which is given by

$$s \ = \ \int_{\Gamma_{yx}} ds, \tag{1.96}$$

where

$$ds \ = \ \sqrt{\dot{z}^2(\sigma)}\, d\sigma \tag{1.97}$$

and $\sigma \in [\sigma_0, \sigma_1]$ is some parametrization. For obvious reasons the parametrization

$$t \ = \ \frac{1}{m}s \tag{1.98}$$

with s given by Eq. (1.96) is called the *proper-time* parametrization. Note that the dimension of t is $[\text{length}]^2$ according to Eq. (1.98).

---

* See, for example, the books [Kac59, Sch81, Wie86, Roe94] for a description of the path-integral approach to Brownian motion.

Let us denote*

$$S[\Gamma_{yx}] \equiv \frac{m^2 \tau}{2} + \frac{1}{2} \int_0^\tau \mathrm{d}t\, \dot{z}^2(t) .$$ (1.99)

The sense of this notation is that the RHS coincides with the classical action of a relativistic free (scalar) particle in the proper-time parametrization (1.98) when

$$\int_0^\tau \mathrm{d}t\, \dot{z}^2(t) = m \int_0^\tau \mathrm{d}s = m\, \text{Length}[\Gamma]$$ (1.100)

since

$$\left(\frac{\mathrm{d}z_\mu(s)}{\mathrm{d}s} \right)^2 = 1$$ (1.101)

and $m\tau = \text{Length}[\Gamma]$ by the definition of the proper time.

Therefore, the path-integral representation (1.70) is nothing but the sum over trajectories with the weight being an exponential of (minus) the classical action:

$$G(x - y) = \sum_{\Gamma_{yx}} e^{-S[\Gamma_{yx}]} .$$ (1.102)

This sum is split in Eq. (1.70) into the trajectories along which the particle propagates during the proper time τ and the integral over τ.

Equation (1.102) implies that the transition amplitude in quantum mechanics is a sum over all paths which connects x and y. In other words, a particle propagates from x to y along all paths Γ_{yx}, including the ones which are forbidden by the free classical equation of motion

$$\ddot{z}_\mu(t) = 0 .$$ (1.103)

Only the classical trajectory (1.103) survives the path integral in the classical limit $\hbar \to 0$. The reason for this is that if the dependence on Planck's constant is restored, it appears in the exponent:

$$G(x - y) = \sum_{\Gamma_{yx}} e^{-S[\Gamma_{yx}]/\hbar} .$$ (1.104)

As $\hbar \to 0$ the path integral is dominated by a saddle point, which is given in the free case by the classical equation of motion (1.103).

---

* The notation $S[\Gamma]$ with square brackets means that S is a functional of Γ, while $f(x)$ with parentheses stands for functions.

It is worth noting that the sum-over-path representation (1.102) is written entirely in terms of trajectories as geometric objects and does not refer to a concrete parametrization. For the free theory $S[\Gamma]$ is proportional to the length of the trajectory Γ:

$$S_{\text{free}}[\Gamma] = m \, \text{Length}[\Gamma], \qquad (1.105)$$

where the length is given for some parametrization σ of the trajectory Γ by

$$\text{Length}[\Gamma] = \int_{\sigma_0}^{\sigma_1} d\sigma \sqrt{\dot{z}^2(\sigma)}. \qquad (1.106)$$

The sum-over-path representation (1.102) with $S[\Gamma]$ given by the classical action (Eq. (1.105) in the free case) is often considered as a first principle of constructing quantum mechanics given the classical action $S[\Gamma]$.

Problem 1.9 Represent the matrix element of the (Euclidean) evolution operator $\langle y | \exp(-H\tau)| x \rangle$ for the nonrelativistic Hamiltonian

$$H = -\frac{\partial^2}{2m} + V(x) \qquad (1.107)$$

as a path integral.

Solution The calculation is similar to that already done in Sect. 1.4. It is most convenient to use the path integral over velocity which was considered in Problem 1.6 on p. 16. The appropriate disentangling formula is given as

$$\langle y | e^{-H\tau} | x \rangle$$
$$= \int \mathcal{D}v_\mu(t) \, e^{-\frac{m}{2} \int_0^\tau dt \, v_\mu^2(t)} \, \boldsymbol{P} \, e^{-\int_0^\tau dt \, v^\mu(t) \partial_\mu(t) - \int_0^\tau dt \, V(x;t)} \, \delta^{(d)}(x - y). \qquad (1.108)$$

Here the argument t in $V(x;t)$ is just the ordering parameter, while the same formula holds when the potential is explicitly time-dependent.

In contrast to Eq. (1.71), we have put the minus sign in front of the linear-in-v term in the exponent in Eq. (1.108), so that it agrees with Appendix B of Feynman's paper [Fey51]. In fact, it does not matter what sign is used since the integral over $v(t)$ is Gaussian, so only even powers of v survive after the integration.

The path-ordered exponential in Eq. (1.108) reads explicitly as

$$\boldsymbol{P} \, e^{-\int_0^\tau dt \, v^\mu(t) \partial_\mu(t) - \int_0^\tau dt \, V(x;t)} = \lim_{\epsilon \to 0} \prod_{i=1}^{M} \left[1 - \epsilon v_i^\mu \frac{\partial}{\partial x^\mu} - \epsilon V(x;i\epsilon) \right], \qquad (1.109)$$

which can be rewritten as

$$\boldsymbol{P}\,\mathrm{e}^{-\int_0^\tau \mathrm{d}t\, v^\mu(t)\partial_\mu(t)-\int_0^\tau \mathrm{d}t\, V(x;t)} \;=\; \lim_{\epsilon\to 0}\prod_{i=1}^{M}\left[1-\epsilon v_i^\mu\frac{\partial}{\partial x^\mu}\right][1-\epsilon V(x;i\epsilon)]\,,$$

(1.110)

if terms which vanish as $\epsilon\to 0$ are neglected, or equivalently as

$$\boldsymbol{P}\,\mathrm{e}^{-\int_0^\tau \mathrm{d}t\, v^\mu(t)\partial_\mu(t)-\int_0^\tau \mathrm{d}t\, V(x;t)} \;=\; \prod_{t=0}^{\tau}\left[1-\mathrm{d}t\, v^\mu(t)\frac{\partial}{\partial x^\mu}\right][1-\mathrm{d}t\, V(x;t)]\,.$$

(1.111)

There is no need to write down the t-dependence of $\partial_\mu(t)$ in these formulas since the order of the operators is explicit.

To disentangle the operator expression (1.111), let us note that

$$[1-\mathrm{d}t\, v^\mu(t)\partial_\mu] \;=\; U^{-1}(t+\mathrm{d}t)\,U(t)$$

(1.112)

with

$$U(t) \;=\; \exp\left[\int_0^t \mathrm{d}t'v^\mu(t')\partial_\mu\right]$$

(1.113)

being the shift operator. It obviously obeys the differential equation

$$\frac{\mathrm{d}}{\mathrm{d}t}U(t) \;=\; v^\mu(t)\,\partial_\mu\, U(t)\,.$$

(1.114)

Now since

$$U(t)\,[1-\mathrm{d}t\, V(x;t)]\,U^{-1}(t) \;=\; \left[1-\mathrm{d}t\, V\!\left(x+\int_0^t \mathrm{d}t'v(t');t\right)\right]\,,$$

(1.115)

the RHS of Eq. (1.111) can be written in the form

$$\prod_{t=0}^{\tau}\left[1-\mathrm{d}t\, v^\mu(t)\frac{\partial}{\partial x^\mu}\right][1-\mathrm{d}t\, V(x;t)]$$

$$=\; U^{-1}(\tau)\prod_{t=0}^{\tau}\left[1-\mathrm{d}t\, V\!\left(x+\int_0^t \mathrm{d}t'v(t');t\right)\right]$$

$$=\; U^{-1}(\tau)\,\exp\left[-\int_0^\tau \mathrm{d}t\, V\!\left(x+\int_0^t \mathrm{d}t'v(t');t\right)\right]\,,$$

(1.116)

which is completely disentangled.

The operator $U^{-1}(\tau)$ is now in the proper order to be applied to the variable y in the argument of the delta-function, which results in the shift

$$\delta^{(d)}(x-y) \;\Longrightarrow\; \delta^{(d)}\!\left(x+\int_0^\tau \mathrm{d}t\, v(t)-y\right).$$

(1.117)

This will be explained in more detail in the next paragraphs.

Passing to the variable (1.76), we get finally

$$\langle y \,|\, \mathrm{e}^{-\boldsymbol{H}\tau} \,|\, x \rangle \;\; = \;\; \int_{\substack{z_\mu(0)=x_\mu \\ z_\mu(\tau)=y_\mu}} \mathcal{D}z_\mu(t)\,\mathrm{e}^{-\int_0^\tau \mathrm{d}t\,\mathcal{L}(t)}, \tag{1.118}$$

where

$$\mathcal{L}(t) \;\; = \;\; \frac{m}{2}\dot{z}_\mu^2(t) + V(z(t)) \tag{1.119}$$

is the Lagrangian associated with the Hamiltonian $\boldsymbol{H}$. The unusual plus sign in this formula is due to the Euclidean-space formalism. It is clear from the derivation that Eq. (1.118) holds for time-dependent potentials as well.

Notice that the path integral in Eq. (1.118) is now over trajectories along which the particle propagates in the fixed proper time τ with no integration over τ.

A special comment about the operator $U^{-1}(\tau)$ in Eq. (1.116) is required. In the Schrödinger representation of quantum mechanics, one is interested in the matrix elements of the evolution operator between some vectors $\langle g|$ and $|f\rangle$ in the Hilbert space. According to Eq. (1.23), in the coordinate representation one has

$$\langle g \,|\, \mathrm{e}^{-\boldsymbol{H}\tau} \,|\, f \rangle \;\; = \;\; \int \mathrm{d}^d x \, g(x)\,\mathrm{e}^{-\boldsymbol{H}\tau} f(x)\,. \tag{1.120}$$

Integrating by parts, the operator $U^{-1}(\tau)$ can then be applied to $g(x)$ which results in the shift

$$g(x) \;\; \Longrightarrow \;\; U(\tau)\,g(x)\,U^{-1}(\tau) \;\; = \;\; g\!\left(x + \int_0^\tau \mathrm{d}t\, v(t)\right). \tag{1.121}$$

Passing to the variable (1.76), Eq. (1.120) becomes

$$\langle g \,|\, \mathrm{e}^{-\boldsymbol{H}\tau} \,|\, f \rangle \;\; = \;\; \int \mathcal{D}z_\mu(t)\,\mathrm{e}^{-\int_0^\tau \mathrm{d}t\,\mathcal{L}(t)} g(z(\tau))f(z(0))\,. \tag{1.122}$$

There are no restrictions on the initial and final points of the trajectories $z_\mu(t)$ in this formula.

Problem 1.10 Calculate the diagonal resolvent of the Schrödinger operator in the potential $V(x)$:

$$R_\omega(x, x; V) \;\; = \;\; \left\langle x \,\left|\, \frac{1}{-\mathcal{G}\partial^2 + \omega^2 + V} \,\right|\, x \right\rangle, \tag{1.123}$$

in the limit $\mathcal{G} \to 0$ for $d = 1$.

Solution Using the formula of the type (1.118), we represent $R_\omega(x, x; V)$ as the path integral

$$R_\omega(x, x; V) = \frac{1}{2} \int\limits_0^\infty d\tau\, e^{-\frac{1}{2}\tau\omega^2} \int\limits_{\substack{z_\mu(0)=x_\mu \\ z_\mu(\tau)=x_\mu}} \mathcal{D}z_\mu(t)\, e^{-\frac{1}{2\mathcal{G}} \int_0^\tau dt\, \dot{z}_\mu^2(t) - \int_0^\tau dt\, V(z(t))}.$$
(1.124)

As $\mathcal{G} \to 0$ this path integral is dominated by the t-independent saddle-point trajectory

$$z(t) = x,$$
(1.125)

which is associated with a particle standing at the point x. Substituting V at this saddle point, i.e. replacing $V(z(t))$ by $V(x)$, and calculating the Gaussian integral over quantum fluctuations around the trajectory (1.125) using Eqs. (1.88) and (1.90), one finds

$$R_\omega(x, x; V) = \frac{1}{2\sqrt{\omega^2 + V(x)}}$$
(1.126)

in $d = 1$.

Equation (1.126) can be alternatively derived by applying the Gel'fand–Dikii technique [GD75] which says that $R_\omega(x, x; V)$ obeys the third-order linear differential equation

$$\frac{1}{2} \left[\frac{\mathcal{G}}{2} \partial^3 - \partial V(x) - V(x)\partial \right] R_\omega(x, x; V) = \omega^2 \partial R_\omega(x, x; V).$$
(1.127)

$R_\omega(x, x; V)$ given by Eq. (1.126) obviously satisfies this equation as $\mathcal{G} \to 0$.

One more way to derive Eq. (1.126) is to perform a semiclassical Wantzel–Kramers–Brillouin (WKB) expansion of $R_\omega(x, y; V)$ in the parameter $\mathcal{G}$. This is explained in Chapter 7 of the book [LL74].

Problem 1.11 Derive Eq. (1.127).

Solution The resolvent

$$R_\omega(x, y; V) = \left\langle y \left| \frac{1}{-\mathcal{G}\partial^2 + \omega^2 + V} \right| x \right\rangle$$
(1.128)

obeys the equations

$$\left. \begin{aligned} \left[-\mathcal{G}\frac{\partial^2}{\partial x^2} + \omega^2 + V(x) \right] R_\omega(x, y; V) &= \delta^{(1)}(x - y), \\ \left[-\mathcal{G}\frac{\partial^2}{\partial y^2} + \omega^2 + V(y) \right] R_\omega(x, y; V) &= \delta^{(1)}(x - y). \end{aligned} \right\}$$
(1.129)

It can be expressed via the two solutions $f_\pm(x)$ of the homogeneous equation

$$\left[-\mathcal{G}\frac{\partial^2}{\partial x^2} + \omega^2 + V(x) \right] f_\pm(x) = 0,$$
(1.130)

where f_+ or f_- are regular at $+\infty$ or $-\infty$, respectively. Then the full solution is

$$R_\omega(x, y; V) = \frac{f_+(x)f_-(y)\theta(x-y) + f_-(x)f_+(y)\theta(y-x)}{\mathcal{G}\, W_\omega} \qquad (1.131)$$

with

$$W_\omega = f_+(x)f'_-(x) - f'_+(x)f_-(x) \qquad (1.132)$$

being the Wronskian of these solutions. Applying $\partial/\partial x$ to Eq. (1.132), it is easy to show that W_ω is an x-independent function of ω.

The simplest way to prove Eq. (1.127) is to differentiate

$$R_\omega(x, x; V) = \frac{f_+(x)f_-(x)}{\mathcal{G}\, W_\omega} \qquad (1.133)$$

using Eq. (1.130), in order to verify that it satisfies the nonlinear differential equation

$$-2\mathcal{G}R_\omega R''_\omega + \mathcal{G}\left(R'_\omega\right)^2 + 4\left(\omega^2 + V\right)R_\omega^2 = 1. \qquad (1.134)$$

One more differentiation of Eq. (1.134) with respect to x results in Eq. (1.127).

It is worth noting that Eq. (1.134) is very convenient for calculating the semi-classical expansion of $R_\omega(x, x; V)$ in $\mathcal{G}$. In particular, the leading order (1.126) is obvious.

Remark on parametric invariant representation

The Green function $G(x - y)$ can alternatively be calculated from the parametric invariant representation

$$G(x-y) \propto \int_{\substack{z_\mu(\sigma_0)=x_\mu \\ z_\mu(\sigma_1)=y_\mu}} \mathcal{D}z_\mu(\sigma)\, e^{-m_0 \int_{\sigma_0}^{\sigma_1} d\sigma \sqrt{\dot{z}^2(\sigma)}} \qquad (1.135)$$

as prescribed by Eqs. (1.105) and (1.106). In contrast to (1.70), this path integral is not easy to calculate. The integration over $\mathcal{D}z_\mu(\sigma)$ in Eq. (1.135) involves integration over the reparametrization group, which gives the proper group-volume factor since the exponent is parametric invariant. Eq. (1.70) is recovered after fixing parametrization to be proper time. How this calculation can be performed is explained in Chapter 9 of the book by Polyakov [Pol87].

If one makes a naive discretization of the parameter σ using equidistant intervals, the exponent in Eq. (1.135) is highly nonlinear in the variables z_i, leading to complicated integrals. In contrast, the discretization (1.91) of the path integral in Eq. (1.70), where the parametric invariance is fixed, results in a Gaussian integral which is easily calculable.

Problem 1.12 Calculate the path integral in Eq. (1.135), discretizing the measure by

$$\mathcal{D}z_\mu \quad \rightarrow \quad \sum_{M=1}^{\infty} \prod_{i=1}^{M} \frac{\mathrm{d}^d z_i}{(2\pi\epsilon)^{d/2}} \tag{1.136}$$

and applying the central limit theorem as $M \to \infty$.

Solution By making the discretization, we represent the RHS of Eq. (1.135) as the probability integral

$$G_\epsilon(x - y) = \frac{1}{(2\pi\epsilon)^{d/2}} \sum_{M=1}^{\infty} \int \prod_{i=1}^{M-1} \frac{\mathrm{d}^d z_i}{(2\pi\epsilon)^{d/2}}$$
$$\times \, \rho(x \to z_1) \, \rho(z_1 \to z_2) \cdots \rho(z_{M-1} \to y) \tag{1.137}$$

with

$$\rho(z_{i-1} \to z_i) = \mathrm{e}^{-m_0 |z_i - z_{i-1}|} \tag{1.138}$$

being an (unnormalized) probability function and ϵ is a parameter with the dimension of $[\text{length}]^2$. The probability interpretation of each term in the sum is standard for random walk models, and means, as usual, that a particle propagates via independent intermediate steps. The discretization of the measure given by Eq. (1.137) looks like that in Eq. (1.66), but the summation over M is now added.

Since the integral in Eq. (1.137) is a convolution, the central limit theorem states that

$$G_\epsilon(x - y) = \frac{1}{(2\pi\epsilon)^{d/2}} \sum_{M} \left[\frac{c_0}{(2\pi\epsilon m_0^2)^{d/2}} \right]^M$$
$$\times \, \frac{1}{(2\pi\sigma^2 M)^{d/2}} \, \mathrm{e}^{-m_0^2 (x-y)^2 / (2\sigma^2 M) + \mathcal{O}(M^{-2})} \tag{1.139}$$

at large M, where c_0 and σ^2 are the zeroth and (normalized) second moments of ρ:

$$\left. \begin{aligned} c_0 &= \int \mathrm{d}^d x \, \mathrm{e}^{-|x|} = 2\pi^{d/2} \frac{\Gamma(d)}{\Gamma(d/2)}, \\ \sigma^2 &= \frac{1}{c_0} \int \mathrm{d}^d x \, x^2 \, \mathrm{e}^{-|x|} = d(d+1). \end{aligned} \right\} \tag{1.140}$$

The sum over M in Eq. (1.139) is convergent for

$$m_0 > m_\mathrm{c} = \frac{c_0^{1/d}}{\sqrt{2\pi\epsilon}} \tag{1.141}$$

and is divergent for $m_0 < m_\mathrm{c}$. Choosing $m_0 > m_\mathrm{c}$, but $m_0^2 - m_\mathrm{c}^2 \sim 1$ in the limit $\varepsilon \to 0$, the sum over M will be convergent, while dominated by terms with large

$$M \sim m_\mathrm{c}^2 \sim \frac{1}{\epsilon}. \tag{1.142}$$

This is easily seen by rewriting Eq. (1.139) as

$$G_\epsilon(x - y) = \sum_M \left(\frac{m_c^2}{2\pi\sigma^2 M}\right)^{d/2}$$

$$\times e^{-dM \ln(m_0/m_c) - m_0^2(x-y)^2/(2\sigma^2 M) + \mathcal{O}(M^{-2})} . \qquad (1.143)$$

Each term with $M \sim m_c^2$ contributes $\mathcal{O}(1)$ to the sum, so that

$$G_\epsilon(x - y) \sim m_c^2 . \qquad (1.144)$$

This justifies the using of the central limit theorem in this case. The typical distances between the z_i, which are essential in the integral on the RHS of Eq. (1.137), are

$$|z_i - z_{i-1}| \sim \frac{1}{m_0} \sim \sqrt{\epsilon} \qquad (1.145)$$

as in Eq. (1.93). The relation (1.142) between the essential values of M and ϵ is also similar to what we had in Sect. 1.4.

The sum over M in Eq. (1.143) can be replaced by a continuous integral over the variable

$$\tau = \frac{\sigma^2 M}{m_c^2} , \qquad (1.146)$$

which is $\mathcal{O}(1)$ for $M \sim m_c^2$. Also introducing the variable m by

$$m^2 \equiv \frac{d}{\sigma^2}\left(m_0^2 - m_c^2\right) > 0 , \qquad (1.147)$$

we rewrite Eq. (1.143) as

$$G_\epsilon(x - y) \overset{\epsilon \to 0}{\longrightarrow} \frac{m_c^2}{\sigma^2} \int_0^\infty d\tau \frac{1}{(2\pi\tau)^{d/2}} e^{-\frac{1}{2}m^2\tau - (x-y)^2/2\tau} , \qquad (1.148)$$

the RHS of which is proportional to that in Eq. (1.89) for the Euclidean propagator.

Remark on discretized path-ordered exponential

As is discussed in Sect. 1.3, the order of operators A_i and B_i with the same index i is not essential in the path-ordered exponential (1.52) as $M \to \infty$. If Eq. (1.52) is promoted to be valid at finite M (or at least to the order of $\mathcal{O}(M^{-1})$), this specifies the commutator of A_i and B_i. Analogously, a discretization of Eq. (1.118) specifies in which order the product of $x_i p_i$ in the classical theory should be substituted by the operators x_i and p_i in the operator formalism. For details see the books by Berezin [Ber86] (Chapter 1 of Part II) and Sakita [Sak85] (Chapter 6).

1.7 Propagators in external field

Let us now consider a (quantum) particle in a classical electromagnetic field. The standard way of introducing an external electromagnetic field is to substitute the (operator of the) four-momentum p^μ by

$$p^\mu \longrightarrow p^\mu - eA^\mu(x). \tag{1.149}$$

Recalling the definition (1.26) of the Euclidean four-momentum, ∂_μ needs to be replaced by the covariant derivative

$$\partial_\mu \longrightarrow \nabla_\mu = \partial_\mu - ieA_\mu(x). \tag{1.150}$$

Inverting the operator ∇_μ^2 using the disentangling procedure, one finds

$$
\begin{aligned}
G(x, y; A) \\
\equiv \left\langle y \left| \frac{1}{-\nabla_\mu^2 + m^2} \right| x \right\rangle \\
= \frac{1}{2} \int_0^\infty d\tau\, e^{-\frac{1}{2}\tau m^2} \int_{\substack{z_\mu(0)=x_\mu \\ z_\mu(\tau)=y_\mu}} \mathcal{D}z_\mu(t)\, e^{-\frac{1}{2}\int_0^\tau dt\, \dot{z}_\mu^2(t) + ie\int_0^\tau dt\, \dot{z}^\mu(t)A_\mu(z(t))}.
\end{aligned}
\tag{1.151}
$$

Note that the exponent is just the classical (Euclidean) action of a particle in an external electromagnetic field. Therefore, this expression is again of the type in Eq. (1.102).

The path-integral representation (1.151) for the propagator of a scalar particle in an external electromagnetic field is due to Feynman [Fey50] (Appendix A).

Problem 1.13 Derive Eq. (1.151) using Eq. (1.71) with $D_\mu = -\nabla_\mu$.

Solution The calculation is analogous to that of Problem 1.9 on p. 22. We have

$$D_\mu(t) = -\nabla_\mu(t) \equiv -\partial_\mu(t) + ieA_\mu(x; t) \tag{1.152}$$

so that explicitly

$$
\begin{aligned}
\boldsymbol{P}e^{-\int_0^\tau dt\, v^\mu(t)\nabla_\mu(t)} &= \prod_{t=0}^{\tau} \left[1 - dt\, v^\mu(t)\frac{\partial}{\partial x^\mu} + ie\, dt\, v^\mu(t)A_\mu(x; t) \right] \\
&= \prod_{t=0}^{\tau} \left[1 - dt\, v^\mu(t)\frac{\partial}{\partial x^\mu} \right] \left[1 + ie\, dt\, v^\mu(t)A_\mu(x; t) \right].
\end{aligned}
\tag{1.153}
$$

This looks exactly like the expression (1.111) with

$$V(x; t) = -ie\, v^\mu(t)A_\mu(x; t). \tag{1.154}$$

Substituting this potential into Eq. (1.118) and remembering the additional integration over τ, we obtain the path-integral representation (1.151).

We can alternatively rewrite Eq. (1.151) in the spirit of Sect. 1.6 as

$$G(x, y; A) \;=\; {\sum_{\Gamma_{yx}}}' \, \mathrm{e}^{\mathrm{i}e \int_{\Gamma_{yx}} \mathrm{d}z^\mu A_\mu(z)}, \tag{1.155}$$

where we have included the free action in the definition of the sum over trajectories:

$${\sum_{\Gamma_{yx}}}' \;\overset{\text{def}}{=}\; \sum_{\Gamma_{yx}} \mathrm{e}^{-S_{\text{free}}[\Gamma_{yx}]}, \tag{1.156}$$

and represented the (parametric invariant) integral over $\mathrm{d}t$ as the contour integral over

$$\mathrm{d}z^\mu \;=\; \mathrm{d}t \, \dot{z}^\mu(t) \tag{1.157}$$

along the trajectory Γ_{yx}.

The meaning of Eq. (1.155) is that the transition amplitude of a quantum particle in a classical electromagnetic field is the sum over paths of the Abelian *phase factor*

$$U[\Gamma_{yx}] \;=\; \mathrm{e}^{\mathrm{i}e \int_{\Gamma_{yx}} \mathrm{d}z^\mu A_\mu(z)}. \tag{1.158}$$

Under the gauge transformation

$$A_\mu(z) \;\overset{\text{g.t.}}{\longrightarrow}\; A_\mu(z) + \frac{1}{e} \partial_\mu \alpha(z), \tag{1.159}$$

the Abelian phase factor transforms as

$$U[\Gamma_{yx}] \;\overset{\text{g.t.}}{\longrightarrow}\; \mathrm{e}^{\mathrm{i}\alpha(y)} \, U[\Gamma_{yx}] \, \mathrm{e}^{-\mathrm{i}\alpha(x)}. \tag{1.160}$$

Noting that a wave function at the point x is transformed under the gauge transformation (1.159) as

$$\varphi(x) \;\overset{\text{g.t.}}{\longrightarrow}\; \mathrm{e}^{\mathrm{i}\alpha(x)} \, \varphi(x), \tag{1.161}$$

we conclude that the phase factor is transformed as the product $\varphi(y)\varphi^\dagger(x)$:

$$U[\Gamma_{yx}] \;\overset{\text{g.t.}}{\sim}\; \text{``}\varphi(y)\,\varphi^\dagger(x)\text{''}, \tag{1.162}$$

where "$\cdots$" means literally "transforms as $\ldots$".

As a consequence of Eqs. (1.160) and (1.161), a wave function at the point x transforms like one at the point y after multiplication by the phase factor:

$$U[\Gamma_{yx}]\,\varphi(x) \;\overset{\text{g.t.}}{\sim}\; \text{``}\varphi(y)\text{''}, \tag{1.163}$$

and analogously

$$\varphi^\dagger(y) \, U[\Gamma_{yx}] \overset{\text{g.t.}}{\sim} \text{``}\varphi^\dagger(x)\text{''}. \qquad (1.164)$$

Equations (1.163) and (1.164) show that the phase factor plays the role of a *parallel transporter* in an electromagnetic field, and that in order to compare phases of a wave function at points x and y, one should first make a parallel transport along some contour Γ_{yx}. The result is, generally speaking, Γ-dependent except when $A_\mu(z)$ is a pure gauge. The sufficient and necessary condition for the phase factor to be Γ-independent is the vanishing of the field strength, $F_{\mu\nu}(z)$, which is a consequence of the Stokes theorem when applied to the Abelian phase factor.[*]

Below we shall deal with determinants of various operators. Analogous to Eq. (1.151), one finds

$$\ln \det \nabla_\mu^2 \;=\; \frac{1}{2} \int_0^\infty \frac{d\tau}{\tau} \, \text{Tr} \, e^{\frac{1}{2}\tau \nabla_\mu^2}$$

$$=\; \frac{1}{2} \int_0^\infty \frac{d\tau}{\tau} \int_{z_\mu(0)=z_\mu(\tau)} \mathcal{D}z_\mu(t) \, e^{-\frac{1}{2}\int_0^\tau dt \, \dot{z}_\mu^2(t) + ie \oint_\Gamma dz^\mu A_\mu(z)}, \qquad (1.165)$$

where the path integral goes over trajectories which are closed owing to the periodic boundary condition $z_\mu(0) = z_\mu(\tau)$. To derive Eq. (1.165), we have used the formula

$$\ln \det \boldsymbol{D} \;=\; \text{Tr} \ln \boldsymbol{D}, \qquad (1.166)$$

which relates the determinant and the trace of a Hermitian operator (or a matrix) $\boldsymbol{D}$.

Problem 1.14 Prove Eq. (1.166).

Solution Let $\boldsymbol{D}$ be positive definite. We first reduce $\boldsymbol{D}$ to a diagonal form by a unitary transformation and denote (positive) eigenvalues as D_i. Then Eq. (1.166) can be written as

$$\ln \prod_i D_i \;=\; \sum_i \ln D_i \qquad (1.167)$$

which is obviously true.

---

[*] Strictly speaking, this statement holds for the case when Γ can be chosen everywhere in space-time, i.e. which is simply connected. However, there exist situations when Γ cannot penetrate into some regions of space as for the Aharonov–Bohm experiment which is discussed below in Sect. 5.4.

The phase factor for a closed contour Γ enters Eq. (1.165). It describes parallel transportation along a closed loop, and is gauge invariant as a consequence of Eq. (1.160):

$$e^{ie \oint_\Gamma dz^\mu A_\mu(z)} \xrightarrow{\text{g.t.}} e^{ie \oint_\Gamma dz^\mu A_\mu(z)}. \qquad (1.168)$$

This quantity, which plays a crucial role in modern formulations of gauge theories, will be discussed in more detail in Chapter 5.

Problem 1.15 Show how the path-integral representation (1.151) recovers for $G(x, y; A)$ the diagrammatic expansion of propagator in an external field A_μ:

$$G(x, y; A) \quad = \quad \underset{x \qquad y}{\underline{\qquad}} \; + \; \underset{x \qquad y}{\wr} \; + \; \underset{x \qquad y}{\wr\wr} \; + \; \underset{x \qquad y}{\curlyvee} \; + \cdots.$$

$$(1.169)$$

Solution Let us expand the phase factor in Eq. (1.155) in e. The linear-in-e term can be transformed using the formula

$$\sum_{\Gamma_{yx}}' \int_{\Gamma_{yx}} d\xi_\mu\, \delta^{(d)}(\xi - z) \cdots \quad = \quad \sum_{\Gamma_{yz}}' \overset{\leftrightarrow}{\frac{\partial}{\partial z_\mu}} \sum_{\Gamma_{zx}}' \cdots, \qquad (1.170)$$

where

$$\overset{\leftrightarrow}{\partial}_\mu \quad = \quad -\overset{\leftarrow}{\partial}_\mu + \overset{\rightarrow}{\partial}_\mu, \qquad (1.171)$$

to reproduce the second diagram on the RHS of Eq. (1.169). Equation (1.170) can be proven by varying both sides of Eq. (1.155) with respect to $A_\mu(z)$.

Equation (1.170) can be rewritten using the formula

$$\partial_\mu^x \sum_{\Gamma_{zx}}' \cdots \quad = \quad -\sum_{\Gamma_{zx}}' v_\mu(x) \cdots, \qquad (1.172)$$

where $v_\mu(x) = \dot{\xi}_\mu(0)$ is the velocity at the point x of the trajectory Γ. Using Eq. (1.172), we find

$$\sum_{\Gamma_{yx}}' \int_{\Gamma_{yx}} d\xi_\mu\, \delta^{(d)}(\xi - z) \cdots \quad = \quad \sum_{\Gamma_{yz}}' v_\mu(z) \sum_{\Gamma_{zx}}' \cdots + \sum_{\Gamma_{yz}}' \sum_{\Gamma_{zx}}' v_\mu(z) \cdots.$$

$$(1.173)$$

Equation (1.172) can be proven by shifting variable in the path integral, while Eq. (1.173) holds, strictly speaking, only if an integrand (denoted by $\cdots$) does not include velocities. Otherwise, additional contact terms might appear.

They can be obtained by noting that the velocity $v_\nu(x)$ corresponds to the covariant derivative (1.150), where $A_\nu(x)$ is also to be varied. Doing so, we arrive at

$$\sum_{\Gamma_{yx}}' v_\nu(x) \int_{\Gamma_{yx}} \mathrm{d}\xi_\mu \, \delta^{(d)}(\xi - z) \cdots$$

$$= \sum_{\Gamma_{yz}}' v_\nu(x) \, v_\mu(z) \sum_{\Gamma_{zx}}' \cdots + \sum_{\Gamma_{yz}}' v_\nu(x) \sum_{\Gamma_{zx}}' v_\mu(z) \cdots + \delta_{\mu\nu} \sum_{\Gamma_{yx}}' \cdots . \tag{1.174}$$

For the case of a more complicated integrand, each velocity produces the same type of contact terms since the variation $\delta/\delta A_\mu(z)$ acts linearly. This reproduces the contact terms as in the fourth term on the RHS of Eq. (1.169).

This Problem is based on Appendix A of the paper [MM81].

Problem 1.16 Establish the equivalence of the path-integral representation (1.165) of $\ln \det \nabla_\mu^2$ and the sum of one-loop diagrams in an external field A_μ:

$$\tag{1.175}$$

Solution The derivation is the same as in the previous Problem. The combinatoric factor of $1/2$ in the third diagram on the RHS of Eq. (1.175) is associated with a symmetry factor.

Remark on analogy with statistical mechanics

A formula of the type (1.165), which represents the trace of an operator via a path integral over closed trajectories, is known as the Feynman–Kac formula. The terminology comes from statistical mechanics where the partition function (or equivalently the statistical sum) is given by the Boltzmann formula

$$Z = \mathrm{Tr}\, e^{-\beta \boldsymbol{H}} \tag{1.176}$$

(with β being the inverse temperature) whose path-integral representation is of the type given in Eq. (1.165). The expression which is integrated on the RHS of Eq. (1.165) over $\mathrm{d}\tau/\tau$ is associated, in statistical-mechanical language, with the partition function of a closed elastic string, the energy of which is proportional to its length, that interacts with an external electromagnetic field. This shows an analogy between Euclidean quantum mechanics in d dimensions and statistical mechanics in d (spatial) and one (temporal) dimensions whose time-dependence disappears, since nothing depends on time at equilibrium. We shall explain this analogy in more detail in Part 2 (Chapter 9) when discussing quantum field theory at finite temperature.

2
Second quantization

In the previous chapter we considered first quantization of particles, where the operators of coordinate and momentum, x and p respectively, are represented in the coordinate space by

$$x = x, \qquad p = -i\frac{\partial}{\partial x} \qquad \boxed{= \text{ first quantization}}. \qquad (2.1)$$

In the language of path integrals, first quantization is associated with integrals over trajectories in the coordinate space.

While propagators can be easily represented as path integrals, it is very difficult to describe, in this language, a (nongeometric) self-interaction of a particle, since this would correspond to extra weights for self-intersecting paths. For the free case (or a particle in an external gauge field) there are no such extra weights, and the transition amplitude is completely described by the classical action of the particle.

In the operator formalism, self-interactions of a particle are described using second quantization – this is where the transition from quantum mechanics to quantum field theory begins. Second quantization is a quantization of fields, and is associated with path integrals over fields, which is the subject of this chapter. We demonstrate how perturbation theory and the Schwinger–Dyson equations can be derived using path integrals.

2.1 Integration over fields

Let us define the following (Euclidean) *partition function*:

$$Z = \int \mathcal{D}\varphi(x)\,\mathrm{e}^{-S}, \qquad (2.2)$$

where the action in the exponent is given for the free case by

$$S_{\text{free}}[\varphi] \;=\; \frac{1}{2} \int \mathrm{d}^d x \left((\partial_\mu \varphi)^2 + m^2 \varphi^2 \right). \tag{2.3}$$

The measure $\mathcal{D}\varphi(x)$ is defined analogously to Eq. (1.66):

$$\int \mathcal{D}\varphi(x) \cdots \;=\; \prod_x \int_{-\infty}^{+\infty} \mathrm{d}\varphi(x) \cdots, \tag{2.4}$$

where the product runs over all space points x and the integral over $\mathrm{d}\varphi$ is the Lebesgue one.

The propagator is given by the average

$$G(x, y) \;=\; \langle\, \varphi(x)\,\varphi(y) \,\rangle, \tag{2.5}$$

where a generic average is defined by the formula

$$\langle\, F[\varphi] \,\rangle \;=\; Z^{-1} \int \mathcal{D}\varphi(x)\, \mathrm{e}^{-S[\varphi]} F[\varphi]. \tag{2.6}$$

The notation is obvious since on the RHS of Eq. (2.6) we average over all field configurations with the same weight as in the partition function (2.2). The normalization factor of Z^{-1} provides the necessary property of an average

$$\langle 1 \rangle \;=\; 1. \tag{2.7}$$

Since the free action (2.3) is Gaussian, the average (2.5) equals

$$G(x - y) \;=\; \left\langle y \left| \frac{1}{-\partial^2 + m^2} \right| x \right\rangle \tag{2.8}$$

which is identical to (1.61). Therefore, we have obtained the same propagator (1.89) as in the previous chapter.

Problem 2.1 By discretizing the (Euclidean) space, derive

$$Z^{-1} \int \prod_x \mathrm{d}\varphi_x \exp\left(-\tfrac{1}{2} \sum_{x,y} \varphi_x D_{xy} \varphi_y \right) \varphi_x \varphi_y \;=\; D_{xy}^{-1}. \tag{2.9}$$

Solution The Gaussian integral can be calculated using the change of variable

$$\varphi_x \;\rightarrow\; \varphi_x' \;=\; \sum_y \left(D^{-1/2} \right)_{xy} \varphi_y \tag{2.10}$$

which results in Eq. (2.9).

Note that the integrals over $\varphi(x)$ are convergent in Euclidean space. If a discretization of space is introduced, the path integrals in Eqs. (2.2) or (2.6) are defined rigorously.

Remark on Minkowski-space formulation

In Minkowski space, perturbation theory is well-defined since the Gaussian path integral, which determines the propagator

$$\int \mathcal{D}\varphi \, e^{iS} = \int \mathcal{D}\varphi \, e^{i\int d^d x \, \varphi D\varphi}, \qquad (2.11)$$

equals

$$\langle \varphi_x \varphi_y \rangle = \left\langle y \left| \frac{1}{iD} \right| x \right\rangle, \qquad (2.12)$$

where D is a proper operator in Minkowski space.

It cannot be said *a priori* whether a nonperturbative formulation of a given (interacting) theory via the path integral in Minkowski space exists since the weight factor is complex and the integral may be divergent.

2.2 Grassmann variables

Path integrals over anticommuting Grassmann variables are used to describe fermionic systems.

The Grassmann variables ψ_x and $\bar{\psi}_y$ obey the anticommutation relations

$$\{\psi_y, \psi_x\} = 0, \qquad \{\bar{\psi}_y, \bar{\psi}_x\} = 0, \qquad \{\bar{\psi}_y, \psi_x\} = 0. \qquad (2.13)$$

Consequently, the square of a Grassmann variable vanishes

$$\psi_x^2 = 0 = \bar{\psi}_x^2. \qquad (2.14)$$

The path integral over the Fermi fields equals

$$\int \mathcal{D}\bar{\psi} \, \mathcal{D}\psi \, e^{-\int d^d x \, \bar{\psi} D\psi} = \det D \qquad (2.15)$$

while an analogous integral over the Bose fields is

$$\int \mathcal{D}\varphi^\dagger \, \mathcal{D}\varphi \, e^{-\int d^d x \, \varphi^\dagger D\varphi} = (\det D)^{-1}. \qquad (2.16)$$

Problem 2.2 Define integrals over Grassmann variables.

Solution Assuming that ψ and $\bar{\psi}$ belong to the same Grassmann algebra, the Berezin integrals are defined by

$$\left.\begin{aligned}
\int d\psi_x &= 0 = \int d\bar{\psi}_x, \\
\int d\psi_x \, \psi_y &= \delta_{xy} = \int d\bar{\psi}_x \, \bar{\psi}_y, \\
\int d\psi_x \, \bar{\psi}_y &= 0 = \int d\bar{\psi}_x \, \psi_y.
\end{aligned}\right\} \qquad (2.17)$$

The simplest interesting integral is

$$\int \mathrm{d}\bar{\psi}_x \, \mathrm{d}\psi_x \, \mathrm{e}^{-\bar{\psi}_x \psi_x} \;\; = \;\; 1 . \tag{2.18}$$

Equation (2.15) can now be easily derived by representing $\langle y \,|\, D \,|\, x \rangle$ in the diagonal form, expanding the exponential up to a term which is linear in all the Grassmann variables and calculating the integrals of this term according to Eq. (2.17). See more details in the book by Berezin [Ber86] (§3 of Part I).

The average over the Fermi fields, defined with the same weight as in Eq. (2.15), equals

$$\langle \psi(x) \, \bar{\psi}(y) \rangle \;\; = \;\; \langle y \,|\, D^{-1} \,|\, x \rangle \tag{2.19}$$

which is the same as for bosons.

Note that the fermion partition function (2.15) can be rewritten according to Eq. (1.166) as

$$\det D \;\; = \;\; \mathrm{e}^{\mathrm{Tr}\ln D} . \tag{2.20}$$

The analogous formula for bosons (2.16) is rewritten as

$$(\det D)^{-1} \;\; = \;\; \mathrm{e}^{-\mathrm{Tr}\ln D} . \tag{2.21}$$

The relative difference of sign between the exponents on the RHS of Eqs. (2.20) and (2.21) is a famous minus sign that emerges for closed fermionic loops which contribute to the (logarithm of the) partition function.

2.3 Perturbation theory

The cubic self-interaction of the scalar field is described by the action

$$S[\varphi] \;\; = \;\; \int \mathrm{d}^d x \left(\frac{1}{2} (\partial_\mu \varphi)^2 + \frac{1}{2} m^2 \varphi^2 + \frac{\lambda}{3!} \varphi^3 \right) , \tag{2.22}$$

where λ is the coupling constant.

To construct perturbation theory, we expand the exponential in λ and calculate the Gaussian averages with the free action (2.3).

To order λ^2, the expansion is

$$\begin{aligned}
\langle \varphi(x) \, \varphi(y) \rangle \;\; = \;\; & \langle \varphi(x) \, \varphi(y) \rangle_{\mathrm{free}} \\
& \times \left(1 - \left\langle \frac{\lambda}{3!} \int \mathrm{d}^d x_1 \, \varphi^3(x_1) \, \frac{\lambda}{3!} \int \mathrm{d}^d x_2 \, \varphi^3(x_2) \right\rangle_{\mathrm{free}} \right) \\
& + \left\langle \varphi(x) \, \frac{\lambda}{3!} \int \mathrm{d}^d x_1 \, \varphi^3(x_1) \, \frac{\lambda}{3!} \int \mathrm{d}^d x_2 \, \varphi^3(x_2) \, \varphi(y) \right\rangle_{\mathrm{free}} \\
& + \cdots .
\end{aligned} \tag{2.23}$$

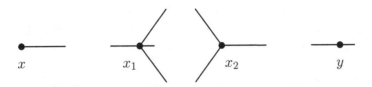

Fig. 2.1.   Diagrammatic representation of the term in the third line of Eq. (2.23).

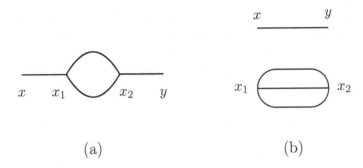

(a)   (b)

Fig. 2.2.   Some of the Feynman diagrams which appear from (2.23) after the Wick pairing.

The term which is linear in λ (as well as the Gaussian average of any odd power of φ) vanishes owing to the reflection symmetry $\varphi \to -\varphi$ of the Gaussian action. The term displayed in the third line of Eq. (2.23) is depicted graphically in Fig. 2.1.

Further calculation of the RHS of Eq. (2.23) is based on the free average

$$\langle \varphi(x_i)\, \varphi(x_j) \rangle_{\text{free}} \;=\; G(x_i - x_j) \tag{2.24}$$

and the rules of Wick pairing of Gaussian averages, which allow us to represent the average of a product as the sum of all possible products of pair averages. Some of the diagrams which emerge after the Wick contraction are depicted in Fig. 2.2. These diagrams are called *Feynman diagrams*.

The diagram shown in Fig. 2.2b is disconnected. Its disconnected part with two loops cancels with the same contribution from Z^{-1} (which yields the factor in the first term on the RHS of Eq. (2.23)). It is a general property that only connected diagrams are left in $\langle \varphi(x_i)\, \varphi(x_j) \rangle$.

Let us note finally that the combinatoric factor of $1/2$ is reproduced correctly for the diagram of Fig. 2.2a. For an arbitrary diagram, this procedure of pairing reproduces the usual combinatoric factor, which is equal to the number of automorphisms of the diagram (i.e. the symmetries of a given graph).

2.4 Schwinger–Dyson equations

Feynman diagrams can be derived alternatively by iterating the coupling constant of the set of Schwinger–Dyson equations which is a quantum analog of the classical equation of motion.

To derive the Schwinger–Dyson equations, let us utilize the fact that the measure (2.4) is invariant under an arbitrary shift of the field

$$\varphi(x) \quad \rightarrow \quad \varphi(x) + \delta\varphi(x). \tag{2.25}$$

This invariance is obvious since the functional integration goes over all the fields, while the shift (2.25) is just a transformation from one field configuration to another.

Since the measure is invariant, the path integral in the average (2.6) does not change under the shift (2.25):

$$\int d^d x \, \delta\varphi(x) \int D\varphi \, e^{-S[\varphi]} \left[-\frac{\delta S[\varphi]}{\delta\varphi(x)} F[\varphi] + \frac{\delta F[\varphi]}{\delta\varphi(x)} \right] = 0. \tag{2.26}$$

Since $\delta\varphi(x)$ is arbitrary, Eq. (2.26) results in the following *quantum equation of motion*

$$\frac{\delta S[\varphi]}{\delta\varphi(x)} \overset{\text{w.s.}}{=} \hbar \frac{\delta}{\delta\varphi(x)}, \tag{2.27}$$

where we have written explicitly the dependence on Planck's constant $\hbar$. It appears this way since the action S is divided by $\hbar$ in Eq. (2.2) when $\hbar$ is restored.

We have put the symbol "w.s." on the top of the equality sign in Eq. (2.27) to emphasize that it is to be understood in the weak sense, i.e. it is valid under averaging when applied to a functional $F[\varphi]$. In other words, the variation of the action on the LHS of Eq. (2.27) can always be substituted by the variational derivative on the RHS when integrated over fields with the same weight as in Eq. (2.6). Therefore, one arrives at the following functional equation:

$$\left\langle \frac{\delta S[\varphi]}{\delta\varphi(x)} F[\varphi] \right\rangle = \hbar \left\langle \frac{\delta F[\varphi]}{\delta\varphi(x)} \right\rangle. \tag{2.28}$$

This equation is quite similar to that which Schwinger considered within the framework of his variational technique.

2.5 Commutator terms

In order to show how Eq. (2.28) reproduces Eq. (1.34) for the free propagator, let us choose

$$F[\varphi] = \varphi(y). \tag{2.29}$$

Substituting into Eq. (2.28) and calculating the variational derivative, one obtains

$$\left(-\partial^2 + m^2\right) \langle \varphi(x)\,\varphi(y)\rangle \;=\; \hbar \left\langle \frac{\delta\varphi(y)}{\delta\varphi(x)} \right\rangle \;=\; \hbar\,\delta^{(d)}(x-y)\,, \qquad (2.30)$$

which coincides with Eq. (1.34).

The LHS of Eq. (2.30) emerges from the variation of the free classical action (2.3)

$$\frac{\delta S_{\text{free}}}{\delta\varphi(x)} \;=\; \left(-\partial^2 + m^2\right)\varphi(x) \qquad (2.31)$$

while the RHS, which results from the variational derivative, emerges in the operator formalism from the canonical commutation relations

$$\delta(x_0 - y_0)\left[\varphi(x_0,\vec{x}),\dot{\varphi}(y_0,\vec{y})\right] \;=\; i\,\delta^{(d)}(x-y) \qquad (2.32)$$

as is explained in Sect. 1.1.

For this reason, the RHS of Eq. (2.30) and, more generally, the RHS of Eq. (2.28) are called *commutator terms*. The variational derivative on the RHS of Eq. (2.27) plays the role of the conjugate momentum in the operator formalism. The calculation of this variational derivative in Euclidean space is equivalent to differentiating the T-product and using canonical commutation relations in Minkowski space.

When Planck's constant vanishes, $\hbar \to 0$, the RHS of Eq. (2.27) (or Eq. (2.28)) vanishes. Therefore it reduces to the classical equation of motion for the field φ:

$$\frac{\delta S[\varphi]}{\delta\varphi(x)} \;=\; 0\,. \qquad (2.33)$$

This implies that the path integral over fields has a saddle point as $\hbar \to 0$ which is given by Eq. (2.33).

Another lesson we have learned is that the average (2.5), which is defined via the Euclidean path integral, is associated with the Wick-rotated T-product. We have already seen this property in the previous chapter in the language of first quantization. More generally, the Euclidean average (2.6) is associated with the vacuum expectation value of $\langle 0\,|\,\boldsymbol{T}F[\varphi]\,|\,0\rangle$ in Minkowski space.

2.6 Schwinger–Dyson equations (continued)

The set of the Schwinger–Dyson equations for an interacting theory can be derived analogously to the free case.

Let us consider the cubic interaction which is described by the action (2.22). Choosing again $F[\varphi]$ to be given by Eq. (2.29) and calculating the variation of the action (2.22), one obtains

$$\left(-\partial^2 + m^2\right) \langle \varphi(x)\, \varphi(y) \rangle + \frac{\lambda}{2} \langle \varphi^2(x)\, \varphi(y) \rangle = \delta^{(d)}(x-y). \qquad (2.34)$$

Problem 2.3 Rederive Eq. (2.34) by analyzing Feynman diagrams.

Solution Let us introduce the Fourier-transformed two- and three-point Green functions

$$G(p) = \int d^d x\, e^{-ipx} \langle \varphi(x)\, \varphi(0) \rangle, \qquad (2.35)$$

and

$$G_3(p, q, -p-q) = \int d^d x\, d^d y\, e^{-ipx-iqy} \langle \varphi(x)\, \varphi(y)\, \varphi(0) \rangle. \qquad (2.36)$$

Let us also denote the free momentum-space propagator as

$$G_0(p) = \frac{1}{p^2 + m^2}. \qquad (2.37)$$

The perturbative expansion of G_3 starts from

$$G_3(p, q, -p-q) = -\lambda\, G_0(p)\, G_0(q)\, G_0(p+q) + \cdots. \qquad (2.38)$$

It is standard to truncate three external legs, introducing the vertex function

$$\Gamma(p, q, -p-q) = G_3(p, q, -p-q)\, G^{-1}(p)\, G^{-1}(q)\, G^{-1}(p+q) \qquad (2.39)$$

with a perturbative expansion which starts from λ:

$$\Gamma(p, q, -p-q) = -\lambda - \lambda^3 \int \frac{d^d k}{(2\pi)^d} G_0(k-p)\, G_0(k)\, G_0(k+q) + \cdots. \qquad (2.40)$$

This expansion can be represented diagrammatically as

$$(2.41)$$

where the filled circle on the LHS represents the exact vertex and the thin lines are associated with the bare propagator (2.37).

An analogous expansion of the propagator is

$$\text{(2.42)}$$

where the bold line represents the exact propagator. It is commonly rewritten as an equation for the self-energy $G_0^{-1}(p) - G^{-1}(p)$, which involves only the one-particle irreducible (1PI) diagrams. The last diagram shown on the RHS of Eq. (2.42) is not 1PI.

Resumming the diagrams according to definition (2.41) of the exact vertex Γ and the exact propagator G, the propagator equation can be represented graphically as

$$G_0(p) - G_0(p)\, G^{-1}(p)\, G_0(p) \;=\; \quad\quad\quad , \qquad \text{(2.43)}$$

where the bold lines represent the exact propagator G, while the external (thin) ones are associated with the bare propagator G_0. One vertex on the RHS of Eq. (2.43) is exact and the other is bare. It does not matter which one is exact and which one is bare since we can collect the diagrams of Eq. (2.42) into the exact vertex either on the LHS or on the RHS. Equation (2.43) can be written analytically as

$$G_0^{-1}(p) - G^{-1}(p) \;=\; -\frac{\lambda}{2} \int \frac{d^d q}{(2\pi)^d} G(q)\, \Gamma(-q, p, q - p)\, G(p - q). \qquad \text{(2.44)}$$

Multiplying Eq. (2.44) by $G(p)$ and using the definition (2.36), we obtain the Fourier transform of Eq. (2.34).

Note that Eq. (2.34) is not closed. It relates the two-point average (propagator) to the three-point average (which is associated with a vertex). The closed set of the Schwinger–Dyson equations can be obtained for the n-point averages

$$G_n(x_1, x_2, \cdots, x_n) \;=\; \langle \varphi(x_1)\, \varphi(x_2) \cdots \varphi(x_n) \rangle. \qquad \text{(2.45)}$$

They are also called the correlators, in analogy with statistical mechanics, or the n-point Green functions, in analogy with the Green functions in Minkowski space.

Choosing

$$F[\varphi] \;=\; \varphi(x_2)\cdots\varphi(x_n) \tag{2.46}$$

and calculating the variational derivative, one finds from Eq. (2.28) the following chain of equations:

$$\left(-\partial^2 + m^2\right)\langle\,\varphi(x)\,\varphi(x_2)\cdots\varphi(x_n)\,\rangle + \frac{\lambda}{2}\langle\,\varphi^2(x)\,\varphi(x_2)\cdots\varphi(x_n)\,\rangle$$

$$= \sum_{j=2}^{n}\delta^{(d)}(x - x_j)\left\langle\,\varphi(x_2)\cdots\underline{\varphi(x_j)}\cdots\varphi(x_n)\,\right\rangle, \tag{2.47}$$

where $\underline{\varphi(x_j)}$ denotes that the corresponding term $\varphi(x_j)$ is missing in the product. Using the notation (2.45), Eq. (2.47) can be rewritten as

$$\left(-\partial^2 + m^2\right)G_n(x, x_2, \ldots, x_n) + \frac{\lambda}{2}G_{n+1}(x, x, x_2, \ldots, x_n)$$

$$= \sum_{j=2}^{n}\delta^{(d)}(x - x_j)\,G_{n-2}(x_2, \ldots, \underline{x_j}, \ldots, x_n) \tag{2.48}$$

with $\underline{x_j}$ again denoting the missing argument.

Remark on connected correlators

The n-point correlators (2.45) include both connected and disconnected parts. The presence of disconnected parts is most easily seen in the free case when all connected parts disappear, while G_n for even n is given by the Wick pairing as is discussed in Sect. 2.3.

The correlators can also be defined by introducing the generating functional, which is a functional of an external source $J(x)$:

$$Z[J] \;=\; \int \mathcal{D}\varphi\,(x)\;e^{-S+\int d^d x\, J(x)\,\varphi(x)}, \tag{2.49}$$

and varying with respect to the source. The n-point correlators (2.45) are then given by

$$G_n(x_1, \ldots, x_n) \;=\; \frac{1}{Z[J]}\frac{\delta}{\delta J(x_1)}\cdots\frac{\delta}{\delta J(x_n)}Z[J], \tag{2.50}$$

while the connected parts are given by

$$\langle\,\varphi(x_1)\cdots\varphi(x_n)\,\rangle_{\text{conn}} \;=\; \frac{\delta}{\delta J(x_1)}\cdots\frac{\delta}{\delta J(x_n)}\ln Z[J]. \tag{2.51}$$

This is because

$$W[J] = \ln Z[J] \tag{2.52}$$

involves only a set of connected diagrams, while disconnected ones emerge in $Z[J]$ after the exponentiation. We have already touched on this property in Sect. 2.3 to order λ^2. The functional $W[J]$ is called, for this reason, the generating functional for connected diagrams.

Remark on the LSZ reduction formula

The correlators $G_n(x_1, \ldots, x_n)$ (analytically continued to Minkowski space) determine the amplitude of the process when n, generally speaking, virtual particles produce k on-mass-shell particles. Let us denote this amplitude as $A_{n \to k}(q_1, \ldots, q_n; p_1, \ldots, p_k)$, where $q_1, \ldots, q_n$ and $p_1, \ldots, p_k$ are the four-momenta of the incoming and outgoing particles, respectively. Four-momentum conservation requires

$$q_1 + \cdots + q_n = p_1 + \cdots + p_k. \tag{2.53}$$

The Lehman–Symanzik–Zimmerman (LSZ) reduction formula reads as

$$
\begin{aligned}
&A_{n \to k}(q_1, \ldots, q_n; p_1, \ldots, p_k) \\
&= \prod_{j=1}^{n} (q_j^2 + m^2) \prod_{i=1}^{k} \lim_{p_i^2 \to -m^2} (p_i^2 + m^2) \int \prod_{i=1}^{k} \frac{d^d p_i}{(2\pi)^d} \int \prod_{j=1}^{n-1} \frac{d^d q_j}{(2\pi)^d} \\
&\quad \times \exp\left(-i \sum_{i=1}^{k} p_i x_i + i \sum_{j=1}^{n-1} q_j x_{k+j}\right) G_n(x_1, \ldots, x_{n+k-1}, 0).
\end{aligned}
\tag{2.54}
$$

The unusual sign of the square of the particle mass m arises from the Euclidean metric.

Equation (2.54) makes sense for timelike p_i, when $p_i^2 < 0$, while q_j^2 is arbitrary. The amplitude for the case of on-mass-shell incoming particles is given by Eq. (2.54) with $q_j^2 \to -m^2$.

2.7 Regularization

The ultraviolet divergences (i.e. those at small distances or large momenta) are an intrinsic property of quantum field theory which makes it different from the quantum mechanics of a finite number of degrees of freedom. The divergences emerge, roughly speaking, because of the delta-function in the canonical commutation relations.

The idea of regularization is to somehow smooth the effect of the delta-function. The usual procedures of regularization are to:

(1) smear the delta-function by

      – point splitting,
      – Schwinger proper-time regularization,
      – latticizing;

(2) add a negative-norm regulating term to the action

      – Pauli–Villars regularization;

(3) introduce higher derivatives in the kinetic term;[*]
(4) change the dimension, $4 \to 4-\epsilon$;
(5) regularize the measure in the path integral.

As an example of point splitting, let us consider the regularization when the delta-function in the commutator term is replaced by

$$\delta^{(d)}(x-y) \ \stackrel{\text{reg.}}{\Longrightarrow} \ \boldsymbol{R}\,\delta^{(d)}(x-y) \ = \ R(x,y). \qquad (2.55)$$

The regularizing operator $\boldsymbol{R}$ is, for instance,

$$\boldsymbol{R} \ = \ \mathrm{e}^{a^2(\partial^2-m^2)}, \qquad (2.56)$$

where the parameter a with the dimension of length plays the role of an ultraviolet cutoff. The cutoff disappears as $a \to 0$ when

$$\left.\begin{array}{rcl} \boldsymbol{R} & \to & 1, \\[4pt] R(x,y) & \to & \delta^{(d)}(x-y). \end{array}\right\} \qquad (2.57)$$

It is easy to calculate how the regularization (2.55) modifies the propagator. The result is

$$\begin{aligned} G_R(x-y) & = \ \frac{1}{-\partial^2+m^2}\,\boldsymbol{R}\,\delta^{(d)}(x-y) \\[6pt] & = \ \frac{1}{2}\int\limits_{a^2}^{\infty} \mathrm{d}\tau \ \mathrm{e}^{-\frac{1}{2}\tau m^2}\,\mathrm{e}^{-(x-y)^2/2\tau}\,\frac{1}{(2\pi\tau)^{d/2}}. \end{aligned} \qquad (2.58)$$

The lower limit in the integral over the proper time τ is now a^2 rather than 0 as in the nonregularized expression (1.89). This particular method of point splitting coincides with the Schwinger proper-time regularization.

A regularization via point splitting can be performed nonperturbatively, while the dimensional regularization (which is listed in item (4)) is defined only within the framework of perturbation theory. The regularization of the measure listed in item (5) will be considered in the next chapter.

When a regularization is introduced, some of the first principles (called the axioms), on which quantum field theory is constructed, are violated. For instance, the regularization via the point splitting (2.55) violates locality.

---

[*] That is in the quadratic-in-fields part of the action.

3

Quantum anomalies from path integral

As is well-known, the Lagrangian approach in classical field theory is very useful for constructing conserved currents associated with symmetries of the Lagrangian. Noether's theorems[*] describe how to construct corresponding currents and when they are conserved.

An analogous approach in quantum field theory is based on path integrals over fields. It naturally incorporates the classical results since the weight in the path integral is given by the classical action.

However, anomalous terms (i.e. those in addition to the classical ones) in the divergences of currents can appear in the quantum case owing to a contribution from regulators which make the theory finite in the ultraviolet limit. They are called *quantum anomalies.*

In this chapter we first consider the chiral anomaly, i.e. the quantum anomaly in the divergence of the axial current, which appears in the path-integral approach as a result of the noninvariance of the regularized measure. Then we briefly repeat the analysis for the scale anomaly, i.e. the quantum anomaly in the divergence of the dilatation current.

3.1 QED via path integral

Let us restrict ourselves to the case of *quantum electrodynamics* (QED), though most of the formulas will be valid for a non-Abelian Yang–Mills theory as well.

QED is described by the following partition function:

$$Z = \int \mathcal{D}A_\mu \int \mathcal{D}\bar{\psi}\mathcal{D}\psi \, e^{-S[A,\psi,\bar{\psi}]}, \qquad (3.1)$$

where A_μ is the vector-potential of the electromagnetic field, ψ_i and $\bar{\psi}_i$ are the Grassmann variables which describe the electron–positron field with

---

[*] See, for example, §2 of the book [BS76].

i being the spinor index. They are independent but are interchangeable under involution

$$\psi \overset{\text{inv.}}{\longleftrightarrow} \bar{\psi}, \tag{3.2}$$

which is defined such that*

$$\psi_1\psi_2 \overset{\text{inv.}}{\longrightarrow} \bar{\psi}_2\bar{\psi}_1. \tag{3.3}$$

In particular, $\bar{\psi}\psi$ is invariant under involution. Therefore, $\bar{\psi}$ is an analog of $i\bar{\psi} = i\psi^\dagger\gamma_0$ in the operator formalism, while involution is analogous to Hermitian conjugation.

The Euclidean QED action in Eq. (3.1) is given by

$$S[A, \psi, \bar{\psi}] = \int \mathrm{d}^d x \left(\bar{\psi}\gamma_\mu\nabla_\mu\psi + m\bar{\psi}\psi + \frac{1}{4}F^2_{\mu\nu} \right), \tag{3.4}$$

where $\nabla_\mu = \partial_\mu - ieA_\mu(x)$ is the covariant derivative as before,

$$F_{\mu\nu} = \partial_\mu A_\nu - \partial_\nu A_\mu \tag{3.5}$$

is the field strength, and γ_μ are the Euclidean γ-matrices which are discussed in Sect. 1.2.

3.2 Chiral Ward identity

Let us perform the local *chiral transformation* (c.t.)

$$\left.\begin{array}{l} \psi(x) \overset{\text{c.t.}}{\longrightarrow} \psi'(x) = e^{i\alpha(x)\gamma_5}\psi(x), \\[2mm] \bar{\psi}(x) \overset{\text{c.t.}}{\longrightarrow} \bar{\psi}'(x) = \bar{\psi}(x)\,e^{i\alpha(x)\gamma_5}. \end{array}\right\} \tag{3.6}$$

Here the parameter of the transformation $\alpha(x)$ is a function of x and γ_5 is the Hermitian matrix

$$\gamma_5 = \gamma_1\gamma_2\gamma_3\gamma_4 \tag{3.7}$$

in $d = 4$ dimensions. Note that both ψ and $\bar{\psi}$ have the same transformation law since in Minkowski space

$$\bar{\psi} = \psi^\dagger\gamma_0, \qquad \gamma_5^\dagger = \gamma_5, \qquad \gamma_0^\dagger = \gamma_0, \tag{3.8}$$

while γ_5 and γ_0 anticommute.

---

* See the book by Berezin [Ber86] (§3.5 of Part I). Sometimes involution is defined with an opposite sign (i.e. $\bar{\psi}$ is substituted by $i\bar{\psi}$) which results in a multiplication of the fermionic part of the action (3.4) by an extra factor of i.

The variation of the classical action (3.4) under the chiral transformation (3.6) reads as

$$\delta S = \int d^d x \left[\partial_\mu \alpha(x) \, J_\mu^A(x) + 2im\alpha(x)\bar{\psi}(x)\gamma_5\psi(x) \right], \qquad (3.9)$$

where the axial current

$$J_\mu^A = i\bar{\psi}\gamma_\mu\gamma_5\psi \qquad (3.10)$$

is the Noether current associated with the chiral transformation.

It follows from Eq. (3.9) that the divergence of the axial current (3.10) is given by

$$\partial_\mu J_\mu^A = 2im\,\bar{\psi}\gamma_5\psi, \qquad (3.11)$$

so that it is conserved in the massless case ($m = 0$) at the classical level:

$$\partial_\mu J_\mu^A \stackrel{m=0}{=} 0. \qquad (3.12)$$

Problem 3.1 Verify Eq. (3.11) using the classical Dirac equation

$$\left(\widehat{\nabla} + m\right)\psi(x) = 0, \qquad \widehat{\nabla} = \gamma_\mu\nabla_\mu. \qquad (3.13)$$

Solution Calculate the divergence of the axial current (3.10) using Eq. (3.13) and the conjugate one

$$\bar{\psi}(x)\left(\overleftarrow{\widehat{\nabla}} - m\right) = 0 \qquad (3.14)$$

with

$$\overleftarrow{\nabla}_\mu = \overleftarrow{\partial}_\mu + ieA_\mu(x). \qquad (3.15)$$

Let us now discuss how the measure in the path integral changes under the chiral transformation (3.6). The old and new measures are related by

$$\mathcal{D}\bar{\psi}\mathcal{D}\psi = \mathcal{D}\bar{\psi}'\mathcal{D}\psi' \det\left[e^{2i\alpha(x)\gamma_5} \delta^{(d)}(x-y) \right], \qquad (3.16)$$

where the determinant is over the space indices x and y, as well as over the γ-matrix indices i and j. Note that the determinant, which is nothing but the Jacobian of the transformation (3.6), emerges for the Grassmann variables to the positive rather than the negative power as for commuting variables. This is a known property of the integrals (2.17) over Grassmann variables [Ber86] which look more like derivatives.

The logarithm of the Jacobian in Eq. (3.16) can be calculated as

$$
\ln \det \left[e^{2i\alpha(x)\gamma_5} \delta^{(d)}(x - y) \right]
$$

$$
= \ \mathrm{Tr} \ln \left(e^{2i\alpha\gamma_5} \right) \ = \ \mathrm{Tr} \left(2i\,\alpha\,\gamma_5 \right)
$$

$$
= \ 2i \int d^d x \, \alpha(x) \, \delta^{(d)}(0) \ \mathrm{sp}\,\gamma_5 \ = \ 0 \,, \tag{3.17}
$$

where sp is the trace only over the γ-matrix indices i and j. The RHS vanishes naively since the trace vanishes. A subtlety with the appearance of the infinite factor of $\delta^{(d)}(0)$ will be discussed in the next section.

Note that the infinitesimal version of the transformation (3.6) is a particular case of the more general one

$$
\left.\begin{array}{l}
\psi(x) \ \longrightarrow \ \psi'(x) = \psi(x) + \delta\psi(x) \,, \\[4pt]
\bar{\psi}(x) \ \longrightarrow \ \bar{\psi}'(x) = \bar{\psi}(x) + \delta\bar{\psi}(x) \,,
\end{array}\right\} \tag{3.18}
$$

which is an analog of the transformation (2.25) and leaves the measure invariant. The calculation given in Eq. (3.17) is an explicit illustration of this fact.

The general transformation (3.18) leads, when applied to the path integral in Eq. (3.1), to the Schwinger–Dyson equations

$$
\left.\begin{array}{l}
\left(\hat{\nabla} + m \right) \psi(x) \ \stackrel{\mathrm{w.s.}}{=} \ \dfrac{\delta}{\delta\bar{\psi}(x)} \,, \\[14pt]
\bar{\psi}(x) \left(\overleftarrow{\hat{\nabla}} - m \right) \ \stackrel{\mathrm{w.s.}}{=} \ \dfrac{\delta}{\delta\psi(x)} \,,
\end{array}\right\} \tag{3.19}
$$

which hold in the weak sense, i.e. under the averaging over $\bar{\psi}$ and ψ.

More restrictive transformations of the same type as (3.6), which are associated with symmetries of the classical action and result in conserved currents, lead to some (less restrictive) relations between correlators which are called *Ward identities*. This terminology goes back to the 1950s when a proper relation between the two- and three-point Green functions was first derived for the gauge symmetry in QED.

The simplest Ward identity, which is associated with the chiral transformation (3.6), is given as

$$
\langle \partial_\mu J_\mu^A(0) \, \psi_i(x) \, \bar{\psi}_j(y) \rangle
$$

$$
\stackrel{m=0}{=} i\,\delta^{(d)}(x) \langle (\gamma_5 \psi)_i(0) \, \bar{\psi}_j(y) \rangle - i\,\delta^{(d)}(y) \langle \psi_i(x) \, (\bar{\psi}\gamma_5)_j(0) \rangle .
$$

$$
\tag{3.20}
$$

It is clear from the way in which Eq. (3.20) was derived, that it is always satisfied as a consequence of the quantum equations of motion (3.19).

Problem 3.2 Derive Eq. (3.20) in the operator formalism when the averages are substituted by the vacuum expectation values of the T-products.

Solution Equation (3.13) acquires an extra $-i$ in Minkowski space, where the spatial γ-matrices are anti-Hermitian rather than Hermitian as in Euclidean space, and holds in the quantum case in the weak sense, i.e. when applied to a state. Using it and the canonical equal-time anticommutation relations for ψ and $\bar\psi$ with the only nonvanishing anticommutator being

$$\delta(x_0 - y_0)\left\{\bar\psi_i(y), \psi_j(x)\right\} = \delta_{ij}\delta^{(d)}(x - y),\qquad (3.21)$$

we reproduce Eq. (3.20) in the operator formalism.

Remark on γ_5 in d dimensions

Let us recall that

$$\gamma_5 = \gamma_1\gamma_2\cdots\gamma_d \qquad (3.22)$$

only exists for even d when the size of the γ-matrices is $2^{d/2} \times 2^{d/2}$. For this reason the dimensional regularization is not applicable in calculations of the chiral anomaly.

Remark on gauge-fixing

Note that we did not add a gauge-fixing term to the action (3.4). It is harmless to do that since the gauge-fixing term does not contribute to the variation of the action under the chiral transformation. Moreover, all gauge-invariant quantities do not depend on the gauge-fixing. How one can quantize a gauge theory without adding a gauge-fixing term will be explained in Part 2.

3.3 Chiral anomaly

As has already been mentioned, Eq. (3.17) involves the uncertainty

$$\delta^{(d)}(0) \cdot \mathrm{sp}\,\gamma_5 = \infty \cdot 0. \qquad (3.23)$$

To regularize $\delta^{(d)}(0)$, one needs [Ver78, Fuj79] to regularize the measure in the path integral over ψ and $\bar\psi$, since this term comes from the change of the measure under the chiral transformation.

Let us expand the fields ψ and $\bar\psi$ over some set of the orthogonal basis functions, similarly to Eq. (1.82):

$$\psi^i(x) = \sum_n c_n^i \phi_n^i(x), \qquad \bar\psi^i(x) = \sum_n \bar c_n^i \phi_n^{i\dagger}(x), \qquad (3.24)$$

where there is no summation over the spinor index i. Here c_n^i and $\bar{c}_n^i$ are Grassmann variables. The measure is then similar to that of Eq. (1.83) and reads explicitly as

$$\mathcal{D}\bar{\psi}\mathcal{D}\psi \;=\; \prod_{n=1}^{\infty}\prod_{i} \mathrm{d}\bar{c}_n^i \prod_{m=1}^{\infty}\prod_{j} \mathrm{d}c_m^j \,. \tag{3.25}$$

The idea of regularizing the measure is to restrict ourselves to a large but finite number of basis functions. This is analogous to the discretization of Sect. 1.4. We therefore define the regularized measure as

$$(\mathcal{D}\bar{\psi})_R(\mathcal{D}\psi)_R \;=\; \prod_{n=1}^{M}\prod_{i} \mathrm{d}\bar{c}_n^i \prod_{m=1}^{M}\prod_{j} \mathrm{d}c_m^j \,. \tag{3.26}$$

The change of the measure under the chiral transformation is

$$(\mathcal{D}\bar{\psi})_R(\mathcal{D}\psi)_R \;=\; (\mathcal{D}\bar{\psi}')_R(\mathcal{D}\psi')_R \det\left[\int \mathrm{d}^d x\, \phi_n^{k\dagger}(x)\, e^{2i\alpha(x)\gamma_5^{kj}}\, \phi_m^j(x) \right],$$

$$\tag{3.27}$$

where the determinant is over both the n and m indices and the spinor indices k and j. This is the regularized analog of the nonregularized expression (3.16).

Using the orthogonality of the basis functions:

$$\int \mathrm{d}^d x\, \phi_n^{j\dagger}(x)\, \phi_m^i(x) \;=\; \delta_{nm}\delta^{ij}, \tag{3.28}$$

and Eq. (2.20), we rewrite the determinant on the RHS of Eq. (3.27) for an infinitesimal parameter α as

$$\det \int \mathrm{d}^d x\, \phi_n^{k\dagger}(x)\, e^{2i\alpha(x)\gamma_5^{kj}}\, \phi_m^j(x) \;=\; 1 + 2i\sum_{n=1}^{M} \int \mathrm{d}^d x\, \phi_n^\dagger(x)\alpha(x)\gamma_5\phi_n(x)\,,$$

$$\tag{3.29}$$

where the spinor indices are contracted in the usual way.

It is easy to see how this formula recovers Eq. (3.17) since

$$\sum_{n=1}^{\infty} \phi_n^i(x)\, \phi_n^{j\dagger}(y) \;=\; \delta^{(d)}(x-y)\, \delta^{ij} \tag{3.30}$$

in the nonregularized case owing to the completeness of the basis functions.

In the regularized case, the sum over n on the LHS of Eq. (3.30) is restricted by M from above so that the RHS is no longer equal to the delta-function. We substitute

$$\sum_{n=1}^{M} \phi_n^i(x)\,\phi_n^{j\,\dagger}(y) \;=\; R^{ij}(x,y), \tag{3.31}$$

with the RHS being the matrix element of some regularizing operator $\boldsymbol{R}$. It can be chosen in many ways. We shall work with several forms:

$$\boldsymbol{R} \;=\; e^{a^2\widehat{\nabla}^2}, \tag{3.32}$$

or

$$\boldsymbol{R} \;=\; \frac{1}{1 - a^2\widehat{\nabla}^2}, \tag{3.33}$$

or

$$\boldsymbol{R} \;=\; \frac{1}{1 + a\widehat{\nabla}}, \tag{3.34}$$

etc., where again $\widehat{\nabla} = \gamma_\mu \nabla_\mu$. The parameter a is the ultraviolet cutoff. The cutoff disappears as $a \to 0$ when Eq. (2.57) holds.

These regularizations (3.32)–(3.34) are nonperturbative, and preserve gauge invariance since they are constructed from the covariant derivative ∇_μ. A consistent regularization occurs when $\boldsymbol{R}$ commutes with $\widehat{\nabla}$, which is obviously true for the regularizations (3.32)–(3.34).*

Therefore, we find

$$\int d^d x\, \alpha(x)\, \partial_\mu J_\mu^{\mathrm{A}} \;=\; 2\mathrm{i}\,\mathrm{Tr}\,(\alpha\gamma_5 \boldsymbol{R})$$

$$=\; 2\mathrm{i}\int d^d x\, \alpha(x)\, \mathrm{sp}\,[\gamma_5 R(x,x)]. \tag{3.35}$$

It is worth noting that the extra R in Eq. (3.35) is a consequence of the more general formula

$$\mathrm{Tr}\,\boldsymbol{O} \;\longrightarrow\; \mathrm{Tr}\,\boldsymbol{O}\boldsymbol{R}, \tag{3.36}$$

which describes how to regularize the traces of operators.

---

* This can be shown by choosing the basis functions to be eigenfunctions of the Hermitian operator $\mathrm{i}\widehat{\nabla}$ ($\mathrm{i}\widehat{\nabla}\phi_n = E_n\phi_n$) and applying $\widehat{\nabla}^{ki}(x)[\widehat{\nabla}^{-1}(y)]^{jl}$ to Eq. (3.31). Then the LHS does not change (because $E_n E_n^{-1} = 1$), while $\langle x|\widehat{\nabla}\boldsymbol{R}\widehat{\nabla}^{-1}|y\rangle$ appears on the RHS. It coincides with the RHS of Eq. (3.31) when $\widehat{\nabla}$ and $\boldsymbol{R}$ commute.

Remark on regularization of the measure

The regularization of the measure in the path integral using Eq. (3.26) is equivalent to the point-splitting procedure where the delta-function in the commutator term is smeared according to Eq. (2.55).

To show this, let us note that the variational derivative can be approximated for a finite number of basis functions by

$$\frac{\delta}{\delta\psi_R^j(y)} = \sum_{n=1}^{M} \phi_n^{j\,\dagger}(y) \sum_k \frac{\partial}{\partial c_n^k}. \tag{3.37}$$

This definition extends the standard mathematical one* to the case of spinor indices. The sum over k is included in order for the regularized variational derivative to reflect variations of all the spinor components of c_n when the variation is not diagonal in the spinor indices.

When applied to

$$\psi_R^i(x) = \sum_{n=1}^{M} c_n^i \phi_n^i(x), \tag{3.38}$$

it yields

$$\frac{\delta\psi_R^i(x)}{\delta\psi_R^j(y)} = \sum_{n=1}^{M} \phi_n^i(x)\,\phi_n^{j\,\dagger}(y) = R^{ij}(x,y), \tag{3.39}$$

or, equivalently,

$$\delta^{ij}\delta^{(d)}(x-y) \xrightarrow{\text{reg.}} R^{ij}(x,y), \tag{3.40}$$

which is the fermionic analog of Eq. (2.55).

Thus, we conclude that the regularization of the measure in the path integral is equivalent to smearing the delta-function in commutator terms.

Remark on regularized Schwinger–Dyson equations

The procedure from the previous Remark results in the following regularized Schwinger–Dyson equations:

$$\left.\begin{aligned}
\left(\hat{\nabla} + m\right)\psi(x) &\overset{\text{w.s.}}{=} \int d^d y\, R(x,y)\frac{\delta}{\delta\bar\psi(y)}, \\
\bar\psi(x)\left(\overleftarrow{\hat{\nabla}} - m\right) &\overset{\text{w.s.}}{=} \int d^d y\, R(x,y)\frac{\delta}{\delta\psi(y)}.
\end{aligned}\right\} \tag{3.41}$$

These equations are understood again in the weak sense, i.e. under the averaging over $\bar\psi$ and ψ and obviously reproduce Eq. (3.19) as $a \to 0$.

---

* See, for example, the book by Lévy [Lev51].

Problem 3.3 Derive Eq. (3.35) using the regularized Schwinger–Dyson equation (3.41).

Solution The calculation is similar to that of Problem 3.1 except for the additional terms arising from the RHS of Eq. (3.41). For $m = 0$ one finds

$$
\partial_\mu J_\mu^A \stackrel{\text{w.s.}}{=} i \int d^d y \, \frac{\delta}{\delta \psi(y)} R(x,y) \gamma_5 \psi(x) - i \bar{\psi}(x) \gamma_5 \int d^d y \, R(x,y) \frac{\delta}{\delta \bar{\psi}(y)}
$$

$$
= 2i \operatorname{sp} \left[\gamma_5 R(x,x) \right], \tag{3.42}
$$

which is equivalent to Eq. (3.35) since there $\alpha(x)$ is an arbitrary function.

3.4 Chiral anomaly (calculation)

In order to derive an explicit expression for the chiral anomaly, we should calculate the RHS of Eq. (3.35) for some choice of the regularizing operator R. Let us choose R given by Eq. (3.33). The operator $\widehat{\nabla}^2$ in the denominator can be transformed as

$$
\begin{aligned}
\widehat{\nabla}^2 &= \nabla^2 + \frac{1}{2} \left[\gamma_\mu, \gamma_\nu \right] \nabla_\mu \nabla_\nu \\
&= \nabla^2 - \frac{ie}{4} \left[\gamma_\mu, \gamma_\nu \right] F_{\mu\nu} \\
&= \nabla^2 + \frac{e}{2} \Sigma_{\mu\nu} F_{\mu\nu} ,
\end{aligned} \tag{3.43}
$$

where the trace of the spin matrices

$$
\Sigma_{\mu\nu} = \frac{1}{2i} \left[\gamma_\mu, \gamma_\nu \right] \tag{3.44}
$$

is given by

$$
\operatorname{sp} \left(\Sigma_{\mu\nu} \Sigma_{\lambda\rho} \gamma_5 \right) = -4 \epsilon_{\mu\nu\lambda\rho} . \tag{3.45}
$$

Expanding in e,

$$
R = R_0 + R_0 (\cdots) R_0 + \cdots \tag{3.46}
$$

with

$$
R_0 = \frac{1}{1 - a^2 \partial^2} , \tag{3.47}
$$

we find schematically

$$
\begin{aligned}
\operatorname{Tr} \left(\alpha \gamma_5 R \right) &= a^4 \operatorname{Tr} \left[\alpha \gamma_5 R_0 \left(\frac{e \Sigma F}{2} \right) R_0 \left(\frac{e \Sigma F}{2} \right) R_0 \right] \\
&= - \int d^4 x \, \alpha(x) \frac{e^2}{16\pi^2} F_{\mu\nu} \widetilde{F}_{\mu\nu} ,
\end{aligned} \tag{3.48}
$$

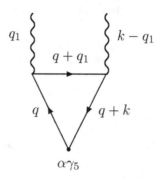

Fig. 3.1. Triangular diagram associated with chiral anomaly in $d = 4$. The solid lines correspond to R_0 given by Eq. (3.50). The wavy lines correspond to the field strength.

where

$$\widetilde{F}_{\mu\nu} = \frac{1}{2}\epsilon_{\mu\nu\lambda\rho}F_{\lambda\rho} \tag{3.49}$$

is the dual field strength.

The calculation described in Eq. (3.48) is most easily performed in momentum space where it is associated with one-loop diagrams. The analytic expression to be calculated can be represented in $d = 4$ graphically as the triangular diagram in Fig. 3.1. The solid lines are associated with R_0 given by Eq. (3.47), which reads in momentum space as

$$R_0(p) = \frac{1}{1 + a^2 p^2}. \tag{3.50}$$

The wavy lines correspond to the field strength. The lower vertex is associated with $\alpha\gamma_5$.

The integral over the four-momentum q, which circulates along the triangular loop, can be easily calculated by introducing $\omega = aq$ and transforming the integral as

$$\int d^4q \, f(q) \quad \rightarrow \quad \frac{1}{a^4}\int d^4\omega \, f\left(\frac{\omega}{a}\right). \tag{3.51}$$

Note that the integral involves a^{-4} which cancels a^4 coming from the expansion in e, for which the proper term is given by the intermediate expression in Eq. (3.48). Therefore, the result is nonvanishing and a-independent as $a \to 0$. Higher terms of the expansion in e are proportional to higher powers in a and vanish as $a \to 0$.

Finally, from Eqs. (3.35) and (3.48) we obtain

$$\partial_\mu J_\mu^{\mathrm{A}} = -\frac{ie^2}{8\pi^2}F_{\mu\nu}\widetilde{F}_{\mu\nu}. \tag{3.52}$$

The anomaly on the RHS is known as the Adler–Bell–Jackiw anomaly. Its appearance is usually related to the fact that any regularization cannot be simultaneously gauge and chiral invariant.

Problem 3.4 Calculate the coefficient in Eq. (3.48) and show that it is regulator-independent.

Solution The contribution of the triangular diagram of Fig. 3.1, which represents the intermediate expression in Eq. (3.48), reads explicitly as

$$2\,\mathrm{Tr}\,(\boldsymbol{\alpha}\gamma_5\boldsymbol{R}) \;=\; -4e^2a^4\int \mathrm{d}^4x \int \mathrm{d}^4y \int \mathrm{d}^4z$$

$$\times\, \alpha(x)R_0(x,y)F_{\mu\nu}(y)R_0(y,z)\tilde{F}_{\mu\nu}(z)R_0(z,x)\,. \qquad (3.53)$$

In momentum space, it becomes

$$-2e^2a^4\int \mathrm{d}^4x\,\alpha(x)\int \frac{\mathrm{d}^4k}{(2\pi)^4}\,\mathrm{e}^{\mathrm{i}kx}\int\frac{\mathrm{d}^4q_1}{(2\pi)^4}F_{\mu\nu}(q_1)\tilde{F}_{\mu\nu}(k-q_1)$$

$$\times \int \frac{\mathrm{d}^4q}{(2\pi)^4}\,\frac{1}{(1+a^2q^2)\,(1+a^2(q+q_1)^2)\,(1+a^2(q+k)^2)}$$

$$\overset{(3.51)}{=}\; -\frac{2e^2}{16\pi^2}\int \mathrm{d}^4x\,\alpha(x)\int\frac{\mathrm{d}^4k}{(2\pi)^4}\,\mathrm{e}^{\mathrm{i}kx}$$

$$\times \int \frac{\mathrm{d}^4q_1}{(2\pi)^4}F_{\mu\nu}(q_1)\tilde{F}_{\mu\nu}(k-q_1)\int\limits_0^\infty\frac{\omega^2\mathrm{d}\omega^2}{(1+\omega^2)^3} \qquad (3.54)$$

which recovers the RHS of Eq. (3.48).

An analogous calculation can be repeated for other regulators (3.32) and (3.34). Let us denote

$$r(a^2p^2) \;\equiv\; R_0(p)\,. \qquad (3.55)$$

Then the only difference with Eq. (3.54) is that the last integral over ω^2 is replaced by

$$\int\limits_0^\infty \mathrm{d}\omega^2\,\omega^2 r''(\omega^2) \;=\; r(0) \;=\; 1 \qquad (3.56)$$

for reasonable functions r which look like those given by Eqs. (3.32)–(3.34).

An anomaly which is analogous to the Adler–Bell–Jackiw anomaly (3.52) exists in $d=2$ where

$$\partial_\mu J_\mu^A \;=\; -\frac{e}{2\pi}\epsilon_{\mu\nu}F_{\mu\nu}\,. \qquad (3.57)$$

This anomaly is given by the diagram depicted in Fig. 3.2. It involves only two lines with the regulator $R_0(p)$ since in $d=2$

$$\int \mathrm{d}^2q\,f(q) \;\to\; \frac{1}{a^2}\int \mathrm{d}^2\omega\,f\!\left(\frac{\omega}{a}\right) \qquad (3.58)$$

so that all terms with more lines vanish as $a\to 0$.

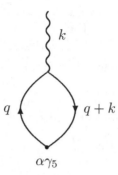

Fig. 3.2.   The diagram associated with the chiral anomaly in $d = 2$. The solid lines correspond to R_0 given by Eq. (3.50). The wavy line corresponds to the field strength.

Problem 3.5   Calculate $2\operatorname{Tr}(\alpha\gamma_5 R)$ in $d = 2$.

Solution   Proceeding as before, we see that only the diagram of Fig. 3.2 is essential in $d = 2$ which yields

$$2iea^2 \int d^2x \int d^2y\, \alpha(x) R_0(x,y) F_{\mu\nu}(y)\epsilon_{\mu\nu} R_0(y,x)$$

$$= 2iea^2 \int d^2x\, \alpha(x) \int \frac{d^2k}{(2\pi)^2}\, e^{ikx} F_{\mu\nu}(k)\epsilon_{\mu\nu} \int \frac{d^2q}{(2\pi)^2} \frac{1}{(1+a^2q^2)\,[1+a^2(q+k)^2]}$$

$$\overset{(3.58)}{=} 2\frac{ie}{4\pi} \int d^2x\, \alpha(x) \int \frac{d^2k}{(2\pi)^2}\, e^{ikx} F_{\mu\nu}(k)\epsilon_{\mu\nu} \int_0^\infty \frac{d\omega^2}{(1+\omega^2)^2}$$

$$= \int d^2x\, \alpha(x) \frac{ie F_{\mu\nu}(x)\epsilon_{\mu\nu}}{2\pi}. \tag{3.59}$$

The linear-in-$F_{\mu\nu}$ term is nonvanishing since

$$\operatorname{sp}(\Sigma_{\mu\nu}\gamma_5) = 2i\epsilon_{\mu\nu} \tag{3.60}$$

in $d = 2$.

The result is again regulator-independent since the integral over ω is replaced for an arbitrary $R_0(p)$ by

$$-\int_0^\infty d\omega^2 r'(\omega^2) = r(0) = 1 \tag{3.61}$$

where Eq. (3.55) has been used.

Remark on the non-Abelian chiral anomaly

Equation (3.52) also holds in the case of a non-Abelian gauge group where $F_{\mu\nu}^a$ is the non-Abelian field strength

$$F_{\mu\nu}^a(x) \;=\; \partial_\mu A_\nu^a(x) - \partial_\nu A_\mu^a(x) + g f^{abc} A_\mu^b(x) A_\nu^c(x) \,. \qquad (3.62)$$

Here f^{abc} are the structure constants of the gauge group and g is the coupling constant. The non-Abelian analog of Eq. (3.52) for the axial current, which is a singlet with respect to the gauge group, is given by*

$$\partial_\mu J_\mu^{\mathrm{A}} \;=\; -\frac{ig^2}{8\pi^2} \sum_a F_{\mu\nu}^a \tilde{F}_{\mu\nu}^a \,. \qquad (3.63)$$

The $d = 2$ anomaly (3.57) exists for the singlet axial current only in the Abelian case.

A description of the chiral anomaly in non-Abelian gauge theories is given, for example, in Chapter 22 of the book by Weinberg [Wei98].

3.5 Scale anomaly

The scale transformation is defined by

$$x_\mu \;\longrightarrow\; x_\mu' = \rho\, x_\mu \,, \qquad (3.64)$$
$$\varphi(x) \;\longrightarrow\; \varphi'(x') = \rho^{l_\varphi} \varphi(x') \,. \qquad (3.65)$$

The index l_φ is called the *scale dimension* of the field φ. The value of l_φ in a free theory is called the canonical dimension, which equals $(d-2)/2$ for bosons (scalar or vector fields) and $(d-1)/2$ for the spinor Dirac field, i.e. 1 and 3/2 in $d = 4$, respectively. Sometimes l_φ is called, for historical reasons, the anomalous dimension. More often the term "anomalous dimension" is used for the difference between l_φ and the canonical value.

The proper Noether current, which is called the *dilatation current*, is expressed via the energy–momentum tensor $\theta_{\mu\nu}$ as

$$D_\mu \;=\; x_\nu \theta_{\mu\nu} \qquad (3.66)$$

so that its divergence equals the trace of the energy–momentum tensor over the spatial indices:

$$\partial_\mu D_\mu \;=\; \theta_{\mu\mu} \,, \qquad (3.67)$$

---

* The coefficient in this formula is the same as in Eq. (3.52) and is twice as large as the conventional one. This is owing to our normalization, which is described in Sect. 5.1.

since the energy–momentum tensor is conserved. For the action (3.4) one finds

$$\theta_{\mu\mu} = -m\bar{\psi}\psi \qquad (3.68)$$

at the classical level.

The above formulas can be obtained from the Noether theorems which state

$$\delta S = \int d^d x \, \rho(x) \, \partial_\mu D_\mu(x) \qquad (3.69)$$

or

$$\partial_\mu D_\mu(x) = \frac{\delta S}{\delta \rho(x)}. \qquad (3.70)$$

In the massless case, $m = 0$, the RHS of Eq. (3.68) vanishes and the dilatation current is conserved. This is a well-known property of electrodynamics with a massless electron that is scale invariant at the classical level. A generic scale-invariant theory does not depend on parameters of the dimension of mass or length. This usual dimension is to be distinguished from the scale dimension which is defined by Eq. (3.65). The dimensional parameters do not change under the scale transformation (3.64).

In the quantum case, the scale invariance is broken by the (dimensional) cutoff a. The energy–momentum tensor is no longer traceless owing to loop effects. The relation (3.67) holds in the quantum case in the weak sense, i.e. for the averages

$$\langle \partial_\mu D_\mu F[A, \psi, \bar{\psi}] \rangle = \langle \theta_{\mu\mu} F[A, \psi, \bar{\psi}] \rangle, \qquad (3.71)$$

where $F[A, \psi, \bar{\psi}]$ is a gauge-invariant functional of A, ψ and $\bar{\psi}$.

For a renormalizable theory such as QED, the RHS of Eq. (3.71) is proportional to the Gell-Mann–Low function $\mathcal{B}(e^2)$ which is defined by

$$-a\frac{de^2}{da} = \mathcal{B}(e^2). \qquad (3.72)$$

A nontrivial property of a renormalizable theory is that the RHS in this formula is a function solely of e^2 – the bare charge.

The meaning of the *renormalizability* is very simple: physical quantities do not depend on the cutoff a, provided the bare charge e is chosen to be cutoff-dependent according to Eq. (3.72). This dependence of e on a effectively accounts for distances smaller than a, which are excluded from the theory.

The precise relation between the trace of the energy–momentum tensor and the Gell-Mann–Low function is given by

$$\theta_{\mu\mu} \overset{\text{w.s.}}{=} \frac{\mathcal{B}(e^2)}{4e^2} F_{\mu\nu}^2, \qquad (3.73)$$

where the equality is understood again in the weak sense. This formula was first obtained in [Cre72, CE72] to leading order in e^2 and proven in [ACD77] to all orders in e^2.

Note that this formula holds in the operator formalism only when applied to a gauge-invariant state. The reason is that otherwise a contribution from a gauge-fixing term in the action would be essential. It does not contribute, however, to gauge-invariant averages which can be formally proven using the gauge Ward identity.

Problem 3.6 Prove the relation (3.73).

Solution Let us absorb the coupling e into A_μ introducing

$$\left.\begin{aligned} \mathcal{A}_\mu &= eA_\mu, \\ \mathcal{F}_{\mu\nu} &= eF_{\mu\nu}. \end{aligned}\right\} \tag{3.74}$$

The Lagrangian density of massless QED then reads as

$$\mathcal{L} = \bar{\psi}\left(\hat{\partial} - i\hat{\mathcal{A}}\right)\psi + \frac{1}{4e^2}\mathcal{F}^2. \tag{3.75}$$

To prove Eq. (3.73), let us use the chain of Eqs. (3.67) and (3.70). It is crucial that in the absence of other dimensional parameters the derivative $\partial/\partial\rho$ can be replaced by $\partial/\partial a$, since all dimensionless quantities in a theory with a cutoff depend only on ratios of the type x/a.* Since the dependence on the cutoff a enters in Eq. (3.75) formally only via e^{-2} in front of $\mathcal{F}_{\mu\nu}^2$, Eq. (3.73) can be proven heuristically by first differentiating with respect to a and then expressing the result via $F_{\mu\nu}$ again. Here we have used the fact that $\mathcal{F}_{\mu\nu}$ is invariant under the renormalization-group transformation and, therefore, does not depend on a.

In the path-integral approach, a contribution to the scale anomaly comes from the regularized quantum measure. Proceeding as in Sect. 3.3, we obtain

$$\partial_\mu D_\mu(x) = -\mathrm{sp}\,[R(x,x)] \tag{3.76}$$

which determines the scale anomaly.

Problem 3.7 Derive Eq. (3.76) using the regularized Schwinger–Dyson equations (3.41).

Solution The energy–momentum tensor of QED is given by

$$\theta_{\mu\nu} = F_{\mu\lambda}F_{\nu\lambda} - \frac{1}{4}\delta_{\mu\nu}F_{\rho\lambda}^2 + \frac{1}{4}\left(\bar{\psi}\gamma_\mu\overleftrightarrow{\nabla}_\nu\psi + \bar{\psi}\gamma_\nu\overleftrightarrow{\nabla}_\mu\psi\right). \tag{3.77}$$

Taking the trace, one obtains

$$\theta_{\mu\mu} = \frac{1}{2}\bar{\psi}\gamma_\mu\overleftrightarrow{\nabla}_\mu\psi. \tag{3.78}$$

---

* This is the reason why the Callan–Symanzik equations, which are nothing but the dilatation Ward identities, coincide with the renormalization-group equations.

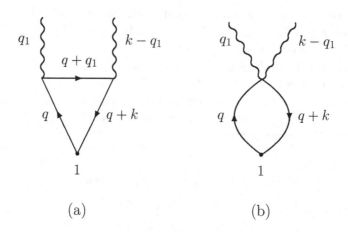

Fig. 3.3.   The diagrams which contribute to the scale anomaly in $d = 4$. The wavy line corresponds to the field strength.

Using Eq. (3.41), it can be transformed as

$$\theta_{\mu\mu} = -m\bar{\psi}\psi + \frac{1}{2}\left[\int d^d y\, R(x,y)\frac{\delta}{\delta\psi(y)}\psi(x) - \bar{\psi}(x)\int d^d y\, R(x,y)\frac{\delta}{\delta\bar{\psi}(y)}\right]$$
$$= -m\bar{\psi}\psi - \mathrm{sp}\,[R(x,x)],\qquad\qquad (3.79)$$

which reproduces Eq. (3.76) as $m \to 0$.

To calculate the scale anomaly we should therefore perform a one-loop calculation of

$$\mathrm{sp}\,[R(x,x)] = \mathrm{sp}\left\langle x\left|\frac{1}{1 + a^2(\mathrm{i}\widehat{\nabla}^2)}\right|x\right\rangle$$
$$= \mathrm{sp}\left\langle x\left|\frac{1}{1 + a^2\,(\mathrm{i}\partial_\mu + eA_\mu)^2 - \frac{1}{2}a^2 e\Sigma_{\mu\nu}F_{\mu\nu}}\right|x\right\rangle \qquad (3.80)$$

which is again most convenient to do in momentum space. The propagator is given by Eq. (3.47), while the vertices, which emerge in the corresponding Feynman rules for the expansion in e, come from the operators

$$-2\mathrm{i}ea^2 A_\mu\partial_\mu\,,\qquad -e^2 a^2 A_\mu^2\,,\qquad \frac{1}{2}ea^2\Sigma_{\mu\nu}F_{\mu\nu}\,.$$

The only diagrams which survive as $a \to 0$ in $d = 4$ are depicted in Fig. 3.3. The calculation of the diagram of Fig. 3.3a is the same as in Sect. 3.3 while the diagram of Fig. 3.3b gives a total derivative which does not contribute to the scale anomaly.

The calculation of the diagram of Fig. 3.3a yields

$$\operatorname{sp}\left[R(x, x)\right] \;=\; -\frac{e^2 F_{\mu\nu}^2(x)}{24\pi^2}. \tag{3.81}$$

The one-loop Gell-Mann–Low function can now be calculated using Eqs. (3.76) and (3.73), which reproduces the known result for QED. The higher-order corrections in e do not vanish for the scale anomaly.

Remark on the non-Abelian scale anomaly

Equation (3.73) holds in the non-Abelian Yang–Mills theory as well if $F_{\mu\nu}$ is substituted by the non-Abelian field strength $F_{\mu\nu}^a$ given by Eq. (3.62). The corresponding formula, is given as

$$\theta_{\mu\mu} \;\overset{\text{w.s.}}{=}\; \frac{\mathcal{B}(g^2)}{4g^2} \sum_a F_{\mu\nu}^a F_{\mu\nu}^a. \tag{3.82}$$

A heuristic proof, presented in Problem 3.6 for the Abelian case, can be repeated. The equality is again understood in the weak sense when averaged between gauge-invariant states. The contribution of gauge-fixing and ghost terms are then canceled owing to the gauge Ward identity which is called in this case the Slavnov–Taylor identity. The proof of Eq. (3.82) was given in [CDJ77, Nie77].

4

Instantons in quantum mechanics

Instantons are solutions of the classical equations of motion with a finite Euclidean action. Such field configurations are not taken into account in perturbation theory. Instantons are characterized by a topological charge which may result in a conserved quantum number and never show up in perturbation theory. In Minkowski space, instantons are associated with tunneling processes between vacua labeled by a distinct topological charge.

Instantons first appear in Yang–Mills theory [BPS75], although this kind of classical solution was known long before in statistical physics [Lan67].

In this chapter we consider instantons in quantum mechanics as an illustration of path-integral calculations. We follow the original paper by Polyakov [Pol77] except for technical details.

4.1 Double-well potential

Let us consider a one-dimensional quantum-mechanical system with the double-well potential

$$V(x) \;=\; \frac{\lambda}{4} \left(x^2 - \frac{\mu^2}{\lambda} \right)^2 \;=\; -\frac{1}{2}\mu^2 x^2 + \frac{1}{4}\lambda x^4 + \frac{\mu^4}{4\lambda}. \qquad (4.1)$$

This is nothing but an anharmonic oscillator with the opposite sign for the coefficient of the quadratic term,* which usually appears with a positive

---

* It is often called the mass term. This terminology comes from quantum field theory, where the potential (4.1) is considered in the context of a spontaneous breaking of the reflection symmetry $x \rightarrow -x$. In our quantum-mechanical problem, defined by the Euclidean action (4.3), the mass of the nonrelativistic particle is absorbed in τ which has, therefore, the dimension of [length]^2. This has already been explained in Sect. 1.6.

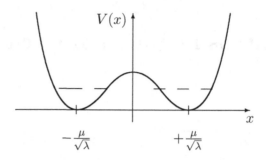

Fig. 4.1. The double-well potential (4.1). The short vertical lines represent the position of the minima (4.4). The dashed lines correspond to the energy E_0 of the lowest state in a single well, i.e. to that in the limit $\lambda \to 0$.

coefficient $\omega^2/2$. We have introduced

$$\mu^2 = -\omega^2 \tag{4.2}$$

in order to work with real numbered values. The constant term is added for later convenience. The potential (4.1) as a function x is depicted in Fig. 4.1.

The (Euclidean) action is defined by

$$S[x] = \int d\tau \left[\frac{1}{2}\dot{x}^2(\tau) + V(x(\tau)) \right] \tag{4.3}$$

with $V(x)$ given by Eq. (4.1). The plus sign between the kinetic and potential energies is because we are in Euclidean space.

It follows from Eqs. (4.1) and (4.3) that the parameter μ has the dimension of [length]$^{-2}$ or, in other words, the dimensions of x and τ are $[\mu]^{-1/2}$ and $[\mu]^{-1}$, respectively. Analogously, the dimension of the constant λ is $[\mu]^3$.

For $\lambda \ll \mu^3$, the potential (4.1) has superficially two degenerate vacua

$$x_0^\pm = \pm\frac{\mu}{\sqrt{\lambda}}, \tag{4.4}$$

the positions of which coincide with the minima of the potential in Fig. 4.1.

The degeneracy between the two minima is preserved at all orders of perturbation theory, where an expansion near one of the minima of the potential (either the left- or right-hand one) is carried out:

$$x(\tau) = \pm\frac{\mu}{\sqrt{\lambda}} + \chi(\tau) \tag{4.5}$$

with $\chi(\tau) \ll \mu/\sqrt{\lambda}$. The correlator at asymptotically large τ is

$$\langle x(0)\, x(\tau) \rangle \;\to\; \frac{\mu^2}{\lambda} + \cdots . \tag{4.6}$$

Its nonvanishing value means that a particle is localized at one of the two vacua.

The next terms of the perturbative expansion in λ do not spoil this result since the potential (4.1) becomes

$$V = \mu^2\chi^2 \mp \sqrt{\lambda}\mu\chi^3 + \frac{\lambda}{4}\chi^4 \qquad (4.7)$$

after the shift (4.5), and has a positive sign for the quadratic term. Therefore, a perturbation theory constructed around the vacuum $x_0^\pm$ is a normal one, and the particle lives perturbatively in one of the two vacua.

However, we know from quantum mechanics that (nonperturbatively)

$$\langle x(0)\, x(\tau) \rangle = \sum_n |x_{n0}|^2\, e^{-(E_n-E_0)\tau} \qquad (4.8)$$

at imaginary time $\tau = it$, where E_n is the energy of the nth eigenstate of the Hamiltonian and x_{n0} is the proper matrix element. Therefore,

$$\langle x(0)\, x(\tau) \rangle \sim e^{-\Delta E\tau} \qquad (4.9)$$

for large τ, where

$$\Delta E = \mu\sqrt{\frac{48}{\pi}}\sqrt{\frac{2\sqrt{2}\,\mu^3}{3\lambda}}\, \exp\left(-\frac{2\sqrt{2}\,\mu^3}{3\lambda}\right) \qquad (4.10)$$

is the energy splitting between the two lowest states (symmetric and antisymmetric) for $\lambda \ll \mu^3$, which vanishes exponentially as $\lambda \to 0$.

The appearance of imaginary time in Eq. (4.8) is because under a barrier particles live in imaginary time. We may say that imaginary time is an appropriate language for describing tunneling through a barrier.

Since the RHS of Eq. (4.9) vanishes as $\tau \to \infty$, the reflection symmetry $x \to -x$, which is broken in perturbation theory, is restored nonperturbatively as $\tau \to \infty$.

Problem 4.1 Derive Eq. (4.10) modulo a constant factor within standard quantum mechanics.

Solution Let us use the semiclassical formula [LL74] (Problem 3 in §50)

$$\Delta E = \frac{\sqrt{2}\,\mu}{\pi}\, e^{-\int_{-a}^{+a} dx\sqrt{2[V(x)-E_0]}}, \qquad (4.11)$$

where $\pm a$ are the classical turning points, which are determined by

$$V(\pm a) = E_0, \qquad (4.12)$$

and

$$E_0 = \sqrt{2}\,\mu \qquad (4.13)$$

is the lowest energy for the oscillator potential (4.7) as $\lambda \to 0$. Denoting

$$h = \sqrt{\frac{\lambda}{\sqrt{2}\,\mu^3}}, \qquad z = \frac{\sqrt{\lambda}}{\mu}\,x\,, \tag{4.14}$$

the integral in the exponent on the RHS of Eq. (4.11) can be calculated using an expansion in h which gives

$$\frac{1}{2h^2} \int\limits_{-1+h}^{1-h} dz \sqrt{(1-z^2)^2 - 4h^2} = \frac{2}{3h^2} + \ln h + \mathcal{O}(1)\,. \tag{4.15}$$

Substituting into Eq. (4.11), one recovers Eq. (4.10) modulo a constant factor.

4.2 The instanton solution

In the path-integral approach, the correlator (4.8) is given by

$$\langle\, x(0)\,x(\tau)\,\rangle = \frac{\displaystyle\int \mathcal{D}x\; e^{-S[x]}\, x(0)\, x(\tau)}{\displaystyle\int \mathcal{D}x\; e^{-S[x]}} \tag{4.16}$$

with no restrictions on the integration over x. This is a quantum-mechanical analog of the path integrals defined in Sect. 2.1.

At small λ, the path integral (4.16) can be evaluated using the saddle-point method. The reason for this is that for x given by Eq. (4.4) (i.e. the minima of the action (4.3)), the Gaussian fluctuations around (4.4) are not essential as $\lambda \to 0$. This is most easily seen by making the shift (4.5) and noting that $\chi(\tau)$ is $\mathcal{O}(1)$ at the saddle points according to Eq. (4.7), the RHS of which is quadratic in $\chi(\tau)$ as $\lambda \to 0$.

Performing the saddle-point evaluation of the path integral (4.16), one obtains

$$\langle\, x(0)\,x(\tau)\,\rangle = \frac{\mu^2}{\lambda} + \cdots\,. \tag{4.17}$$

Note that $x(0)$ and $x(\tau)$ in the integrand can be substituted using the saddle-point values after which the integral over Gaussian fluctuations cancels with the same integral in the denominator. In other words, we have reproduced the fact that each of the trivial minima (4.4) results in Eq. (4.6).

Minima of the action (4.3) can also be obtained from the classical equation of motion

$$-\ddot{x} - \mu^2 x + \lambda x^3 = 0\,. \tag{4.18}$$

The trivial minima (4.4) obviously satisfy this equation.

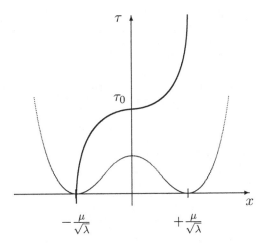

Fig. 4.2. Graphical representation of the one-kink solution (4.19) as a function of τ.

However, another solution of the classical equation of motion (4.18) exists:

$$x_{\text{inst}}(\tau - \tau_0) = \frac{\mu}{\sqrt{\lambda}} \tanh \frac{\mu (\tau - \tau_0)}{\sqrt{2}}, \qquad (4.19)$$

which is associated with another (local) minimum of the classical action. This solution is called an *instanton* or a *pseudoparticle*. The arbitrary constant τ_0 in Eq. (4.19) is the position of the center of the instanton.

The solution (4.19) is also known as a *kink* in this quantum-mechanical problem. It interpolates between the two minima (4.4) when τ changes from $-\infty$ to $+\infty$ as depicted in Fig. 4.2. Also shown in this figure is the double-well potential, $V(x)$, from Fig. 4.1.

An analogous solution which interpolates between $\mu/\sqrt{\lambda}$ at $\tau = -\infty$ and $-\mu/\sqrt{\lambda}$ at $\tau = +\infty$ is called an *anti-instanton*. It differs from Eq. (4.19) by an overall minus sign:

$$x_{\text{ainst}}(\tau - \tau_0) = -\frac{\mu}{\sqrt{\lambda}} \tanh \frac{\mu (\tau - \tau_0)}{\sqrt{2}}, \qquad (4.20)$$

and is obviously also a solution of the classical equation (4.18).

Problem 4.2 Find all solutions of Eq. (4.18) with the boundary conditions $x(-\infty) = -\mu/\sqrt{\lambda}$ and $x(+\infty) = \mu/\sqrt{\lambda}$.

Solution Equation (4.18) looks like Newton's equation for a classical particle, with unit mass, in the upside-down potential $-V(x)$ (its shape can be obtained from that depicted in Fig. 4.1 by reflecting with respect to the horizontal axis

$V = 0$). The first integral of motion is the energy

$$\mathcal{E} \;=\; \frac{1}{2}\dot{x}^2 - \frac{\lambda}{4}\left(x^2 - \frac{\mu^2}{\lambda}\right)^2 \tag{4.21}$$

which is obviously conserved owing to Eq. (4.18).

Equation (4.21) can easily be solved for the velocity

$$\dot{x} \;=\; \sqrt{2[\mathcal{E} + V(x)]}, \tag{4.22}$$

where we have chosen the positive sign according to the boundary condition. It also says that $\mathcal{E} = 0$ in order for the particle to stay at $x = \mu/\sqrt{\lambda}$ for $\tau \to \infty$, since this point is associated with the maximum of $-V(x)$. Therefore, we find

$$\dot{x} \;=\; \sqrt{\frac{\lambda}{2}}\left(\frac{\mu^2}{\lambda} - x^2\right), \tag{4.23}$$

which after integration results in Eq. (4.19) with τ_0 being the integration constant. It is evident that the solution is unique.

For the instanton (or anti-instanton) minimum, one finds, substituting in Eq. (4.3),

$$S\left[x_{\text{inst}}\right] \;=\; \frac{2\sqrt{2}\,\mu^3}{3\lambda}, \tag{4.24}$$

which differs only by sign from the exponent in Eq. (4.10) for the energy splitting ΔE.

4.3 Instanton contribution to path integral

The contribution of the instanton configuration looks as if it is suppressed in the path integral by a factor of $\exp\left(-S\left[x_{\text{inst}}\right]\right)$, but, in fact, this exponential is multiplied by τ since the instanton has a zero mode. This factor of τ appears after an integration over the collective coordinate τ_0 – the instanton center. The explicit result for the one-kink contribution to the correlator (4.16) may be written as [Pol77]

$$\langle x(0)\,x(\tau)\rangle \;=\; \frac{\mu^2}{\lambda}\left[1 - C\tau\sqrt{\frac{2\sqrt{2}\,\mu^3}{3\lambda}}\exp\left(-\frac{2\sqrt{2}\,\mu^3}{3\lambda}\right)\right], \tag{4.25}$$

where C is a (dimensional) constant.

Problem 4.3 Derive Eq. (4.25) using the Faddeev–Popov method to deal with the collective coordinate τ_0.

Solution Let us approximate the path integrals in the numerator and denominator of Eq. (4.16) for small λ by the sum of the contributions from the trivial minima (4.4) and the one-kink minima (4.19) and (4.20). Since the one-kink contribution is suppressed by $\exp\left(-S[x_{\text{inst}}]\right)$, we can expand the denominator to give

$$\langle\, x(0)\,x(\tau)\,\rangle \;\;=\;\; \frac{\mu^2}{\lambda} + e^{-S[x_{\text{inst}}]}\, \frac{\displaystyle\int \mathcal{D}x(\tau)\left[x(0)x(\tau) - \frac{\mu^2}{\lambda}\right] e^{-(S[x]-S[x_{\text{inst}}])}}{\displaystyle\int \mathcal{D}\chi(\tau)\, e^{-\int d\tau\left(\frac{1}{2}\dot{\chi}^2 + \mu^2\chi^2\right)}}\,,$$

(4.26)

where the path integral in the numerator is over fluctuations around the instanton solution (4.19). The normalizing factor in the denominator is associated with averaging over the Gaussian fluctuations around the trivial minima (4.4), the potential energy of which is described by the quadratic term in Eq. (4.7). There are two such trivial minima (x_+ and x_-) and two one-kink minima (instanton and anti-instanton) so these factors of 2 cancel.

Keeping the quadratic term in the expansion around the instanton:

$$x(\tau) \;\;=\;\; x_{\text{inst}}(\tau - \tau_0) + \chi(\tau - \tau_0)\,,$$

(4.27)

one obtains

$$S[x] - S[x_{\text{inst}}] \;\;=\;\; \frac{1}{2}\int d\tau\,\left(\dot{\chi}^2 - \mu^2\chi^2 + 3\lambda x_{\text{inst}}^2\chi^2\right)\,.$$

(4.28)

The fluctuations around the instanton are Gaussian except for one mode, which is associated with a translation of the instanton center, τ_0. This zero mode is given by

$$\chi_0(\tau) \;\propto\; \dot{x}_{\text{inst}}(\tau)\,.$$

(4.29)

This is obvious because

$$\left(-\frac{d^2}{d\tau^2} - \mu^2 + 3\lambda x_{\text{inst}}^2\right)\dot{x}_{\text{inst}} \;\;=\;\; 0$$

(4.30)

as a result of differentiating Eq. (4.18) with respect to τ_0.

To deal with the zero mode, let us insert

$$1 \;\;=\;\; \int_{-\infty}^{+\infty} d\tau\,\delta(u[x] - \tau)$$

(4.31)

into the path integral in the numerator on the RHS of Eq. (4.26). Here $u[x]$ is determined by the equation

$$\int_{-\infty}^{+\infty} d\tau\, y(\tau - u[x])\,x(\tau) \;\;=\;\; 0$$

(4.32)

with

$$y(\tau) \;\;=\;\; \frac{\dot{x}(\tau)}{\left[\displaystyle\int_{-\infty}^{+\infty} dt\,\dot{x}^2(t)\right]^{1/2}}$$

(4.33)

which is the normalized derivative of $x(\tau)$.

Under the translation,

$$\tau \quad \rightarrow \quad \tau' = \tau - \tau_0, \tag{4.34}$$

one obtains

$$x(\tau) \quad \rightarrow \quad x(\tau') = x(\tau - \tau_0). \tag{4.35}$$

This leaves the measure and the action in the path integral (4.26) invariant, while

$$u[x] \quad \rightarrow \quad u[x] + \tau_0. \tag{4.36}$$

Therefore, the integration over the instanton center, τ_0, in the numerator of Eq. (4.26) factorizes and we find

$$\int \mathcal{D}x(\tau) \left(x(0) x(\tau) - \frac{\mu^2}{\lambda} \right) e^{-(S[x] - S[x_{\text{inst}}])}$$

$$= \int_{-\infty}^{+\infty} d\tau_0 \left[x_{\text{inst}}(-\tau_0) x_{\text{inst}}(\tau - \tau_0) - \frac{\mu^2}{\lambda} \right]$$

$$\times \int \mathcal{D}\chi(\tau) \, \delta(u[x_{\text{inst}}(\tau) + \chi(\tau)]) \, e^{-\frac{1}{2} \int d\tau (\dot{\chi}^2 - \mu^2 \chi^2 + 3\lambda x_{\text{inst}}^2 \chi^2)}. \tag{4.37}$$

We have substituted the integration over the zero mode χ_0 by integration over the collective coordinate τ_0. The remaining path integral is finite since the integration runs over directions which are orthogonal to the zero mode.

The integral over τ_0 is equal to

$$\int_{-\infty}^{+\infty} d\tau_0 \left[x_{\text{inst}}(-\tau_0) x_{\text{inst}}(\tau - \tau_0) - \frac{\mu^2}{\lambda} \right] = -\frac{2\mu^2}{\lambda} \tau \tag{4.38}$$

as $\lambda \to 0$. This is because

$$x_{\text{inst}}(\tau - \tau_0) = \frac{\mu}{\sqrt{\lambda}} \text{sign}(\tau - \tau_0) \tag{4.39}$$

as $\lambda \to 0$.

Expanding the delta-function in χ:

$$\delta(u[x]) = \left| \int_{-\infty}^{+\infty} d\tau \, \dot{y}_{\text{inst}}(\tau) x_{\text{inst}}(\tau) \right| \delta \left(\int_{-\infty}^{+\infty} d\tau \, y_{\text{inst}}(\tau) \chi(\tau) \right), \tag{4.40}$$

and noting that

$$\int_{-\infty}^{+\infty} d\tau \, \dot{x}_{\text{inst}}^2(\tau) = \frac{2\sqrt{2} \, \mu^3}{3\lambda}, \tag{4.41}$$

we obtain

$$\int \mathcal{D}\chi(\tau)\, \delta(u[x_{\text{inst}}(\tau) + \chi(\tau)])\, e^{-\frac{1}{2}\int d\tau(\dot{\chi}^2 - \mu^2\chi^2 + 3\lambda x_{\text{inst}}^2\chi^2)}$$

$$= \sqrt{\frac{2\sqrt{2}\,\mu^3}{3\lambda}} \int \mathcal{D}\chi(\tau)\, \delta\left(\int_{-\infty}^{+\infty} d\tau\, y_{\text{inst}}(\tau)\chi(\tau)\right)\, e^{-\frac{1}{2}\int d\tau(\dot{\chi}^2 - \mu^2\chi^2 + 3\lambda x_{\text{inst}}^2\chi^2)}.$$

(4.42)

Note the appearance of the factor of $\sqrt{S[x_{\text{inst}}]}$.

We have thus obtained Eq. (4.25) with

$$C = 2 \frac{\int \mathcal{D}\chi(\tau)\, \delta\left(\int_{-\infty}^{+\infty} d\tau\, y_{\text{inst}}(\tau)\chi(\tau)\right)\, e^{-\frac{1}{2}\int d\tau(\dot{\chi}^2 - \mu^2\chi^2 + 3\lambda x_{\text{inst}}^2\chi^2)}}{\int \mathcal{D}\chi(\tau)\, e^{-\int d\tau(\frac{1}{2}\dot{\chi}^2 + \mu^2\chi^2)}}.$$

(4.43)

Problem 4.4 Calculate the ratio of determinants in Eq. (4.43).

Solution Let us introduce the notation

$$z = \frac{\mu\tau}{\sqrt{2}}, \qquad D = \frac{d}{dz}.$$

(4.44)

Noting that

$$\lambda x_{\text{inst}}^2(\tau) = \mu^2\left(1 - \frac{1}{\cosh^2 z}\right),$$

(4.45)

we can rewrite the ratio of determinants as

$$B^{-2} = \frac{4\pi}{\mu^2} \frac{\det'\left[-D^2 + 4 - 6/\cosh^2 z\right]}{\det\left[-D^2 + 4\right]}.$$

(4.46)

The notation $\det'$ means that the zero eigenvalue is excluded. An extra factor of 2π comes from the normalization of the Gaussian integral in the denominator which involves one further integral.

The RHS of Eq. (4.46) can be calculated using the limiting procedure

$$\frac{\det'\left[-D^2 + 4 - 6/\cosh^2 z\right]}{\det\left[-D^2 + 4\right]} = \lim_{\omega \to 2} \frac{\det\left[-D^2 + \omega^2 - 6/\cosh^2 z\right]}{(\omega^2 - 4)\det\left[-D^2 + \omega^2\right]}.$$

(4.47)

To compute the ratio of the Fredholm determinants

$$\mathcal{R}_\omega[v] \equiv \frac{\det\left[-D^2 + \omega^2 + v(z)\right]}{\det\left[-D^2 + \omega^2\right]}$$

(4.48)

for the potential

$$v(z) = -\frac{6}{\cosh^2 z},$$

(4.49)

let us note that

$$\frac{\partial}{\partial \omega^2} \ln \mathcal{R}_\omega[v] = \text{Tr}\left[\frac{1}{-D^2 + \omega^2 + v(z)}\right] - \text{Tr}\left[\frac{1}{-D^2 + \omega^2}\right]$$

$$= \int_{-\infty}^{+\infty} dz \left[R_\omega(z, z; v) - \frac{1}{2\omega}\right], \qquad (4.50)$$

where the diagonal resolvent $R_\omega(z, z; v)$ is defined by Eq. (1.123) with $G = 1$ and $V \equiv v$. The term $1/2\omega$ on the RHS, which equals the diagonal resolvent in the free case when $v = 0$ (see Eq. (1.38)), comes from the free determinant in the denominator on the RHS of Eq. (4.48).

A crucial observation is that the diagonal resolvent for the potential (4.49) is given by the simple formula

$$R_\omega(z, z; v) = \frac{1}{2\omega} - \frac{v(z)}{4\omega(\omega^2 - 1)} + \frac{v^2(z)}{8\omega(\omega^2 - 1)(\omega^2 - 4)}, \qquad (4.51)$$

which can easily be verified by substituting into the Gel'fand–Dikii equation (1.127) with $\mathcal{G} = 1$. The reason for this is that the potential (4.49) is integrable and possesses two bound states (see, for example, §23 of [LL74]).

Calculating the integral over z on the RHS of Eq. (4.50), using the formulas

$$\int_{-\infty}^{+\infty} \frac{dz}{\cosh^2 z} = 2, \qquad \int_{-\infty}^{+\infty} \frac{dz}{\cosh^4 z} = \frac{4}{3}, \qquad (4.52)$$

we obtain

$$\frac{\partial}{\partial \omega^2} \ln \mathcal{R}_\omega[v] = \frac{1}{\omega}\left(\frac{1}{\omega^2 - 1} + \frac{2}{\omega^2 - 4}\right), \qquad (4.53)$$

which is easily integrated over ω to give

$$\frac{\det\left[-D^2 + \omega^2 - 6/\cosh^2 z\right]}{\det\left[-D^2 + \omega^2\right]} = \frac{(\omega - 2)(\omega - 1)}{(\omega + 2)(\omega + 1)}. \qquad (4.54)$$

The integration constant has been determined by requiring that

$$\lim_{\omega \to \infty} \mathcal{R}_\omega[v] = 1. \qquad (4.55)$$

Substituting into Eq. (4.47), we obtain

$$C = 2B = \sqrt{\frac{48}{\pi}}\mu \qquad (4.56)$$

which coincides with the constant in Eq. (4.10).

For other methods of calculating the ratio of determinants in the one-instanton contribution, see the original papers [Lan67, Pol77], the reviews [Col77, VZN82] or Chapter 4 of the book [Pol87].

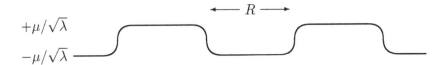

Fig. 4.3. The many-kink configuration $x_{M\text{-kink}}(\tau)$ which is combined from the solution (4.19).

4.4 Symmetry restoration by instantons

At $\tau \sim 1/\Delta E$, many kinks become essential. A many-kink "solution" can be approximately constructed from several single kinks and antikinks, which are separated along the τ-axis by the some distance $R \gg 1/\mu$, since the interaction between kinks would be $\sim \exp(-\mu R)$. Such a configuration is depicted in Fig. 4.3 for the case when the number of kinks is equal to the number of antikinks. An analogous configuration with the number of kinks being one more greater than the number of antikinks connects the $-\mu/\sqrt{\lambda}$ and $\mu/\sqrt{\lambda}$ vacua.

It is not an exact solution of Eq. (4.18) since the kink and the antikink attract and have a tendency to annihilate. However, it is an approximate solution as $\lambda \to 0$.

Analytically, the M-kink configuration can be represented as

$$x_{M\text{-kink}}(\tau) \;=\; \frac{\mu}{\sqrt{\lambda}} \prod_{i=1}^{M} \operatorname{sign}(\tau - \tau_i) , \tag{4.57}$$

where τ_i are the centers of the instantons (or anti-instantons), from which the M-kink configuration is built out, and

$$\tau_1 \leq \tau_2 \leq \cdots \leq \tau_M . \tag{4.58}$$

Equation (4.57) assumes that the kinks do not interact and are infinitely thin as $\lambda \to 0$. The action of the configuration (4.57) is therefore given by

$$S[x_{M\text{-kink}}] \;=\; \frac{2\sqrt{2}\mu^3}{3\lambda} M , \tag{4.59}$$

i.e. it equals M times the action for the one-kink case.

Summing over many-kink configurations, one finds [Pol77]

$$\langle\, x(0)\, x(\tau)\,\rangle \;=\; \frac{\mu^2}{\lambda}\, \mathrm{e}^{-\tau \Delta E} , \tag{4.60}$$

where ΔE is given by Eq. (4.10). The $x \to -x$ symmetry is now restored as $\tau \to \infty$. This restoration is produced by instantons = classical trajectories with a finite (Euclidean) action.

Fig. 4.4.   Graphical representation of a periodic potential.

Problem 4.5 Obtain the exponentiation of the one-kink contribution (4.25) after summing over the M-kink configurations (4.57) in the dilute gas approximation when the interaction between kinks is disregarded.

Solution The calculation of the contribution of the M-kink configuration (4.57) to the path integral is quite analogous to that for the one-kink case which is described in Problem 4.3. One finds

$$\langle x(0)\,x(\tau)\rangle \;=\; \frac{\mu^2}{\lambda}\sum_{M=0}^{\infty}(-\Delta E)^M\int_0^{\tau}\mathrm{d}\tau_1\int_0^{\tau_1}\mathrm{d}\tau_2\cdots\int_0^{\tau_{M-1}}\mathrm{d}\tau_M\,, \qquad (4.61)$$

which reproduces Eq. (4.60) by noting that the ordered integral is equal to

$$\int_0^{\tau}\mathrm{d}\tau_1\int_0^{\tau_1}\mathrm{d}\tau_2\cdots\int_0^{\tau_{M-1}}\mathrm{d}\tau_M \;=\; \frac{\tau^M}{M!}\,. \qquad (4.62)$$

This calculation is very similar to that in statistical mechanics for the exponentiation of a single-particle contribution to the partition function in the case of an ideal gas.

4.5 Topological charge and θ-vacua

Let us consider a periodic potential whose period equals 1, which is depicted in Fig. 4.4. It can be viewed as being defined on a circle S^1 of unit length. The boundary conditions are

$$\left.\begin{array}{rcl} x(1) &=& x(0) \quad \boxed{\text{in perturbation theory}}\,, \\[2mm] x(1) &=& x(0)+n \quad \boxed{\text{for } n\text{-instanton solution}}\,. \end{array}\right\} \qquad (4.63)$$

The multi-instanton solution always exists because of the topological formula[*]

$$\pi_1(S^1) \;=\; \mathbb{Z}\,, \qquad (4.64)$$

where $\pi_k(M)$ is the kth homotopy group with elements that are classes of continuous maps of the k-sphere S^k onto M. Equation (4.64) describes the fact that an (integer) winding number $n\in\mathbb{Z}$ is associated with the mapping $S^1\to S^1$, which counts how many times the target is covered.

---

[*] See, for example, the book [DNF86] (§17.5 of Part II).

We see the difference between the M-kink configuration for the double-well potential and the multi-instanton solution for the periodic potential. The former was not an exact solution of the classical field equation (4.18). Only a single instanton or anti-instanton was a solution that connects the two vacua. This is why we need a periodic potential for the multi-instanton solution to exist owing to the topological argument.

The value of n in the boundary condition (4.63) is called the *topological charge* of the instantons, while $n < 0$ is associated with anti-instantons. The vacuum states are labeled by n: $|n\rangle$. The n-instanton configuration connects the $|m\rangle$ and $\langle m+n|$ states. Therefore, instantons are associated in Minkowski space with the process of tunneling between topologically distinct vacua* rather than with real particles. For this reason, they are sometimes called pseudoparticles in Euclidean space.

It is convenient to consider another representation of vacuum states

$$|\theta\rangle \;=\; \sum_{n=-\infty}^{\infty} e^{i\theta n}\,|n\rangle,\qquad (4.65)$$

which are called the θ-vacua. The θ-vacua are orthogonal

$$\langle \theta \mid \theta'\rangle \;=\; \sum_{m=-\infty}^{\infty}\sum_{n=-\infty}^{\infty} e^{i(\theta n - \theta' m)}\,\langle m \mid n\rangle \;=\; \delta_{2\pi}\left(\theta - \theta'\right),\qquad (4.66)$$

where $\delta_{2\pi}$ is a periodic delta-function with period 2π. Here we have used the orthogonality of the n-states:

$$\langle m \mid n\rangle \;=\; \delta_{mn}.\qquad (4.67)$$

The θ-vacuum partition function is given by

$$Z(\theta) \;=\; \int \mathcal{D}x\, e^{-S[x]+i\theta\int_0^1 d\tau\, \dot{x}(\tau)}.\qquad (4.68)$$

Here in the exponent θ is multiplied by the topological charge

$$\int_0^1 d\tau\, \dot{x}(\tau) \;=\; x(1) - x(0)\qquad (4.69)$$

---

* The Minkowski-space interpretation of instantons is attributed to V.N. Gribov (unpublished). It is based on the fact that when the particle is localized in one of the two wells its momentum is indefinite and can sometimes be very large so that the proper energy is above the barrier between the two wells. Such a particle jumps from the given well to the other one. The characteristic time of this process is small in the typical units given by μ. In other words, this process is instantaneous, which explains the term "instanton" as introduced by 't Hooft. The exponential suppression with λ of the one-instanton contribution (4.25) represents quantitatively the fact that the probability of having large momentum is small.

which never appears in perturbation theory. Therefore, the partition function (4.68) can be alternatively represented as

$$Z(\theta) \;=\; \sum_n \int\limits_{x(1)=x(0)+n} \mathcal{D}x\,e^{-S[x]+i\theta n}\,. \tag{4.70}$$

The second term in the exponent in Eq. (4.68) is known as the θ-*term*. The parameter θ plays the role of a new fundamental constant that does not show up in perturbation theory. The amplitude of physical processes generated by instantons may depend on θ.

Remark on description of instantons

A description of instantons in the first-quantized language can only be given in quantum mechanics (where the first and second quantizations do not differ essentially). The path-integral representation (4.16) is more in the spirit of second quantization, which is discussed in Chapter 2, where $x(\tau)$ plays the role of a field that depends on the one-dimensional coordinate τ.

Remark on instantons in Yang–Mills theory

In the Yang–Mills theory, instantons are conveniently described by a (Euclidean) path integral over fields. The saddle-point equation, which describes instantons in the $SU(2)$ Yang–Mills theory, is given by [BPS75]

$$F_{\mu\nu}^a(x) \;=\; \tilde{F}_{\mu\nu}^a(x)\,, \tag{4.71}$$

for which nontrivial solutions exist owing to the fact that the mapping of the asymptotic boundary S^3 of four-dimensional Euclidean space onto $SU(2)$ is nontrivial:

$$\pi_3\left(SU(2)\right) \;=\; \mathbb{Z}\,. \tag{4.72}$$

Correspondingly, the topological charge is given by[*]

$$n \;=\; \frac{g^2}{16\pi^2}\int d^4x \sum_{a=1}^{3} F_{\mu\nu}^a(x)\tilde{F}_{\mu\nu}^a(x)\,, \tag{4.73}$$

which equals one-half of the nonconservation of the axial charge given by the Minkowski-space integral of the chiral anomaly (3.63). This expression is also known in topology as the Pontryagin index or the second Chern class. See, for example, the lectures/reviews [Col77, VZN82, SS98] and the book [Shi94] for an introduction to instantons in Yang–Mills theory.

---

[*] Concerning the coefficient, see the footnote on p. 59.

Bibliography to Part 1

Reference guide

The operator formalism in quantum field theory is described in the canonical books [AB69, BS76, BD65] which were written in the 1950s or at the beginning of the 1960s. Modern textbooks on this subject include those by Brown [Bro92] and Weinberg [Wei98].

Feynman disentangling is contained in the original paper [Fey51], the appendices of which are especially relevant. A classic book on path integrals in quantum mechanics is that by Feynman and Hibbs [FH65]. The path-integral approach to the very closely related problem of Brownian motion is discussed in the books [Kac59, Sch81, Wie86, Roe94]. Many information on path integrals can be found in the book by Kleinert [Kle95].

An introduction to path integrals in quantum mechanics and quantum field theory can be found in many books. I shall list some of those that I have on my bookshelf: [Ber86, Pop91, FS80, IZ80, Ram89, Sak85, Riv88]. The ordering is according to the appearance of the first edition. The book by Berezin [Ber86], which is mathematically more rigorous, contains an excellent description of operations with Grassmann variables.

An introduction to path integrals in statistical mechanics can be found in the books [Kac59, Fey72, Pop91, Wie86, ID91, Roe94]. The well-written book by Parisi [Par88] describes a modern view of the relation between statistical mechanics and quantum field theory. A very good, while slightly more advanced, book where contemporary problems of quantum field theory and statistical mechanics are discussed using the unified language of Euclidean path integrals is that by Polyakov [Pol87].

The derivation of quantum anomalies from the noninvariance of the measure in the path integral is contained in the original papers [Ver78, Fuj79, Fuj80] (see also the review [Mor86]). It can also be found in Chapter 22 of the book by Weinberg [Wei98].

Instantons in the Yang–Mills theory were discovered by Belavin, Polyakov, Schwartz and Tyupkin [BPS75]. The role of instantons in quantum mechanics is clarified in the original paper by Polyakov [Pol77]. Their description is given in the books by Sakita [Sak85] and Polyakov [Pol87]. The review articles [Col77, VZN82, SS98] are also useful for an introduction to the subject. The original papers on instantons in quantum field theory are collected in the book edited by Shifman [Shi94].

References

[AB69] AKHIEZER A.I. AND BERESTETSKII V.B. *Quantum electrodynamics* (Oldbourne Press, London, 1962).

[ACD77] ADLER S.L., COLLINS J.C., AND DUNCAN A. 'Energy–momentum-tensor trace anomaly in spin-1/2 quantum electrodynamics'. *Phys. Rev.* **D15** (1977) 1712.

[BD65] BJORKEN J.D. AND DRELL S.D. *Relativistic quantum fields* (McGraw-Hill, New York, 1965).

[Ber86] BEREZIN F.A. *The method of second quantization* (Academic Press, New York, 1966).

[BPS75] BELAVIN A.A., POLYAKOV A.M., SCHWARTZ A.S., AND TYUPKIN YU.S. 'Pseudoparticle solutions of the Yang–Mills equations'. *Phys. Lett.* **B59** (1975) 85.

[Bro92] BROWN L.S. *Quantum field theory* (Cambridge Univ. Press, 1992).

[BS76] BOGOLIUBOV N.N. AND SHIRKOV D.V. *Introduction to the theory of quantized fields* (Interscience, New York, 1959).

[CDJ77] COLLINS J.C., DUNCAN A., AND JOGLEKAR S.D. 'Trace and dilatation anomalies in gauge theories'. *Phys. Rev.* **D16** (1977) 438.

[CE72] CHANOWITZ M.S. AND ELLIS J. 'Canonical anomalies and broken scale invariance'. *Phys. Lett.* **B40** (1972) 397.

[Col77] COLEMAN S. 'The uses of instantons', in Proc. of Erice Int. School of subnuclear physics 1977 (Plenum, New York, 1979) p. 805. Reprinted in COLEMAN S. *Aspects of symmetry* (Cambridge Univ. Press, 1985) p. 265.

[Col86] COLLINS J.C. *Renormalization* (Cambridge Univ. Press, 1986).

[Cre72] CREWTHER R. 'Nonperturbative evaluation of the anomalies in low-energy theorems'. *Phys. Rev. Lett.* **28** (1972) 1421.

[Dir58] DIRAC P.A.M. *The principles of quantum mechanics* (Oxford Univ. Press, 1958).

[DNF86] DUBROVIN B.A., NOVIKOV C.P. AND FOMENKO A.T. *Modern geometry. Methods and applications* (Springer-Verlag, Berlin, Part I 1984, Part II 1985).

[Dys49] DYSON F.J. 'The radiation theories of Tomonaga, Schwinger, and Feynman'. *Phys. Rev.* **75** (1949) 486.

[Fey50] FEYNMAN R.P. 'Mathematical formulation of the quantum theory of electromagnetic interaction'. *Phys. Rev.* **80** (1950) 440.

[Fey51] FEYNMAN R.P. 'An operator calculus having applications in quantum electrodynamics'. *Phys. Rev.* **84** (1951) 108.

[Fey72] FEYNMAN R.P. *Statistical mechanics. A set of lectures* (Benjamin, Reading, 1972).

[FH65] FEYNMAN R.P. AND HIBBS A.R. *Quantum mechanics and path integrals* (McGraw-Hill, New York, 1965).

[FS80] FADDEEV L.D. AND SLAVNOV A.A. *Gauge fields. Introduction to quantum theory* (Benjamin, Reading, 1980).

[Fuj79] FUJIKAWA K. 'Path-integral measure for gauge-invariant fermion theories'. *Phys. Rev. Lett.* **42** (1979) 1195.

[Fuj80] FUJIKAWA K. 'Path integral for gauge theories with fermions'. *Phys. Rev.* **D21** (1980) 2848, **D22** (1980) 1499(E).

[GD75] GEL'FAND I.M. AND DIKII L.A. 'Asymptotic behavior of the resolvent of Sturm–Liouville equations and the algebra of the Korteweg–de Vries equations'. *Russ. Math. Surv.* **30** (1975) v. 5, p. 77.

[ID91] ITZYKSON C. AND DROUFFE J.-M. *Statistical field theory*, vol. 1, 2 (Cambridge Univ. Press, 1991).

[IZ80] ITZYKSON C. AND ZUBER J.-B. *Quantum field theory* (McGraw-Hill, New York, 1980).

[Kac59] KAC M. *Probability and related topics in the physical science* (Interscience, New York, 1959).

[Kle95] KLEINERT H. *Path integrals in quantum mechanics, statistics, and polymer physics*, 2nd edn (World Scientific, Singapore, 1995).

[Lan67] LANGER J.S. 'Theory of the condensation point'. *Ann. Phys.* **41** (1967) 108.

[Lev51] LÉVY P. *Problèmes concrets d'analyse fonctionnelle* (Gauthier-Villars, Paris, 1951).

[LL74] LANDAU L.D. AND LIFSHITS E.M. *Quantum mechanics. Nonrelativistic theory* (Pergamon, New York, 1977).

[MM81] MAKEENKO YU.M. AND MIGDAL A.A. 'Quantum chromodynamics as dynamics of loops'. *Nucl. Phys.* **B188** (1981) 269.

[Mor86] MOROZOV A.YU. 'Anomalies in gauge theories'. *Sov. Phys. Usp.* **29** (1986) 993.

[Nie77] NIELSEN N.K. 'Energy–momentum tensor in a non-Abelian quark–gluon theory'. *Nucl. Phys.* **B120** (1977) 212.

[Par88] PARISI G. *Statistical field theory* (Addison-Wesley, New York, 1988).

Bibliography to Part 1

[Pol77]   POLYAKOV A.M. 'Quark confinement and topology of gauge groups'. *Nucl. Phys.* **B120** (1977) 429.

[Pol87]   POLYAKOV A.M. *Gauge fields and strings* (Harwood Academic Pub., Chur, 1987).

[Pop91]   POPOV V.N. *Functional integrals and collective excitations* (Cambridge Univ. Press, 1991).

[Ram89]   RAMOND P. *Field theory. A modern primer*, 2nd edn (Addison-Wesley, New York, 1989).

[Riv88]   RIVERS R. *Path-integral methods in quantum field theory* (Cambridge Univ. Press, 1988).

[Roe94]   ROEPSTORFF G. *Path-integral approach in quantum physics* (Springer-Verlag, Berlin, 1994).

[Sak85]   SAKITA B. *Quantum theory of many-variable systems and fields* (World Scientific, Singapore, 1985).

[Sch58]   SCHWINGER J. *Selected papers on quantum electrodynamics* (Dover, New York, 1958).

[Sch81]   SCHULMAN L.S. *Techniques and applications of path integration* (Wiley, New York, 1981).

[Shi94]   SHIFMAN M. *Instantons in gauge theories* (World Scientific, Singapore, 1994).

[SS98]   SCHAFER T. AND SHURYAK E.V. 'Instantons in QCD'. *Rev. Mod. Phys.* **70** (1998) 323.

[Ver78]   VERGELES S.N. unpublished, as quoted in MIGDAL A.A. 'Effective low-energy Lagrangian for QCD'. *Phys. Lett.* **B81** (1979) 37.

[VZN82]   VAINSHTEIN A.I., ZAKHAROV V.I., NOVIKOV V.A., AND SHIFMAN M.A. 'ABC of instantons'. *Sov. Phys. Usp.* **25** (1982) 195.

[Wei98]   WEINBERG S. *Quantum fields* (Cambridge Univ. Press, 1998).

[Wie23]   WIENER N. 'Differential space'. *J. Math. Phys.* **2** (1923) v. 3, p. 131.

[Wie86]   WIEGEL F.W. *Introduction to path-integral methods in physics and polymer science* (World Scientific, Singapore, 1986).

Part 2
Lattice Gauge Theories

> "I never said it."
> "Now you are telling us when you did say it. I'm asking you to tell us when you didn't say it."
>
> J. HELLER, *Catch-22*

Lattice gauge theories in their modern form were proposed in 1974 by Wilson [Wil74] in connection with the problem of quark confinement in quantum chromodynamics (QCD).

Lattice gauge theories are a nonperturbative regularization of a gauge theory. The lattice formulation is a nontrivial definition of a gauge theory beyond perturbation theory. The problem of nonperturbative quantization of gauge fields is solved in a simple and elegant way on a lattice.

The use of the lattice formulation clarifies an analogy between quantum field theory and statistical mechanics. It offers the possibility of applying nonperturbative methods, such as the strong-coupling expansion or the numerical Monte Carlo method, to quantum chromodynamics and to other gauge theories, which provide evidence for quark confinement.

However, the lattice in QCD is no more than an auxiliary tool in obtaining results for the continuum limit. In order to pass to the continuum, the lattice spacing should be many times smaller than the characteristic scale of the strong interaction.

We shall start this part with a description of the continuum formulation of non-Abelian gauge theories, and will return to this from time to time when discussing the lattice approach. The point is that some ideas, e.g. concerning the possibility of reformulating gauge theories in terms of gauge-invariant variables, which were originally introduced by Wilson [Wil74] on a lattice, are applicable for the continuum theory as well.

5

Observables in gauge theories

Modern theories of fundamental interactions are gauge theories. The principle of local gauge invariance was introduced by H. Weyl for the electromagnetic interaction in analogy with general covariance in Einstein's theory of gravitation. An extension to non-Abelian gauge groups was given by Yang and Mills [YM54].

A crucial role in gauge theories is played by the phase factor which is associated with parallel transport in an external gauge field. The phase factors are observable in quantum theory, in contrast with the classical theory. For the electromagnetic field, this is known as the Aharonov–Bohm effect.

In this chapter we initially consider the matrix notation for the non-Abelian gauge fields and introduce proper non-Abelian phase factors. Then we discuss the relation between observables in classical and quantum theories.

5.1 Gauge invariance

The principle of local gauge invariance deals with the gauge transformation (g.t.) of a matter field ψ, which is given by

$$\psi(x) \xrightarrow{\text{g.t.}} \psi'(x) = \Omega(x)\,\psi(x)\,. \tag{5.1}$$

Here $\Omega(x) \in G$ with G being a semisimple Lie group which is called the *gauge group* ($G = SU(3)$ for QCD). Equation (5.1) demonstrates that ψ belongs to the fundamental representation of G.

We restrict ourselves to a unitary gauge group when

$$\Omega^{-1}(x) = \Omega^{\dagger}(x)\,, \tag{5.2}$$

while an extension to other Lie groups is obvious. Then we have

$$\psi^\dagger(x) \xrightarrow{\text{g.t.}} \psi'^\dagger(x) = \psi^\dagger(x)\,\Omega^\dagger(x). \tag{5.3}$$

In analogy with QCD, the gauge group $G = SU(N)$ is usually associated with *color*, while the proper index of ψ is called the color index.

The gauge transformation (5.1) of the matter field ψ can be compensated by a transformation of the non-Abelian gauge field $\mathcal{A}_\mu$ which belongs to the adjoint representation of G:

$$\mathcal{A}_\mu(x) \xrightarrow{\text{g.t.}} \mathcal{A}'_\mu(x) = \Omega(x)\,\mathcal{A}_\mu(x)\,\Omega^\dagger(x) + \mathrm{i}\,\Omega(x)\,\partial_\mu\Omega^\dagger(x). \tag{5.4}$$

We have introduced in Eq. (5.4) the Hermitian matrix $\mathcal{A}_\mu(x)$ with the elements

$$[\mathcal{A}_\mu(x)]^{ij} = g\sum_a A^a_\mu(x)\,[t^a]^{ij}. \tag{5.5}$$

Here $[t^a]^{ij}$ are the generators of G ($a = 1,\dots,N^2 - 1$ for $SU(N)$) which are normalized such that[*]

$$\operatorname{tr} t^a t^b = \delta^{ab}, \tag{5.6}$$

where tr is the trace over the matrix indices i and j, while g is the gauge coupling constant.

Equation (5.5) can be inverted to give

$$A^a_\mu(x) = \frac{1}{g}\operatorname{tr}\mathcal{A}_\mu(x)\,t^a. \tag{5.7}$$

Substituting

$$\Omega(x) = \mathrm{e}^{\mathrm{i}\alpha(x)}, \tag{5.8}$$

we obtain for an infinitesimal α:

$$\delta\,\mathcal{A}_\mu(x) \overset{\text{g.t.}}{=} \nabla^{\text{adj}}_\mu\,\alpha(x). \tag{5.9}$$

Here

$$\nabla^{\text{adj}}_\mu\,\alpha \equiv \partial_\mu\alpha - \mathrm{i}\,[\mathcal{A}_\mu,\alpha] \tag{5.10}$$

is the covariant derivative in the adjoint representation of G, while

$$\nabla^{\text{fun}}_\mu\,\psi \equiv \partial_\mu\psi - \mathrm{i}\,\mathcal{A}_\mu\psi \tag{5.11}$$

---

[*] Quite often another normalization of the generators with an extra factor of $1/2$, $\operatorname{tr}\tilde{t}^a\tilde{t}^b = \frac{1}{2}\delta^{ab}$, is used for historical reasons, in particular, $\tilde{t}^a = \sigma^a/2$ for the $SU(2)$ group, where σ^a are the Pauli matrices. This results in the redefinition of the coupling constant, $\tilde{g}^2 = 2g^2$.

is that in the fundamental representation. It is evident that

$$\nabla_\mu^{\text{adj}} B(x) = [\nabla_\mu^{\text{fun}}, B(x)], \tag{5.12}$$

where $B(x)$ is a matrix-valued function of x.

The QCD action is given in the matrix notation as

$$S[\mathcal{A}, \psi, \bar{\psi}] = \int \mathrm{d}^4 x \left[\bar{\psi} \gamma_\mu (\partial_\mu - \mathrm{i}\mathcal{A}_\mu) \psi + m\bar{\psi}\psi + \frac{1}{4g^2} \operatorname{tr} \mathcal{F}_{\mu\nu}^2 \right], \tag{5.13}$$

where

$$\mathcal{F}_{\mu\nu} = \partial_\mu \mathcal{A}_\nu - \partial_\nu \mathcal{A}_\mu - \mathrm{i}[\mathcal{A}_\mu, \mathcal{A}_\nu] \tag{5.14}$$

is the (Hermitian) matrix of the non-Abelian field strength.

The action (5.13) is manifestly invariant under the local gauge transformation (5.1) and (5.4) since

$$\mathcal{F}_{\mu\nu}(x) \overset{\text{g.t.}}{\longrightarrow} \Omega(x)\,\mathcal{F}_{\mu\nu}(x)\,\Omega^\dagger(x) \tag{5.15}$$

or

$$\delta\mathcal{F}_{\mu\nu}(x) \overset{\text{g.t.}}{=} -\mathrm{i}[\mathcal{F}_{\mu\nu}(x), \alpha(x)] \tag{5.16}$$

for the infinitesimal gauge transformation.

For the Abelian group $G = U(1)$, the above formulas recover those of the previous part for QED where we have already used the calligraphic notation in Problem 3.6 on p. 61.

Problem 5.1 Rewrite classical equations of motion in the matrix notation.

Solution The non-Abelian Maxwell equation and the Bianchi identity are given, respectively, as

$$\nabla_\mu^{\text{adj}} \mathcal{F}_{\mu\nu} = 0 \tag{5.17}$$

and

$$\nabla_\mu^{\text{adj}} \tilde{\mathcal{F}}_{\mu\nu} = 0, \tag{5.18}$$

where the dual field strength is defined by Eq. (3.49). Rewriting Eq. (5.14) as

$$\mathcal{F}_{\mu\nu} = \mathrm{i}[\nabla_\mu^{\text{fun}}, \nabla_\nu^{\text{fun}}] \tag{5.19}$$

and using Eq. (5.12), we represent the Bianchi identity as

$$\epsilon_{\mu\nu\lambda\rho}[\nabla_\mu^{\text{fun}}, [\nabla_\nu^{\text{fun}}, \nabla_\lambda^{\text{fun}}]] = 0 \tag{5.20}$$

which is obviously satisfied owing to the Jacobi identity.

We have thus proven the well-known fact that the Bianchi identity is satisfied explicitly in the second-order formalism, where $\mathcal{F}_{\mu\nu}$ is expressed via $\mathcal{A}_\mu$ by virtue of Eq. (5.14). In contrast, $\mathcal{A}_\mu$ and $\mathcal{F}_{\mu\nu}$ are considered to be independent variables in the first-order formalism, where both equations (5.17) and (5.18) are essential. The concept of the first- and second-order formalisms comes from the theory of gravity.

5.2 Phase factors (definition)

In order to compare the phases of wave functions at distinct points, one needs a non-Abelian extension of the parallel transporter that was considered in Sect. 1.7. The proper extension of the Abelian formula (1.158) is written as

$$U[\Gamma_{yx}] \;=\; \boldsymbol{P}\, e^{i \int_{\Gamma_{yx}} dz^\mu \mathcal{A}_\mu(z)}. \tag{5.21}$$

Although the matrices $\mathcal{A}_\mu(z)$ do not commute, the path-ordered exponential on the RHS of Eq. (5.21) is defined unambiguously by the general method of Sect. 1.3. This is obvious after rewriting the phase factor in an equivalent form

$$\boldsymbol{P}\, e^{i \int_{\Gamma_{yx}} dz^\mu \mathcal{A}_\mu(z)} \;=\; \boldsymbol{P}\, e^{i \int_0^1 d\sigma\, \dot{z}^\mu(\sigma)\, \mathcal{A}_\mu(z(\sigma))}. \tag{5.22}$$

Therefore, the path-ordered exponential in Eq. (5.21) can be understood as*

$$U[\Gamma_{yx}] \;=\; \prod_{t=0}^{\tau} \left[1 + i\, dt\, \dot{z}^\mu(t)\, \mathcal{A}_\mu(z(t))\right]. \tag{5.23}$$

We have already used this notation for the product on the RHS in Problem 1.9 on p. 22. Using Eq. (1.157), Eq. (5.23) can also be written as

$$U[\Gamma_{yx}] \;=\; \prod_{z \in \Gamma_{yx}} \left[1 + i\, dz^\mu \mathcal{A}_\mu(z)\right]. \tag{5.24}$$

If the contour Γ_{yx} is discretized as is shown in Fig. 1.3, then the non-Abelian phase factor is approximated by

$$U[\Gamma_{yx}] \;=\; \lim_{M \to \infty} \prod_{i=1}^{M} \left[1 + i\,(z_i - z_{i-1})^\mu \mathcal{A}_\mu \left(\frac{z_i + z_{i-1}}{2}\right)\right], \tag{5.25}$$

which obviously reproduces (5.24) in the limit $\epsilon \to 0$.

Note that the non-Abelian phase factor (5.21) is, by construction, an element of the gauge group G itself, while $\mathcal{A}_\mu$ belongs to the Lie algebra of G.

---

* Sometimes the phase factor is defined using a similar formula but with the inverse order of multipliers. Our definition using Eq. (5.23) is exactly equivalent to Dyson's definition of the P-product (see the footnote on p. 3) which can be seen by choosing the contour Γ_{yx} to coincide with the temporal axis.

Problem 5.2 Write down an explicit expansion of the non-Abelian phase factor (5.21) in $\mathcal{A}_\mu$.

Solution Let us use the notation

$$\int_x^y dz^\mu \cdots \equiv \int_{\Gamma_{yx}} dz^\mu \cdots \tag{5.26}$$

for the integral along the contour Γ_{yx}. Then we have

$$P e^{i \int_x^y dz^\mu \mathcal{A}_\mu(z)}$$

$$= \sum_{k=0}^\infty i^n \int_x^y dz_1^{\mu_1} \int_{z_1}^y dz_2^{\mu_2} \cdots \int_{z_{k-1}}^y dz_k^{\mu_k} \mathcal{A}_{\mu_k}(z_k) \cdots \mathcal{A}_{\mu_2}(z_2) \mathcal{A}_{\mu_1}(z_1). \tag{5.27}$$

The ordered integral in this formula can be rewritten in a more symmetric form as

$$\int_0^\tau dt_1 \int_{t_1}^\tau dt_2 \cdots \int_{t_{k-1}}^\tau dt_k \, \dot{z}^{\mu_1}(t_1) \dot{z}^{\mu_2}(t_2) \cdots \dot{z}^{\mu_k}(t_k)$$

$$\times \mathcal{A}_{\mu_k}(z(t_k)) \cdots \mathcal{A}_{\mu_2}(z(t_2)) \mathcal{A}_{\mu_1}(z(t_1))$$

$$= \int_0^\tau dt_1 \int_0^\tau dt_2 \cdots \int_0^\tau dt_k \, \theta(t_k, t_{k-1}, \ldots, t_2, t_1) \, \dot{z}^{\mu_1}(t_1) \dot{z}^{\mu_2}(t_2) \cdots \dot{z}^{\mu_k}(t_k)$$

$$\times \mathcal{A}_{\mu_k}(z(t_k)) \cdots \mathcal{A}_{\mu_2}(z(t_2)) \mathcal{A}_{\mu_1}(z(t_1)), \tag{5.28}$$

where

$$\theta(t_k, t_{k-1}, t_{k-2}, \cdots, t_2, t_1) = \theta(t_k - t_{k-1}) \theta(t_{k-1} - t_{k-2}) \cdots \theta(t_2 - t_1) \tag{5.29}$$

orders the points along the contour. We shall also denote this theta in a parametrization-independent form as

$$\theta(k, k-1, k-2, \ldots, 2, 1) \equiv \theta(t_k, t_{k-1}, t_{k-2}, \ldots, t_2, t_1). \tag{5.30}$$

It satisfies the obvious identity

$$\theta(k, k-1, k-2, \ldots, 2, 1) + \theta(k-1, k, k-2, \ldots, 2, 1)$$
$$+ \text{(other permutations of } k, \ldots, 1)$$
$$= 1. \tag{5.31}$$

For the Abelian case, when $\mathcal{A}_{\mu_i}(z_i)$ commute, Eq. (5.31) results in

$$\int_x^y dz_1^{\mu_1} \int_{z_1}^y dz_2^{\mu_2} \cdots \int_{z_{k-1}}^y dz_k^{\mu_k} \mathcal{A}_{\mu_k}(z_k) \cdots \mathcal{A}_{\mu_2}(z_2) \mathcal{A}_{\mu_1}(z_1) = \frac{1}{k!} \left(\int_x^y dz^\mu \mathcal{A}_\mu(z) \right)^k \tag{5.32}$$

so that the Abelian exponential of the contour integral is reproduced.

Problem 5.3 Disentangle the non-Abelian phase factor using a path integral over Grassmann variables on a contour.

Solution Let us define the average

$$
\left\langle\, F[\psi,\bar\psi]\, \right\rangle_\psi \;=\; \frac{\displaystyle\int \mathcal{D}\bar\psi(t)\,\mathcal{D}\psi(t)\, \mathrm{e}^{-\int_0^\tau \mathrm{d}t\,\bar\psi(t)\dot\psi(t)-\bar\psi(0)\psi(0)}\, F[\psi,\bar\psi]}{\displaystyle\int \mathcal{D}\bar\psi(t)\,\mathcal{D}\psi(t)\, \mathrm{e}^{-\int_0^\tau \mathrm{d}t\,\bar\psi(t)\dot\psi(t)-\bar\psi(0)\psi(0)}}\,. \tag{5.33}
$$

The path integral in this formula looks like those of Chapter 2 with $\bar\psi_i(t)$ and $\psi_j(t)$ being Grassmann variables which depend on the one-dimensional variable $t \in [0,\tau]$ that parametrizes a contour, and i and j are the color indices.

The simplest average, which describes propagation of the color indices along the contour, is

$$
\left\langle\, \psi_i(t_2)\bar\psi_j(t_1)\, \right\rangle_\psi \;=\; \delta_{ij}\,\theta(t_2 - t_1)\,, \qquad 0 \le t_1, t_2 \le \tau\,. \tag{5.34}
$$

This can be easily checked, say, by deriving the Schwinger–Dyson equation

$$
\frac{\partial}{\partial t_2}\left\langle\, \psi_i(t_2)\bar\psi_j(t_1)\, \right\rangle_\psi \;=\; \delta_{ij}\,\delta^{(1)}(t_2 - t_1)\,, \qquad 0 < t_1, t_2 < \tau \tag{5.35}
$$

as was done in Chapter 3. We now see that we need the Grassmann variables because the operator in the action in Eq. (5.33) is $\partial/\partial t$.

A special comment is needed concerning the term $\bar\psi(0)\psi(0)$ in the exponents in Eq. (5.33), the appearance of which in the disentangling procedure is clarified in [HJS77]. The need for this term can be seen from the discretized version of the exponent:

$$
\int_0^\tau \mathrm{d}t\,\bar\psi(t)\dot\psi(t) + \bar\psi(0)\psi(0) \;\longrightarrow\; \sum_{n=1}^M \bar\psi(n\epsilon)\,[\psi(n\epsilon) - \psi(n\epsilon - \epsilon)] + \bar\psi(0)\psi(0)\,. \tag{5.36}
$$

For this discretization we immediately obtain

$$
\left\langle\psi_i(n\epsilon)\bar\psi_j(m\epsilon)\right\rangle_\psi \;=\; \begin{cases} \delta_{ij} & \text{for } n \ge m\,, \\ 0 & \text{for } n < m\,. \end{cases} \tag{5.37}
$$

The term $\bar\psi(0)\psi(0)$ is needed to provide nonvanishing integrals over $\bar\psi(0)$ and $\psi(0)$. It can also be seen from the discretized version that the path integral in the denominator on the RHS of Eq. (5.33) is equal to unity.

The fermionic path-integral representation for the non-Abelian phase factor (see, for example, [GN80]) is given as

$$
\left[\boldsymbol{P}\,\mathrm{e}^{\mathrm{i}\int_0^\tau \mathrm{d}t\,\dot z^\mu(t)\mathcal{A}_\mu(z(t))}\right]_{ij} \;=\; \left\langle\, \mathrm{e}^{\mathrm{i}\int_0^\tau \mathrm{d}t\,\dot z^\mu(t)\bar\psi(t)\mathcal{A}_\mu(z(t))\psi(t)}\,\psi_i(\tau)\bar\psi_j(0)\, \right\rangle_\psi\,. \tag{5.38}
$$

There is no path-ordering sign on the RHS since the matrix indices of $\mathcal{A}_\mu$ are contacted by ψ and $\bar\psi$.

In order to prove Eq. (5.38), one expands the exponential in $\mathcal{A}_\mu$ and calculates the average using Eq. (5.34) and the rules of Wick's pairing, which yields

$$\frac{1}{k!} \left\langle \psi_i(\tau) \left[\int_0^\tau dt\, \dot{z}^\mu(t)\, \bar{\psi}(t)\mathcal{A}_\mu(z(t))\psi(t) \right]^k \bar{\psi}_j(0) \right\rangle_\psi$$

$$= \int_0^\tau dt_1 \int_0^\tau dt_2 \cdots \int_0^\tau dt_k\, \theta(\tau, t_k, \ldots, t_2, t_1, 0)\, \dot{z}^{\mu_1}(t_1)\dot{z}^{\mu_2}(t_2)\cdots \dot{z}^{\mu_k}(t_k)$$

$$\times\, [\mathcal{A}_{\mu_k}(z(t_k))\cdots \mathcal{A}_{\mu_2}(z(t_2))\mathcal{A}_{\mu_1}(z(t_1))]_{ij}\,, \tag{5.39}$$

where $\theta(\tau, t_k, \ldots, t_2, t_1, 0)$ is given by Eq. (5.29). It is crucial in the derivation of this formula that only connected terms contribute to the average (5.33). Equation (5.39) reproduces Eq. (5.27) from the previous Problem, which completes the proof of Eq. (5.38). Moreover, we can say that the path integral (5.33) is nothing but a nice representation of the thetas (5.29).

Problem 5.4 Invert $(-\nabla^2 + m^2)$ when ∇_μ is in the fundamental representation.

Solution The calculation is quite analogous to that of the Problem 1.13 on p. 29. We first use the path-integral representation of the inverse operator:

$$G(x, y; \mathcal{A})$$

$$\equiv \left\langle y \left| \frac{1}{-\nabla_\mu^{\text{fun}}\nabla_\mu^{\text{fun}} + m^2} \right| x \right\rangle$$

$$= \frac{1}{2} \int_0^\infty d\tau\, e^{-\frac{1}{2}\tau m^2} \int_{z_\mu(0)=x_\mu} \mathcal{D}z_\mu(t)\, e^{-\frac{1}{2}\int_0^\tau dt\, \dot{z}_\mu^2(t)} \left\langle y \left| \boldsymbol{P} e^{-\int_x^{z(\tau)} dz^\mu \nabla_\mu^{\text{fun}}} \right| x \right\rangle. \tag{5.40}$$

The integral over $z(\tau)$ – the final point of the trajectory – of the matrix element on the RHS equals

$$\int d^d z(\tau) \left\langle y \left| \boldsymbol{P} e^{-\int_x^{z(\tau)} dz^\mu \nabla_\mu^{\text{fun}}} \right| x \right\rangle = \boldsymbol{P} e^{i\int_x^y dz^\mu \mathcal{A}_\mu(z)}. \tag{5.41}$$

Therefore, the result can be written as

$$G(x, y; \mathcal{A}) = \sum_{\Gamma_{yx}}{}' \boldsymbol{P} e^{i\int_{\Gamma_{yx}} dz^\mu \mathcal{A}_\mu(z)}, \tag{5.42}$$

where $\sum'$ is defined by Eq. (1.156).

Problem 5.5 Invert $(-\nabla^2 + m^2)$ when ∇_μ is in the adjoint representation.

Solution Let us introduce

$$\nabla_\mu^{ab} = \partial_\mu \delta^{ab} - g f^{abc} A_\mu^c \tag{5.43}$$

and the Green function $G^{ab}(x, y; \mathcal{A})$ which obeys

$$\left(-\nabla_\mu^{ac}\nabla_\mu^{cb} + m^2\delta^{ab}\right) G^{bd}(x, y; \mathcal{A}) = \delta^{ad}\delta^{(d)}(x - y). \tag{5.44}$$

Then we obtain

$$G^{ab}(x, y; \mathcal{A}) \;=\; \sum_{\Gamma_{yx}}{}' \operatorname{tr} t^b \, U[\Gamma_{yx}] \, t^a \, U^\dagger[\Gamma_{yx}], \qquad (5.45)$$

where $U[\Gamma_{yx}]$ is given by Eq. (5.21).

Since matrices are rearranged in inverse order under Hermitian conjugation, one has*

$$U^\dagger[\Gamma_{yx}] \;=\; U[\Gamma_{xy}]. \qquad (5.46)$$

In particular, the phase factors obey the *backtracking* condition

$$U[\Gamma_{yx}] \, U[\Gamma_{xy}] \;=\; 1. \qquad (5.47)$$

We have chosen $\mathcal{A}_\mu$ in the discretized phase factor (5.25) at the center on the ith interval in order to satisfy Eq. (5.47) at finite ϵ.

Problem 5.6 Establish the relation between non-Abelian phase factors and the group of paths.

Solution The group of paths (or loops) is defined as follows. The elements of the group are the paths Γ_{yx}. The product of two elements Γ_{zx} and Γ_{yz} is the path Γ_{yx}, which is a composition of Γ_{zx} and Γ_{yz}. In other words, one first passes along the path Γ_{zx} and then the path Γ_{yz}. The product is denoted as

$$\Gamma_{yz} \, \Gamma_{zx} \;=\; \Gamma_{yx}. \qquad (5.48)$$

The multiplication of paths is obviously associative but noncommutative. The inverse element is defined as

$$\Gamma_{yx}^{-1} \;=\; \Gamma_{xy}, \qquad (5.49)$$

i.e. the path with opposite orientation.

It follows from definition (5.24) that

$$U[\Gamma_{yz}] \, U[\Gamma_{zx}] \;=\; U[\Gamma_{yz} \, \Gamma_{zx}]. \qquad (5.50)$$

The backtracking condition (5.47) is then given by

$$U[\Gamma_{yx} \, \Gamma_{xy}] \;=\; 1. \qquad (5.51)$$

In other words, the paths of opposite orientation cancel each other in the phase factors.

---

* The notation Γ_{yx} means that the contour is oriented from x to y, while Γ_{xy} denotes the opposite orientation from y to x. In the path-ordered product (5.24), these two contours result in opposite orders of multiplication for the matrices.

5.3 Phase factors (properties)

Under the gauge transformation (5.4) the non-Abelian phase factor (5.21) transforms as

$$U[\Gamma_{yx}] \xrightarrow{\text{g.t.}} \Omega(y) U[\Gamma_{yx}] \Omega^\dagger(x). \qquad (5.52)$$

This formula stems from the fact that

$$[1 + i\,dz^\mu \mathcal{A}_\mu(z)] \xrightarrow{\text{g.t.}} [1 + i\,dz^\mu \mathcal{A}'_\mu(z)]$$
$$= \Omega(z + dz)\,[1 + i\,dz^\mu \mathcal{A}_\mu(z)]\,\Omega^\dagger(z) \qquad (5.53)$$

under the gauge transformation, which can be proven by substituting Eq. (5.4), so that $\Omega^\dagger(z)$ and $\Omega(z)$ cancel in the definition (5.24) at the intermediate point z.

One of the consequences of Eq. (5.52) is that $\psi(x)$, transported by the matrix $U[\Gamma_{yx}]$ to the point y, transforms under the gauge transformation as $\psi(y)$:

$$U[\Gamma_{yx}]\,\psi(x) \overset{\text{g.t.}}{\sim} \text{``}\psi(y)\text{''}, \qquad (5.54)$$

and, analogously,

$$\bar\psi(y)\,U[\Gamma_{yx}] \overset{\text{g.t.}}{\sim} \text{``}\bar\psi(x)\text{''}. \qquad (5.55)$$

Therefore, $U[\Gamma_{yx}]$ is, indeed, a parallel transporter.

It follows from these formulas that $\bar\psi(y)\,U[\Gamma_{yx}]\,\psi(x)$ is gauge invariant:

$$\bar\psi(y)\,U[\Gamma_{yx}]\,\psi(x) \xrightarrow{\text{g.t.}} \bar\psi(y)\,U[\Gamma_{yx}]\,\psi(x). \qquad (5.56)$$

Another consequence of Eq. (5.52) is that the trace of the phase factor for a closed contour Γ is gauge invariant:

$$\operatorname{tr} \boldsymbol{P}\,e^{i\oint_\Gamma dz^\mu \mathcal{A}_\mu(z)} \xrightarrow{\text{g.t.}} \operatorname{tr} \boldsymbol{P}\,e^{i\oint_\Gamma dz^\mu \mathcal{A}_\mu(z)}. \qquad (5.57)$$

These properties of the non-Abelian phase factor are quite similar to those of the Abelian one which was considered in Sect. 1.7.

Problem 5.7 Calculate $\partial U[\Gamma_{yx}]/\partial x_\mu$ and $\partial U[\Gamma_{yx}]/\partial y_\mu$.

Solution It is convenient to start from Eq. (5.25). Then only $(z_1 - x)$ in the last element of the product should be differentiated with respect to x or $(y - z_{M-1})$ in the first element of the product should be differentiated with respect to y. As $\epsilon \to 0$, we obtain

$$\left.\begin{aligned}
\frac{\partial}{\partial x_\mu} \boldsymbol{P}\,e^{i\int_x^y dz^\mu \mathcal{A}_\mu(z)} &= -i\,\boldsymbol{P}\,e^{i\int_x^y dz^\mu \mathcal{A}_\mu(z)}\,\mathcal{A}_\mu(x), \\
\frac{\partial}{\partial y_\mu} \boldsymbol{P}\,e^{i\int_x^y dz^\mu \mathcal{A}_\mu(z)} &= i\,\mathcal{A}_\mu(y)\,\boldsymbol{P}\,e^{i\int_x^y dz^\mu \mathcal{A}_\mu(z)}.
\end{aligned}\right\} \qquad (5.58)$$

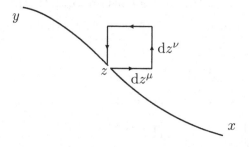

Fig. 5.1. The rectangular loop δC_{zz}, which is added to the contour Γ_{yx} at the intermediate point z in the (μ, ν)-plane.

These formulas are exactly the same as if one were to just differentiate the lower and upper limit in the path-ordered integral, bearing in mind the ordering of matrices.

One can rewrite Eq. (5.58) via the covariant derivatives as

$$\left.\begin{aligned}
\nabla_{\mu}^{\text{fun}}(y)\, U[\Gamma_{yx}] &= 0\,, \\
U[\Gamma_{yx}]\overleftarrow{\nabla}_{\mu}^{\text{fun}}(x) &= 0\,.
\end{aligned}\right\} \tag{5.59}$$

It is the property of the parallel transporter which is annihilated by the covariant derivative.

Problem 5.8 Prove that the sufficient and necessary condition for the phase factor to be independent on a local variation of the path is the vanishing of $\mathcal{F}_{\mu\nu}$.

Solution Let us add to Γ_{yx} at the point $z \in \Gamma_{yx}$ an infinitesimal loop δC_{zz} that lies in the (μ, ν)-plane and encloses the area $\delta\sigma_{\mu\nu}(z)$. Then the variation of the phase factor is

$$\delta U[\Gamma_{yx}] \;\equiv\; U[\Gamma_{yz}\,\delta C_{zz}\,\Gamma_{zx}] - U[\Gamma_{yx}] \;=\; \mathrm{i}\, U[\Gamma_{yz}]\,\mathcal{F}_{\mu\nu}(z)\, U[\Gamma_{zx}]\,\delta\sigma_{\mu\nu}(z)\,. \tag{5.60}$$

We can rewrite Eq. (5.60) as

$$\delta U[\Gamma_{yx}] \;=\; \mathrm{i}\, \boldsymbol{P}\, U[\Gamma_{yx}]\,\mathcal{F}_{\mu\nu}(z)\,\delta\sigma_{\mu\nu}(z) \tag{5.61}$$

since the P-product will automatically put $\mathcal{F}_{\mu\nu}(z)$ at the point z on the contour Γ_{yx}.

A convenient way to prove Eq. (5.60) is to choose δC_{zz} to be a rectangle which is constructed from the vectors dz^{μ} and dz^{ν}, as depicted in Fig. 5.1. Using the representation (5.41), we see that the phase factor acquires the extra factor

$$[1 + dz^{\nu}\nabla_{\nu}]\,[1 + dz^{\mu}\nabla_{\mu}]\,[1 - dz^{\nu}\nabla_{\nu}]\,[1 - dz^{\mu}\nabla_{\mu}] \;=\; 1 - dz^{\mu}dz^{\nu}\,[\nabla_{\mu}, \nabla_{\nu}] \tag{5.62}$$

at the proper order in the path-ordered product. Then Eq. (5.19) results in Eq. (5.61). Alternatively, one can prove Eq. (5.61) using the discretized formula (5.25).

Problem 5.9 Derive a non-Abelian version of the Stokes theorem.

Solution The ordered contour integral can be represented as the double-ordered surface integral [Are80, Bra80]

$$
\boldsymbol{P}\,\mathrm{e}^{\mathrm{i}\oint_{C_{xx}}\mathrm{d}z^{\mu}\mathcal{A}_{\mu}(z)} \;=\; \boldsymbol{P}_{\sigma}\,\boldsymbol{P}_{\tau}\,\mathrm{e}^{\mathrm{i}\int_{S}\mathrm{d}\sigma^{\mu\nu}\,\text{``}\mathcal{F}_{\mu\nu}(x)\text{''}}, \tag{5.63}
$$

where τ and σ parametrize the surface S (spanned by C but arbitrary otherwise), the element of which is given by

$$
\mathrm{d}\sigma^{\mu\nu} \;=\; \mathrm{d}\tau\,\mathrm{d}\sigma\left(\frac{\partial z_{\mu}}{\partial\tau}\frac{\partial z_{\nu}}{\partial\sigma} - \frac{\partial z_{\mu}}{\partial\sigma}\frac{\partial z_{\nu}}{\partial\tau}\right). \tag{5.64}
$$

"$\mathcal{F}_{\mu\nu}(x)$" in Eq. (5.63) means that $\mathcal{F}_{\mu\nu}(z(\tau,\sigma))$ is parallel-transported to the initial point x.

Remark on an analogy with differential geometry

The formulas of the type of Eq. (5.60) are well-known in differential geometry where parallel transport around a small closed contour determines the curvature. Therefore, $\mathcal{F}_{\mu\nu}$ in Yang–Mills theory is the proper curvature in an internal color space while $\mathcal{A}_{\mu}$ is the connection.

A historical remark

An analog of the phase factors was first introduced by Weyl [Wey19] in his attempt to describe the gravitational and the electromagnetic interaction of an electron on an equal footing. What he did is associated in modern language with the scale rather than the gauge transformation, i.e. the vector-potential was not multiplied by i as in Eq. (1.158). This explains the term "gauge invariance" – gauging literally means fixing a scale. The factor of i was inserted by London [Lon27] after the creation of quantum mechanics and the recognition of the fact that the electromagnetic interaction corresponds to the freedom of choice of the phase of a wave function and not to a scale transformation. However, the terminology has remained.

5.4 Aharonov–Bohm effect

The simplest example of a gauge field is the electromagnetic field, for which transverse components describe photons. Otherwise, the longitudinal components of the vector-potential, which are changeable under the gauge transformation, are related to gauging the phase of a wave function, i.e. permit one to compare its values at different space-time points when an electron is placed in an external electromagnetic field.

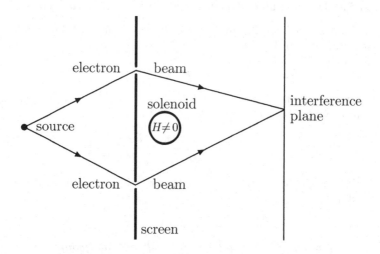

Fig. 5.2. Principal scheme of the experiment that demonstrates the Aharonov–Bohm effect. Electrons do not pass inside the solenoid where the magnetic field is concentrated. Nevertheless, a phase difference arises between the electron beams passing through the two slits. The interference picture changes with the value of the electric current.

As is well-known in quantum mechanics, the wave-function phase itself is unobservable. Only the phase differences are observable, for example via interference phenomena. For the electron in an electromagnetic field, the current (gauged) value of the phase of the wave function ψ at the point y is related, as is discussed in Sect. 1.7, to its value at some reference point x by the parallel transport which is given by Eq. (1.163). Therefore, the phase difference depends on the value of the phase factor for a given path Γ_{yx} along which the parallel transport is performed.

It is essential that the phase factors are observable in quantum theory, in contrast to classical theory. This is seen in the Aharonov–Bohm effect. The corresponding experiment is depicted schematically in Fig. 5.2.

It allows one to measure the phase difference between electrons passing through the two slits and, therefore, going across opposite sides of the solenoid. The fine point is that the magnetic field is nonvanishing only inside the solenoid where electrons do not penetrate. Hence the electrons pass throughout the region of space where the magnetic field strength vanishes! Nevertheless, the vector potential A_μ itself does not vanish which results in observable consequences.

The probability amplitude for an electron to propagate from a source at the point x to the point y in the interference plane is given by the

Minkowski-space analog of Eq. (1.155):

$$\Psi(x, y) = \sum_{\Gamma^+_{yx}}{}' e^{ie \int_{\Gamma^+_{yx}} dz^\mu A_\mu(z)} + \sum_{\Gamma^-_{yx}}{}' e^{ie \int_{\Gamma^-_{yx}} dz^\mu A_\mu(z)}, \qquad (5.65)$$

where the contour Γ^+_{yx} passes through the upper slit, while the contour Γ^-_{yx} passes through the lower one.

The intensity of the interference pattern is given by $|\Psi(x, y)|^2$ which contains, in particular, the term proportional to (the real part of)

$$e^{ie \int_{\Gamma^+_{yx}} dz^\mu A_\mu(z)} \, e^{-ie \int_{\Gamma^-_{yx}} dz^\mu A_\mu(z)} = e^{ie \oint_\Gamma dz^\mu A_\mu(z)}, \qquad (5.66)$$

where the closed contour Γ is composed from Γ^+_{yx} and Γ^-_{xy}. This is nothing but the phase factor associated with a parallel transport along the closed contour Γ.

For the given process this phase factor does not depend on the shape of Γ^+_{yx} and Γ^-_{yx}. Applying the Stokes theorem, one obtains

$$e^{ie \oint_\Gamma dz^\mu A_\mu} = e^{ie \int d\sigma^{\mu\nu} F_{\mu\nu}} = e^{ieHS}, \qquad (5.67)$$

where HS is the magnetic flux through the solenoid. Therefore, the interference picture changes when H changes.*

Remark on quantum vs. classical observables

A moral from the Aharonov–Bohm experiment is that the phase factors are observable in quantum theory while in classical theory only the electric and magnetic field strengths are observable. The vector potential plays, in classical theory, only an auxiliary role in determining the field strength.

For the non-Abelian gauge group $G = SU(N)$, a quark can alter its color under the parallel transport so the non-Abelian phase factor (5.21) is a unitary $N \times N$ matrix. A non-Abelian analog of the quantity, which is measurable in the Aharonov–Bohm experiment, is the trace of the matrix of the parallel transport along a closed path. It is gauge invariant according to Eq. (5.57).

It looks promising to reformulate gauge theories entirely in terms of these observable quantities. How this can be achieved will be explained in Part 3.

---

* A detailed computation of the interference picture for the Aharonov–Bohm experiment is contained, for example, in the review by Kobe [Kob79].

6

Gauge fields on a lattice

The modern formulation of non-Abelian lattice gauge theories is due to Wilson [Wil74]. Independently, gauge theories were discussed on a lattice by Wegner [Weg71] as a gauge-invariant extension of the Ising model and in an unpublished work by A. Polyakov in 1974 which deals mostly with Abelian theories.

Placing gauge fields on a lattice provides, first, a nonperturbative regularization of ultraviolet divergences. Secondly, the lattice formulation of QCD possesses some nonperturbative terms in addition to perturbation theory. A result of this is that one has a nontrivial definition of QCD beyond perturbation theory which guarantees confinement of quarks.

The lattice formulation of gauge theories deals with phase-factor-like quantities, which are elements of the gauge group, and are natural variables for quantum gauge theories.

The gauge group on the lattice is therefore compact, offering the possibility of nonperturbative quantization of gauge theories without fixing the gauge. The lattice quantization of gauge theories is performed in such a way as to preserve the compactness of the gauge group.

The continuum limit of lattice gauge theories is reproduced when the lattice spacing is many times smaller than the characteristic scale. This is achieved when the non-Abelian coupling constant tends to zero as it follows from the renormalization-group equation.

In this chapter we consider the Euclidean formulation of lattice gauge theories. First, we introduce the lattice terminology and discuss the action of lattice gauge theory at the classical level. Then, we quantize gauge fields on the lattice using the path-integral method, where the integration is over the invariant group measure. We explain Wilson's criterion of confinement and demonstrate it using calculations in the strong-coupling limit. Finally, we discuss how to pass to the continuum limit of lattice gauge theories.

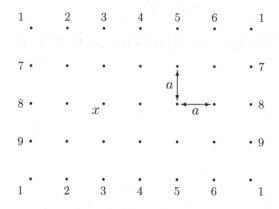

Fig. 6.1. Two-dimensional lattice with periodic boundary conditions. The sites labeled by the same numbers are identified. The lattice spacing equals a, while the spatial size of the lattice corresponds to $L_1 = 6$ and $L_2 = 4$.

6.1 Sites, links, plaquettes and all that

The first step in constructing a lattice gauge theory is to approximate the continuous space by a discrete set of points, i.e. a *lattice*. In the Euclidean formulation, the lattice is introduced along all four coordinates, while the time is left as continuous in the Hamiltonian approach.* We shall discuss only the Euclidean formulation of lattice gauge theories.

The lattice is defined as a set of points of the d-dimensional Euclidean space with the coordinates

$$x_\mu = n_\mu a, \tag{6.1}$$

where the components of the vector

$$n_\mu = (n_1, n_2, \ldots, n_d) \tag{6.2}$$

are integer numbers. The points (6.1) are called the lattice *sites*.

The dimensional constant a, which is equal to the distance between the neighboring sites, is called the *lattice spacing*. Dimensional quantities are usually measured in units of a, therefore setting $a = 1$.

A two-dimensional lattice is depicted in Fig. 6.1. A four-dimensional lattice for which the distances between sites are the same in all directions (as for the lattice in Fig. 6.1) is called a *hypercubic* lattice.

The next concept is the *link* of a lattice. A link is a line which connects two neighboring sites. A link is usually denoted by the letter l and is

---

* A Hamiltonian formulation of lattice gauge theories was developed by Kogut and Susskind [KS75].

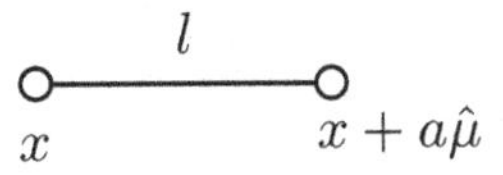

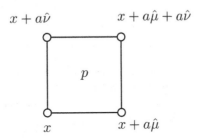

Fig. 6.2.   A link of a lattice. The link connects the sites x and $x + a\hat{\mu}$.

$$x + a\hat{\nu} \qquad\qquad x + a\hat{\mu} + a\hat{\nu}$$

$$p$$

$$x \qquad\qquad x + a\hat{\mu}$$

Fig. 6.3.   A plaquette of a lattice. The plaquette boundary is made of four links.

characterized by the coordinate x of its starting point and its direction $\mu = 1, \ldots, d$:

$$l = \{x; \mu\}. \tag{6.3}$$

The link l connects sites with coordinates x and $x + a\hat{\mu}$, where $\hat{\mu}$ is a unit vector along the μ-direction, as shown in Fig. 6.2. The lengths of all links are equal to a for a hypercubic lattice.

The elementary square enclosed by four links is called the *plaquette*. A plaquette p is specified by the coordinate x of a site and by the two directions μ and ν along which it is constructed:

$$p = \{x; \mu, \nu\}. \tag{6.4}$$

A plaquette is depicted in Fig. 6.3. The set of four links which bound the plaquette p is denoted as ∂p.

If the spatial size of the lattice is infinite, then the number of dynamical degrees of freedom is also infinite (but enumerable). In order to limit the number of degrees of freedom, one deals with a lattice which has a finite size $L_1 \times L_2 \times \cdots \times L_d$ in all directions (see Fig. 6.1).

Usually, one imposes *periodic boundary conditions* to reduce finite-size effects that are due to the finite extent of the lattice. In other words, one identifies pairs of sites which lie on parallel bounding hyperplanes. Usually the sites with the coordinates $(0, n_2, \ldots, n_d)$ and $(L_1, n_2, \ldots, n_d)$ are identified and similarly along other axes.

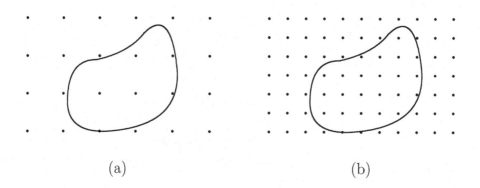

(a)   (b)

Fig. 6.4. Description of continuum field configurations using (a) "coarse" and (b) "fine" lattices. Lattice (a) can represent the given continuum field configuration very roughly, while lattice (b) has a spacing which is small enough.

Problem 6.1 Calculate the numbers of sites, links and plaquettes for a symmetric hypercubic lattice with periodic boundary conditions.

Solution Let us denote $L_1 = L_2 = \cdots = L_d = L$. Then

$$N_s = L^d, \qquad N_l = dL^d, \qquad N_p = \frac{d(d-1)}{2}L^d. \qquad (6.5)$$

Problem 6.2 Label the lattice links by a natural number $l \in [1, N_l]$.

Solution One of the choices is as follows:

$$l = \mu + n_1 d + n_2 dL + \cdots + n_d dL^{d-1}, \qquad (6.6)$$

where $n_\nu = x_\nu/a$ and μ is the direction of the link $\{x; \mu\}$.

6.2 Lattice formulation

The next step is to describe how matter fields and gauge fields are defined on a lattice.

A matter field, say a quark field, is attributed to the lattice sites. One can just think that a continuous field $\varphi(x)$ is approximated by its values at the lattice sites:

$$\varphi(x) \implies \varphi_x. \qquad (6.7)$$

It is clear that, in order for the lattice field φ_x to be a good approximation of a continuous field configuration $\varphi(x)$, the lattice spacing should be much smaller than the characteristic size of a given configuration. This is explained in Fig. 6.4.

The gauge field is attributed to the links of the lattice:

$$\mathcal{A}_\mu(x) \implies U_\mu(x). \qquad (6.8)$$

It looks natural since a link is characterized by a coordinate and a direction (see Eq. (6.3)) – the same as $\mathcal{A}_\mu(x)$. Sometimes the notation $U_{x,\mu}$ is used as an alternative for $U_\mu(x)$ to emphasize that it is attributed to links.

The link variable $U_\mu(x)$ can be viewed as

$$U_\mu(x) \;=\; \boldsymbol{P}\, e^{i \int_x^{x+a\hat\mu} dz^\mu \, \mathcal{A}_\mu(z)}, \tag{6.9}$$

where the integral is along the link $\{x; \mu\}$. As $a \to 0$, this yields

$$U_\mu(x) \;\to\; e^{ia\mathcal{A}_\mu(x)} \tag{6.10}$$

so that $U_\mu(x)$ is expressed via the exponential of the μth component of the vector potential, say, at the center of the link to agree with Eq. (5.25).

Since the path-ordered integral in Eq. (6.9) depends on the orientation, the concept of the orientation of a given link arises. The same link, which connects the points x and $x + a\hat\mu$, can be written either as $\{x; \mu\}$ or as $\{x + a\hat\mu; -\mu\}$. The orientation is positive for $\mu > 0$ in the former case (i.e. the same as the direction of the coordinate axis) and is negative in the latter case.

We have assigned the link variable $U_\mu(x)$ to links with positive orientations. The U-matrices which are assigned to links with negative orientations are given by

$$U_{-\mu}(x + a\hat\mu) \;=\; U_\mu^\dagger(x). \tag{6.11}$$

This is a one-link analog of Eq. (5.46).

It is clear from the relation (6.9) between the lattice and continuum gauge variables how one can construct lattice analogs of the continuum phase factors – one should construct the contours from the links of the lattice.

An important role in the lattice formulation is played by the phase factor for the simplest closed contour on the lattice: the (oriented) boundary of a plaquette, as is shown in Fig. 6.5. The plaquette variable is composed from the link variables (6.9) as

$$U(\partial p) \;=\; U_\nu^\dagger(x)\, U_\mu^\dagger(x + a\hat\nu)\, U_\nu(x + a\hat\mu)\, U_\mu(x). \tag{6.12}$$

The link variable transforms under the gauge transformation, according to Eq. (5.52), as

$$U_\mu(x) \;\xrightarrow{\text{g.t.}}\; \Omega(x + a\hat\mu)\, U_\mu(x)\, \Omega^\dagger(x), \tag{6.13}$$

where the matrix $\Omega(x)$ is attributed to the lattice sites. This defines the lattice gauge transformation.

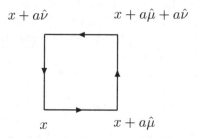

Fig. 6.5.   A contour in the form of an oriented boundary of a plaquette.

The plaquette variable transforms under the lattice gauge transformation as

$$U(\partial p) \xrightarrow{\text{g.t.}} \Omega(x)\, U(\partial p)\, \Omega^\dagger(x)\,. \tag{6.14}$$

Therefore, its trace over the color indices is gauge invariant:

$$\operatorname{tr} U(\partial p) \xrightarrow{\text{g.t.}} \operatorname{tr} U(\partial p)\,. \tag{6.15}$$

The invariance of the trace under the lattice gauge transformation is used in constructing an action of a lattice gauge theory. The simplest (Wilson) action is

$$S_{\text{lat}}[U] \;=\; \sum_p \left[1 - \frac{1}{N}\operatorname{Re}\operatorname{tr} U(\partial p)\right]. \tag{6.16}$$

The summation is over all the elementary plaquettes of the lattice (i.e. over all x, μ, and ν), regardless of their orientations.

Since a reversal of the orientation of the plaquette boundary results, according to Eq. (5.46), in complex conjugation:

$$\operatorname{tr} U(\partial p) \xrightarrow{\text{reor.}} \operatorname{tr} U^\dagger(\partial p) \;=\; [\operatorname{tr} U(\partial p)]^*, \tag{6.17}$$

one can rewrite the action (6.16) in the equivalent form

$$S_{\text{lat}}[U] \;=\; \frac{1}{2}\sum_{\text{orient } p} \left[1 - \frac{1}{N}\operatorname{tr} U(\partial p)\right], \tag{6.18}$$

where the sum is also over the two possible orientations of the boundary of a given plaquette.

In the limit $a \to 0$, the lattice action (6.16) becomes (in $d = 4$) the action of a continuum gauge theory. In order to show this, let us first note that

$$U(\partial p) \;\to\; \exp\left[ia^2 \mathcal{F}_{\mu\nu}(x) + \mathcal{O}(a^3)\right], \tag{6.19}$$

where $\mathcal{F}_{\mu\nu}(x)$ is defined using Eq. (5.14).

In the Abelian theory, the expansion (6.19) is easily found from the Stokes theorem. The commutator of $\mathcal{A}_\mu(x)$ and $\mathcal{A}_\nu(x)$, which arises in the non-Abelian case, complements the field strength to the non-Abelian one, as is ensured by the gauge invariance. Equation (6.19) was, in fact, already derived in Problem 5.8 on p. 94.

The transition to the continuum limit is performed by virtue of

$$a^4 \sum_p \quad \overset{a\to 0}{\longrightarrow} \quad \frac{1}{2} \int d^4x \sum_{\mu,\nu} . \tag{6.20}$$

Expanding the exponential on the RHS of Eq. (6.19) in a, we obtain

$$S_{\text{lat}} \quad \overset{a\to 0}{\longrightarrow} \quad \frac{1}{4N} \int d^4x \sum_{\mu,\nu} \text{tr}\, \mathcal{F}^2_{\mu\nu}(x) , \tag{6.21}$$

which coincides modulo a factor with the action of the continuum gauge theory.

Problem 6.3 Derive the lattice version of the non-Abelian Maxwell equation (5.17).

Solution Let us perform the change of the link variable

$$U_\mu(x) \to U_\mu(x)\left[1 - i\epsilon_\mu(x)\right], \qquad U^\dagger_\mu(x) \to \left[1 + i\epsilon_\mu(x)\right]U^\dagger_\mu(x) , \tag{6.22}$$

where $\epsilon_\mu(x)$ is an infinitesimal traceless Hermitian matrix.

A given link $\{x; \mu\}$ enters $4(d-1)$ plaquettes $p = \{x; \mu, \nu\}$ in the action (6.18). One-half of them have a boundary with a positive orientation and the other half with a negative one. The variation of the action (6.18) under the shift (6.22) is

$$\delta S[U] = \frac{i}{2N} \sum_{\nu \neq \pm\mu} \left[\text{tr}\, U(\partial p)\epsilon_\mu(x) - \text{tr}\, \epsilon_\mu(x)U^\dagger(\partial p)\right] . \tag{6.23}$$

Since $\epsilon_\mu(x)$ is arbitrary, we obtain

$$\sum_{\nu \neq \pm\mu} \left[U(\partial\{x; \mu, \nu\}) - U^\dagger(\partial\{x; \mu, \nu\})\right] = 0 , \tag{6.24}$$

or, graphically,

$$\sum_{\nu \neq \pm\mu} \left(\quad - \quad \right) = 0 . \tag{6.25}$$

In the latter equation we have depicted only plaquettes with positive orientation, while those with negative orientation are recovered by the sum over ν

for $\nu < 0$. Equation (6.24) (or (6.25)) is the lattice analog of the non-Abelian Maxwell equation.

In order to show how this equation reproduces the continuum one (5.17) as $a \to 0$, let us rewrite the second term on the LHS of Eq. (6.25) using (6.11):

$$\sum_{\nu \neq \pm \mu} \left(\quad - \quad \right) = 0 \qquad (6.26)$$

or, analytically,

$$\sum_{\nu \neq \pm \mu} \left[U(\partial\{x; \mu, \nu\}) - U_\nu(x - a\hat{\nu})\, U(\partial\{x - a\hat{\nu}; \mu, \nu\})\, U_\nu^\dagger(x - a\hat{\nu}) \right] = 0. \qquad (6.27)$$

It is now clear that the plaquette boundary in the second term on the LHS, which is the same as the first one but transported by one lattice spacing in the ν-direction, is associated with $\mathcal{F}_{\mu\nu}(x - a\hat{\nu})$. Using Eqs. (6.10) and (6.19), we recover the continuum Maxwell equation (5.17).

Remark on the naive continuum limit

The limit $a \to 0$, when Eqs. (6.10) and (6.19) hold reproducing the continuum action (6.21), is called the *naive continuum limit*. It is assumed in the naive continuum limit that $A_\mu(x)$ is weakly fluctuating at neighboring lattice links. Fluctuations of the order of $1/a$ are not taken into account, since discontinuities of the vector potential in the continuum theory are usually associated with an infinite action.

Another subtlety with the naive continuum limit is that the next order in a terms of the expansion of the lattice action (6.16), say the term $\propto a^2 \operatorname{tr} \mathcal{F}^3$, are associated with nonrenormalizable interactions and the smallness of a^2 can be compensated, in principle, by quadratic divergences.

The actual continuum limit of lattice gauge theories is, in fact, very similar to the naive one modulo some finite renormalizations of the gauge coupling constant. The large fluctuations of $A_\mu(x)$ of the order of $1/a$ become frozen when passing to the continuum limit. How one can pass to the continuum limit of lattice gauge theories is explained in Sect. 6.7.

Remark on ambiguities of the lattice action

The Wilson action (6.16) is the simplest one which reproduces the continuum action in the naive continuum limit. One can alternatively use the characters of $U(\partial p)$ in other representations of $SU(N)$, e.g. in the adjoint representation

$$\chi_{\text{adj}}(U) = |\operatorname{tr} U|^2 - 1, \tag{6.28}$$

to construct the lattice action.

The adjoint-representation lattice action is given as

$$S_{\text{adj}}[U] = \sum_p \left[1 - \frac{1}{N^2} |\operatorname{tr} U(\partial p)|^2\right]. \tag{6.29}$$

The naive continuum limit will be the same as for the Wilson action (6.16).

Moreover, one can define the lattice action as a mixture of the fundamental and adjoint representations [BC81, KM81]:

$$S_{\text{mix}}[U] = \sum_p \left[1 - \frac{1}{N}\operatorname{Re}\operatorname{tr} U(\partial p)\right] + \frac{\beta_A}{2\beta}\sum_p \left[1 - \frac{1}{N^2} |\operatorname{tr} U(\partial p)|^2\right]. \tag{6.30}$$

The ratio β_A/β is a constant ~ 1 which does not affect the continuum limit. This action is called the *mixed action*.

The lattice action (6.29) for $N = 2$ is associated with the action of the $SO(3)$ lattice gauge theory. Since algebras of the $SU(2)$ and $SO(3)$ groups coincide, these two gauge theories coincide in the continuum and differ on the lattice.

One more possibility is to use the phase factor associated, say, with the boundary of two plaquettes having a common link, or the phase factors for more complicated closed contours of finite size on the lattice to construct the action. These actions will also reproduce, in the naive continuum limit, the action of the continuum gauge theory.

The independence of the continuum limit of lattice gauge theories on the choice of lattice actions in called the *universality*. We shall say more about this in Sect. 7.4 when discussing the renormalization group on the lattice.

6.3 The Haar measure

The partition function of a pure* lattice gauge theory is defined by

$$Z(\beta) = \int \prod_{x,\mu} dU_\mu(x) \, e^{-\beta S[U]}, \tag{6.31}$$

where the action is given by Eq. (6.16).

---

* Here "pure" means without matter fields.

This is the analog of a partition function in statistical mechanics at an inverse temperature β given by[*]

$$\beta = \frac{N}{g^2}. \tag{6.32}$$

This formula results from comparing Eq. (6.21) with the gauge-field part of the continuum action (5.13).[**]

A subtle question is what is the measure $dU_\mu(x)$ in Eq. (6.31). To preserve the gauge invariance at finite lattice spacing, the integration is over the *Haar measure* which is an invariant group measure. Invariance of the Haar measure under multiplication by an arbitrary group element from the left or from the right:

$$dU = d(\Omega U) = d(U\Omega'), \tag{6.33}$$

guarantees the gauge invariance of the partition function (6.31).

This invariance of the Haar measure is crucial for the Wilson formulation of lattice gauge theories.

It is instructive to present an explicit expression for the Haar measure in the case of the $SU(2)$ gauge group. An element of $SU(2)$ can be parametrized using the unit four-vector a_μ $(a_\mu^2 = 1)$ as

$$U = a_4\mathbb{1} + i\vec{a}\vec{\sigma}, \tag{6.34}$$

where $\vec{\sigma}$ are the Pauli matrices. The Haar measure for $SU(2)$ then reads

$$dU = \frac{1}{\pi^2} \prod_{\mu=1}^{4} da_\mu\, \delta^{(1)}\left(a_\mu^2 - 1\right), \tag{6.35}$$

since $\det U = a_\mu^2$.

Problem 6.4 Rewrite the Haar measure on $SU(2)$ via a unit three-vector $\vec{n}$ ($\vec{n}^2 = 1$) and an angle φ ($\varphi \in [0, 2\pi]$).

Solution An element of $SU(2)$ reads in this parametrization as

$$U = e^{i\varphi\vec{n}\vec{\sigma}/2} = \cos\frac{\varphi}{2} + i\vec{n}\vec{\sigma}\sin\frac{\varphi}{2}. \tag{6.36}$$

The geometric meaning of this parametrization is simple: the element (6.36) is associated with a rotation through the angle φ around the $\vec{n}$-axis. The Haar measure for the $SU(2)$ group is then

$$dU = \frac{d^2\vec{n}}{4\pi}\frac{d\varphi}{\pi}\sin^2\frac{\varphi}{2}. \tag{6.37}$$

This formula can be obtained from Eq. (6.35) by integrating over $|\vec{a}|$.

---

[*] The standard factor of 2 is missing because of the normalization (5.6).

[**] One has instead $\beta = N/g^2 a^{4-d}$ on a d-dimensional lattice since the Yang–Mills coupling g is dimensional for $d \neq 4$.

Problem 6.5 For the $U(N)$ group represent the Haar measure as a multiple integral over the matrix elements of U.

Solution Elements of a unitary matrix U are complex numbers. The Haar measure can be represented as

$$\int \mathrm{d}U \cdots = \int_{-\infty}^{+\infty} \prod_{i,j} \mathrm{d}\,\mathrm{Re}\,U_{ij}\,\mathrm{d}\,\mathrm{Im}\,U_{ij}\,\delta^{(N^2)}(UU^\dagger - \mathbb{I}) \cdots . \qquad (6.38)$$

The integral in this formula goes over unrestricted U_{ij} as if U were a general complex matrix while the delta-function restricts U to be unitary.

The partition function (6.31) characterizes vacuum effects in the quantum theory. Physical quantities are given by the averages of the same type as Eq. (2.6):

$$\langle F[U] \rangle = Z^{-1}(\beta) \int \prod_{x,\mu} \mathrm{d}U_\mu(x)\,\mathrm{e}^{-\beta S[U]}\,F[U], \qquad (6.39)$$

where $F[U]$ is a gauge-invariant functional of the link variable $U_\mu(x)$. The averages (6.39) become the corresponding expectation values in the continuum theory as $a \to 0$ and β is related to g^2 by Eq. (6.32).

Remark on the lattice quantization

On a lattice of finite size, the integral over the gauge group in Eq. (6.39) is finite since the integration is over a compact group manifold, in contrast to the continuum case, where the volume of the gauge group is infinite. Therefore, the expression (6.39) is a constructive method for calculating averages of gauge-invariant quantities, though the gauge is not fixed.

The gauge can be fixed on the lattice in the standard way by the Faddeev–Popov method. This procedure involves extracting a (finite) common factor, which equals the volume of the gauge group, from the numerator and denominator on the RHS of Eq. (6.39). Therefore, the averages of gauge-invariant quantities coincide for a fixed and unfixed gauge, while the average of a functional which is not gauge invariant vanishes when the gauge is not fixed.

The fixing of gauge is convenient (though not necessary) for calculations in a lattice perturbation theory. A Lorentz gauge cannot be fixed, however, outside perturbation theory because of Gribov copies [Gri78]. In contrast, the lattice path integral (6.39) with an unfixed gauge is a method of nonperturbative quantization.

A price for the compactness of the group manifold on the lattice is the presence of fluctuations $\mathcal{A}_\mu(x) \sim 1/a$ which do not occur in the continuum (say, the values of the vector potential A_μ and $A_\mu + 2\pi/ae$ are identified for the Abelian $U(1)$ group). However, these fluctuations become unimportant when passing to the continuum limit.

6.4 Wilson loops

As has already been mentioned in Sect. 6.2, lattice phase factors are associated with contours which are drawn on the lattice.

In order to write down an explicit representation of the phase factor on the lattice via the link variables, let us specify the (lattice) contour C by its initial point x and by the directions (some of which may be negative) of the links from which the contour is built:

$$C = \{x; \mu_1, \ldots, \mu_n\}. \tag{6.40}$$

Then the lattice phase factor $U(C)$ is given by

$$U(C) = U_{\mu_n}(x + a\hat{\mu}_1 + \cdots + a\hat{\mu}_{n-1}) \cdots U_{\mu_2}(x + a\hat{\mu}_1) U_{\mu_1}(x). \tag{6.41}$$

For the links with a negative direction it is again convenient to use Eq. (6.11).

A closed contour has $\hat{\mu}_1 + \cdots + \hat{\mu}_n = 0$. The trace of the phase factor for a closed contour, which is gauge invariant, is called the *Wilson loop*.

The average of the Wilson loop is determined by the general formula (6.39) to be

$$W(C) \equiv \left\langle \frac{1}{N} \text{tr}\, U(C) \right\rangle$$

$$= Z^{-1}(\beta) \int \prod_{x,\mu} dU_\mu(x)\, e^{-\beta S[U]} \frac{1}{N} \text{tr}\, U(C). \tag{6.42}$$

This average is often called the Wilson loop average.

A very important role in lattice gauge theories is played by the averages of the Wilson loops associated with rectangular contours. Such a contour lying in the (x, t)-plane is depicted in Fig. 6.6.

The Wilson loop average is related for $T \gg R$ to the energy of the interaction of the static (i.e. infinitely heavy) quarks, which are separated by a distance R, by the formula

$$W(R \times T) \stackrel{T \gtrsim R}{=} e^{-E_0(R) \cdot T}. \tag{6.43}$$

Problem 6.6 Derive Eq. (6.43) by fixing the gauge $\mathcal{A}_4 = 0$.

Solution In the axial gauge $\mathcal{A}_4 = 0$, we have $U_4(x) = 1$ so that only vertical segments of the rectangle in Fig. 6.6 contribute to $U(R \times T)$. Denoting

$$\Psi_{ij}(t) \equiv \left[P e^{i \int_0^R dz_1\, A_1(z_1, \ldots, t)} \right]_{ij}, \tag{6.44}$$

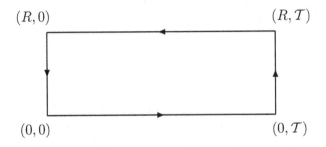

$(R,0)$ (R,T)

$(0,0)$ $(0,T)$

Fig. 6.6.   Rectangular loop of size $R \times T$.

we then have

$$W(R \times T) = \left\langle \frac{1}{N} \mathrm{tr}\, \Psi(0)\, \Psi^\dagger(T) \right\rangle . \tag{6.45}$$

Inserting in Eq. (6.45) a sum over a complete set of intermediate states

$$\sum_n |n\rangle\langle n| = 1, \tag{6.46}$$

we obtain

$$W(R \times T) = \sum_n \frac{1}{N} \langle \Psi_{ij}(0) \mid n\rangle\langle n \mid \Psi_{ji}^\dagger(T)\rangle$$

$$= \sum_n \frac{1}{N} |\langle \Psi_{ij}(0) \mid n\rangle|^2 \, \mathrm{e}^{-E_n T}, \tag{6.47}$$

where E_n is the energy of the state $|n\rangle$. As $T \to \infty$, only the ground state with the lowest energy survives in the sum over states and finally we find

$$W(R \times T) \; \overset{\text{large } T}{\longrightarrow} \; \mathrm{e}^{-E_0 T}, \tag{6.48}$$

which results in Eq. (6.43).

Note that nothing in this derivation relies on the lattice. Therefore, Eq. (6.43) holds for a rectangular loop in the continuum theory as well.

Equation (6.43) can also be understood as follows. Let us consider the Abelian case when the interaction is described by Coulomb's law. The contour integral can then be rewritten as the integral over the whole space

$$e \oint_C \mathrm{d}z^\mu A_\mu(z) = \int \mathrm{d}^d x J^\mu(x) A_\mu(x), \tag{6.49}$$

where

$$J^\mu(x) = e \oint_C \mathrm{d}z^\mu \delta^{(d)}(x - z) \tag{6.50}$$

is a four-vector current of a classical particle moving along the trajectory C which is described by the function $z_\mu(t)$.

It is clear that

$$-\ln W(C) \;=\; -\ln\left\langle e^{i\int d^4x\, J^\mu(x) A_\mu(x)}\right\rangle \tag{6.51}$$

determines the change of action of the classical particle arising from the electromagnetic interaction in accordance with Eq. (6.43). How one may obtain Coulomb's law in this language is shown later in Problem 12.3.

A similar interpretation of Eq. (6.43) in the non-Abelian case is somewhat more complicated. For a heavy particle moving along some trajectory in space-time, color degrees of freedom are quantum and easily respond to changes of the gauge field $A_\mu(x)$, which interacts with them. Let us suppose that a quark and an antiquark are created at the same space-time point in some color state. Then this state must be a singlet with respect to color (or *colorless*) since the average over the gauge field would vanish otherwise. When the quarks separate, their color changes from one point to another simultaneously with the change of color of the gauge field, in order for the system of the quarks plus the gauge field to remain colorless. Therefore, the averaging over the gauge field leads to an averaging over fluctuations of quark color degrees of freedom. $E_0(R)$ in Eq. (6.43) is associated with the interaction energy averaged over color in this way.

Problem 6.7 Derive a non-Abelian analog of Eq. (6.50).

Solution The proper non-Abelian extension of Eq. (6.50) is given by [Won70]

$$\mathcal{J}_\mu^a(x) \;=\; g\int_0^\tau dt\, \dot z_\mu(t)\, \delta^{(d)}(x - z(t))\, I^a(t), \tag{6.52}$$

where $I^a(t)$, which describes the color state of a classical particle moving along the trajectory $z^\mu(t)$ in an external Yang–Mills field $A_\mu(z)$, is a solution of the equation

$$\dot I^a(t) + g f^{abc} \dot z^\mu(t)\, A_\mu^b(z(t))\, I^c(t) \;=\; 0. \tag{6.53}$$

It is convenient to use Grassmann variables again to describe color degrees of freedom as in Problem 5.3 on p. 90. Then [BCL77, BSS77]

$$\mathcal{J}_\mu^a(x) \;=\; \bar\psi(t) t^a \psi(t) \tag{6.54}$$

and $\psi(t)$ is a solution of

$$\dot\psi(t) - i\,\dot z^\mu(t)\, A_\mu(z(t))\, \psi(t) \;=\; 0. \tag{6.55}$$

By definition, $E_0(R)$ in Eq. (6.43) includes a renormalization of the mass of a heavy quark owing to the interaction with the gauge field and which is thus independent of R. To the first order in g^2, it is the same as in QED and is given by

$$\Delta E_{\text{mass}} \;=\; \frac{g^2}{4\pi a}\frac{N^2-1}{N} \tag{6.56}$$

as $a \to 0$. The calculation is presented later in Problem 12.2.

The potential energy of the interaction between the static quarks is therefore defined as the difference

$$E(R) \;=\; E_0(R) - \Delta E_{\text{mass}}\,. \tag{6.57}$$

If $g^2/4\pi a$ in ΔE_{mass} did not become infinite as $a \to 0$, the term resulting from the mass renormalization would not have to be subtracted, since it simply changes the reference level for the potential energy.

6.5 Strong-coupling expansion

We already mentioned in Sect. 6.3 that the path integral (6.39) can be calculated by the lattice perturbation theory in g^2. As was pointed out by Wilson [Wil74], there exists an alternative way of evaluating the same quantity on a lattice by an expansion in $1/g^2$ or in β since they are related by Eq. (6.32). This expansion is called the *strong-coupling* expansion. It is an analog of the high-temperature expansion in statistical mechanics since β is the analog of an inverse temperature.

In order to perform the strong-coupling expansion, we expand the exponential of the lattice action, say in Eq. (6.42), in β. Then the problem is to calculate the integrals over the unitary group of the form

$$I^{i_1\cdots i_m,k_1\cdots k_n}_{j_1\cdots j_m,l_1\cdots l_n} \;=\; \int \mathrm{d}U\, U^{i_1}_{j_1}\cdots U^{i_m}_{j_m} U^{\dagger k_1}_{l_1}\cdots U^{\dagger k_n}_{l_n}\,, \tag{6.58}$$

where the Haar measure (given for $SU(2)$ by Eq. (6.35)) is normalized as

$$\int \mathrm{d}U \;=\; 1\,. \tag{6.59}$$

It is clear from general arguments that the integral (6.58) is nonvanishing only if $n = m$ (mod N), i.e. only if $n = m + kN$, where k is integer.

For the simplest case $m = n = 1$, the answer can easily be found by using the unitarity of U and the orthogonality relation:

$$\int \mathrm{d}U\, U^i_j\, U^{\dagger\,k}_{\;\;l} \;=\; \frac{1}{N}\delta^i_l\delta^k_j\,. \tag{6.60}$$

Problem 6.8 Prove Eq. (6.60) for the $U(N)$ group.

Solution From the general arguments we obtain

$$\int dU \, U_j^i \, {U^\dagger}_l^{\,k} \;=\; A\,\delta_l^i\delta_j^k + B\,\delta_j^i\delta_l^k\,. \tag{6.61}$$

Contracting by δ_i^l, using the unitarity of U, and Eq. (6.59), we have

$$AN + B \;=\; 1\,. \tag{6.62}$$

One more relation between A and B arises from the fact that the character in the adjoint representation is given by Eq. (6.28). Contracting Eq. (6.61) by δ_i^j and δ_k^l, and using the orthogonality of the characters which states

$$\int dU \left(|\operatorname{tr} U|^2 - 1 \right) \;=\; 0\,, \tag{6.63}$$

we find

$$AN + BN^2 \;=\; 1\,. \tag{6.64}$$

Therefore, $A = 1/N$ and $B = 0$ which proves Eq. (6.60).

The simplest Wilson loop average, which is nonvanishing in the strong-coupling expansion, is that for the loop which coincides with the boundary of a plaquette (see Fig. 6.5). It is called the plaquette average and is denoted by

$$W(\partial p) \;=\; \left\langle \frac{1}{N}\operatorname{tr} U(\partial p) \right\rangle\,. \tag{6.65}$$

In order to calculate the plaquette average to order β, it is sufficient to retain only the terms $\mathcal{O}(\beta)$ in the expansion of the exponentials in Eq. (6.42):

$$W(\partial p) \;=\; \frac{\displaystyle\int \prod_{x,\mu} dU_\mu(x)\left[1 + \beta\sum_{p'}\frac{1}{N}\operatorname{Re}\operatorname{tr} U(\partial p')\right]\frac{1}{N}\operatorname{tr} U(\partial p)}{\displaystyle\int \prod_{x,\mu} dU_\mu(x)\left[1 + \beta\sum_{p'}\frac{1}{N}\operatorname{Re}\operatorname{tr} U(\partial p')\right]} + \mathcal{O}(\beta^2)\,. \tag{6.66}$$

The group integration can then be performed by remembering that

$$\int dU_\mu(x)\,[U_\mu(x)]_j^i\,[U_\nu^\dagger(y)]_l^k \;=\; \frac{1}{N}\delta_{xy}\,\delta_{\mu\nu}\,\delta_l^i\,\delta_j^k \tag{6.67}$$

at different links.

Using this property of the group integral in Eq. (6.66), we immediately see that the denominator is equal to 1 (each link is encountered no more than once), while the only nonvanishing contribution in the numerator

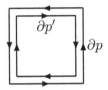

Fig. 6.7. Boundaries of the plaquettes p and p' with opposite orientations ∂p and $\partial p'$, respectively.

is from the plaquette p', which coincides with p but has the opposite orientation as is depicted in Fig. 6.7.

It is convenient to use the graphical notation* for Eq. (6.60) at each link of ∂p:

$$i \xrightarrow{} \genfrac{}{}{0pt}{}{j}{k} = \frac{1}{N} \times \left(\genfrac{}{}{0pt}{}{i}{l} \supset \quad \subset \genfrac{}{}{0pt}{}{j}{k} \right), \tag{6.68}$$

where the semicircles are associated with the Kronecker symbols:

$$\genfrac{}{}{0pt}{}{i}{l}\supset = \delta_l^i. \tag{6.69}$$

This notation is convenient since the lines which denote the Kronecker symbols in the latter equation can be associated with propagation of the color indices. Analogously a closed line represents the contracted Kronecker symbol, which is summed over the color indices,

$$\bigcirc = \delta_i^i = N. \tag{6.70}$$

Using the graphical representation (6.68) for each of the four links depicted in Fig. 6.7, we obtain

$$\int \prod_{x,\mu} dU_\mu(x)\, \mathrm{tr}\, U(\partial p)\, \mathrm{tr}\, U^\dagger(\partial p') = \frac{1}{N^4} \times \begin{matrix} \bigcirc & \bigcirc \\ \bigcirc & \bigcirc \end{matrix} = 1, \tag{6.71}$$

where the contracted Kronecker symbols are associated with the four sites of the plaquette.

---

* A calculation of more complicated group integrals (6.58) using the graphical notation is discussed in the lectures by Wilson [Wil75] and in Chapter 8 of the book by Creutz [Cre83]. An alternative method of calculating the group integrals using the character expansion is described in the review by Drouffe and Zuber [DZ83].

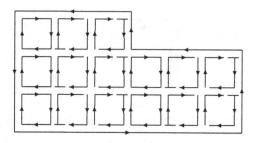

Fig. 6.8.   Filling of a loop with elementary plaquettes.

The final answer for the plaquette average is

$$\left.\begin{array}{rclr} W(\partial p) & = & \dfrac{\beta}{2N^2} & \text{for } SU(N) \text{ with } N \geq 3 \,, \\[2ex] W(\partial p) & = & \dfrac{\beta}{4} & \text{for } SU(2) \,. \end{array}\right\} \qquad (6.72)$$

The result for $SU(2)$ differs by a factor of $1/2$ because $\operatorname{tr} U(\partial p)$ is real for $SU(2)$ so that the orientation of the plaquettes can be ignored.

The graphical representation (6.68) is useful for evaluating the leading order of the strong-coupling expansion for more complicated loops. According to Eq. (6.67), a nonvanishing result emerges only when plaquettes, arising from the expansion of the exponentials of Eq. (6.42) in β, completely cover a surface enclosed by the given loop C as depicted in Fig. 6.8. In this case each link is encountered twice (or never), once in the positive direction and once in the negative direction, so that all the group integrals are nonvanishing. The leading order in β corresponds to filling a *minimal surface*, whose area takes on the smallest possible value. This yields

$$W(C) \;=\; [W(\partial p)]^{A_{\min}(C)} \,, \qquad (6.73)$$

where $W(\partial p)$ is given by Eq. (6.72) and $A_{\min}(C)$ is the area (in units of a^2) of the minimal surface.

For the rectangular loop, which is depicted in Fig. 6.6, the minimal surface is just a piece of the plane bounded by the rectangle. Therefore, we find

$$W(R \times T) \;=\; [W(\partial p)]^{RT} \qquad (6.74)$$

to the leading order in β.

More complicated surfaces, which do not lie in the plane of the rectangle, will give a contribution to $W(C)$ of the order of β^{area}. They are suppressed at small β since their areas are larger than $A_{\min}$.

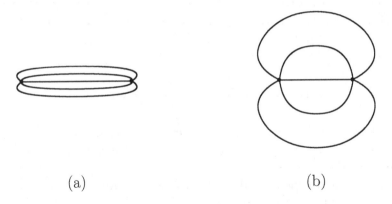

(a)   (b)

Fig. 6.9. Lines of force between static quarks for (a) linear and (b) Coulomb interaction potentials. For the linear potential the lines of force are contracted into a tube, while they are distributed over the whole space for the Coulomb one.

6.6 Area law and confinement

The exponential dependence of the Wilson loop average on the area of the minimal surface (as in Eq. (6.73)) is called the *area law*. It is customarily assumed that if an area law holds for loops of large area in pure gluodynamics (i.e. in the pure $SU(3)$ gauge theory) then quarks are confined. In other words, there are no physical $|in\rangle$ or $\langle out|$ quark states. This is the essence of Wilson's *confinement criterion*. The argument is that physical amplitudes (for example, the polarization operator) do not have quark singularities when the Wilson criterion is satisfied. I refer the reader to the well-written original paper by Wilson [Wil74], where this point is clarified.

Another, somewhat oversimplified, justification for the Wilson criterion is based on the relationship (6.43) between the Wilson loop average and the potential energy of interaction between static quarks. When the area law

$$W(C) \overset{\text{large } C}{\longrightarrow} e^{-KA_{\min}(C)} \tag{6.75}$$

holds for large loops, the potential energy is a linear function of the distance between the quarks:

$$E(R) = KR. \tag{6.76}$$

The coefficient K in these formulas is called the *string tension* because the gluon field between quarks contracts to a tube or string, whose energy is proportional to its length, as is depicted in Fig. 6.9a. The value of K

is the energy of the string per unit length. This string is stretched with the distance between quarks and prevents them from moving apart to macroscopic distances.

Equation (6.74) gives

$$K = \frac{1}{a^2} \ln \frac{2N^2}{\beta} = \frac{1}{a^2} \ln \left(2Ng^2 \right) \tag{6.77}$$

for the string tension to the leading order of the strong-coupling expansion. The next orders of the strong-coupling expansion result in corrections in β to this formula.

Therefore, confinement holds in the lattice gauge theory to any order of the strong-coupling expansion.

Remark on the perimeter law

For the Coulomb potential

$$E(R) = -\frac{g^2}{4\pi R} \frac{N^2 - 1}{N}, \tag{6.78}$$

the gauge field between quarks would be distributed over the whole space as is depicted in Fig. 6.9b. The Wilson loop average would have the behavior

$$W(C) \overset{\text{large } C}{\longrightarrow} e^{-\text{const} \cdot L(C)}, \tag{6.79}$$

where $L(C)$ denotes the length (or perimeter) of the closed contour C.

This behavior of the Wilson loops is called the *perimeter law*. To each order of perturbation theory, it is the perimeter law (6.79), rather than the area law (6.75), that holds for the Wilson loop averages. A perimeter law corresponds to a potential which cannot confine quarks.

Remark on the Creutz ratio

To distinguish between the area and perimeter law behavior of the Wilson loop averages, Creutz [Cre80] proposed to consider the ratio

$$\chi(I, J) = -\ln \frac{W(I \times J) \, W((I - 1) \times (J - 1))}{W((I - 1) \times J) \, W(I \times (J - 1))}, \tag{6.80}$$

where $W(I \times J)$ is as before the average of a rectangular Wilson loop of size $I \times J$. The exponentials of the perimeter, which is equal to

$$L(I \times J) = 2I + 2J, \tag{6.81}$$

cancel out in the ratio (6.80). In particular, the mass renormalization (6.56) cancels out, which is essential for the continuum limit.

The Creutz ratio (6.80) has the meaning of an interaction force between quarks, which can be seen by stretching the rectangle along the "temporal" axis (as illustrated by Fig. 6.6). If the area law (6.75) holds for asymptotically large I and J, then

$$\chi(I, J) \xrightarrow{\text{large } I, J} a^2 K, \tag{6.82}$$

i.e. it does not depend on I or J and coincides with the string tension. This property of the Creutz ratio was used for numerical calculations of the string tension.

6.7 Asymptotic scaling

Equation (6.77) establishes the relationship between values of the lattice spacing a and the coupling g^2 as follows. Let us set K to be equal to its experimental value*

$$K = (400 \text{ MeV})^2 \approx 1 \text{ GeV/fm}. \tag{6.83}$$

Then the renormalizability prescribes that variations of a, which plays the role of a lattice cutoff, and of the bare charge g^2 should be made simultaneously in order that K does not change.

Given Eq. (6.77), this procedure calls for $a \to \infty$ as $g^2 \to \infty$. In other words, the lattice spacing is large in the strong-coupling limit, compared with 1 fm – the typical scale of the strong interaction. This is a situation of the type shown in Fig. 6.4a. Such a coarse lattice cannot describe the continuum limit correctly and, in particular, the rotational symmetry.

In order to pass to the continuum, the lattice spacing a should be decreased to have a picture like that in Fig. 6.4b. Equation (6.77) shows that a decreases with decreasing g^2. However, this formula ceases to be applicable in the intermediate region of $g^2 \sim 1$ and, therefore, $a \sim 1$ fm.

The recipe for further decreasing a is the same as in the strong-coupling region, further decreasing g^2. While no analytic formulas are available at intermediate values of g^2, the expected relation between a and g^2 for small g^2 is predicted by the known two-loop Gell-Mann–Low function of QCD.

---

* This value results from the string model of hadrons where the slope of the Regge trajectory α' and the string tension K are related by $K = 1/2\pi\alpha'$. This formula holds even for a classical string. The slope $\alpha' = 1 \text{ GeV}^{-2}$ say from the $\rho - A_2 - g$ trajectory. A similar value of K is found from the description of mesons made out of heavy quarks using a nonrelativistic potential model.

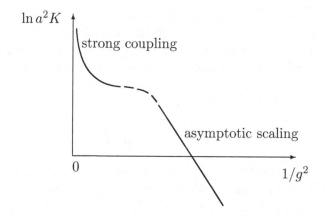

Fig. 6.10.   The dependence of the string tension on $1/g^2$. The strong-coupling formula (6.77) holds for small $1/g^2$. The asymptotic-scaling formula (6.84) sets in for large $1/g^2$. Both formulas are not applicable in the intermediate region of $1/g^2 \sim 1$ which is depicted by the dashed line.

For pure $SU(3)$ gluodynamics, Eq. (6.77) is replaced at small g^2 by

$$K \;=\; \text{const} \cdot \frac{1}{a^2} \left(\frac{8\pi^2}{11g^2} \right)^{\frac{102}{121}} e^{-8\pi^2/11g^2}, \qquad (6.84)$$

where we have used the two-loop Gell-Mann–Low function.

The exponential dependence of K on $1/g^2$ is called *asymptotic scaling*. Asymptotic scaling sets in for some value of $1/g^2$ as depicted in Fig. 6.10. For such values of g^2, where asymptotic scaling holds, the lattice gauge theory has a continuum limit.

The knowledge of the two asymptotic behaviors says nothing about the behavior of a^2K in the intermediate region of $g^2 \sim 1$. There can be either a smooth transition between these two regimes or a phase transition. Numerical methods were introduced to study this problem, some of which are described in the next chapter.

Remark on dimensional transmutation

The QCD action (5.13) does not contain a dimensional parameter of the order of hundreds MeV. The masses of the light quarks are of the order of a few MeV and can be disregarded. The only parameter of the action is the dimensionless bare coupling constant g^2. At the classical level, there is no way to obtain a dimensional parameter of the order of hundreds MeV.

In quantum theory, these is always a dimensional cutoff (such as a for the lattice regularization). The renormalizability says that a and g^2 are not independent but are related by the Gell-Mann–Low equation (3.72). It can be integrated to give the integration constant

$$\Lambda_{\text{QCD}} = \frac{1}{a} \exp \left[- \int \frac{dg^2}{\mathcal{B}(g^2)} \right]. \tag{6.85}$$

Up to this point there has been no difference between QCD and QED. The difference stems from the fact that the Gell-Mann–Low function $\mathcal{B}(g^2)$ is positive for QED and negative for QCD. In QED $e^2(a)$ increases with decreasing a, while in QCD $g^2(a)$ decreases with decreasing a. The latter behavior of the coupling constant is called *asymptotic freedom*. In both cases the Gell-Mann–Low function vanishes when the coupling constant tends to zero. Such values of coupling constants where the Gell-Mann–Low function vanishes are called the *fixed point*. Since the infrared behavior of e^2 in QED is interchangeable with the ultraviolet behavior of g^2 in QCD, the origin is an infrared-stable fixed point in QED and an ultraviolet-stable fixed point in QCD. In QED the fine-structure constant ($\approx 1/137$) is measurable in experiments, while in QCD the constant Λ_{QCD} is measurable.

This phenomenon of the appearance of a dimensional parameter in QCD, which remains finite in the limit of vanishing cutoff, is called *dimensional transmutation*. All observable dimensional quantities, such as the string tension or hadron masses, are proportional to the corresponding powers of Λ_{QCD}. Therefore, their dimensionless ratios, such as the ratio of $\sqrt{K}$ to the hadron masses, are universal numbers which do not depend on g^2. The goal of a nonperturbative approach in QCD is to calculate these numbers but not the overall dimensional parameter.

Remark on second-order phase transition

In statistical physics it is usually said that the continuum limits of a lattice system are reached at the points of second-order phase transitions when the correlation length becomes infinite in lattice units. This statement is in perfect agreement with what has been said above concerning the continuum limit of lattice gauge theories.

A correlation length is inversely proportional to Λ_{QCD} given by Eq. (6.85). The only chance for the RHS of Eq. (6.85) to vanish is to have a zero of the Gell-Mann–Low function $\mathcal{B}(g^2)$ at some fixed point $g^2 = g_*^2$. Therefore, the bare coupling should approach the fixed-point value g_*^2 to describe the continuum.

As we have discussed, $\mathcal{B}(0) = 0$ for a non-Abelian gauge theory so that $g_*^2 = 0$ is a fixed-point value of the coupling constant. Therefore, the continuum limit is associated with $g^2 \to 0$ as mentioned above.

7

Lattice methods

Analytic calculations of observables in the non-Abelian lattice gauge theories are available only in the strong-coupling regime $g^2 \to \infty$, while one needs $g^2 \to 0$ for the continuum limit. When g^2 is decreased, the lattice systems can undergo phase transitions as often happens in statistical mechanics.

To look for phase transitions, the mean-field method was first applied to lattice gauge theories [Wil74, BDI74]. It turned out to be useful for studying the first-order phase transitions which very often happen in lattice gauge systems but do not affect the continuum limit.

The second-order phase transitions are better described by the lattice renormalization group method. The approximate Migdal–Kadanoff recursion relations [Mig75, Kad76] were the first implementation of the renormalization group transformation on a lattice, which indicated the absence of a second-order phase transition in the non-Abelian lattice gauge theories and, therefore, quark confinement.

A very powerful method for practical nonperturbative calculations of observables in lattice gauge theories is the numerical Monte Carlo method. This method simulates statistical processes in a lattice gauge system and for this reason is often called a numerical simulation. The idea of applying it to lattice gauge theories is due to Wilson [Wil77], while the practical implementation was done by Creutz, Jacobs and Rebbi [CJR79] for Abelian gauge groups and by Creutz [Cre79, Cre80] for the $SU(2)$ and $SU(3)$ groups.

In this chapter we briefly describe the mean-field method, the lattice renormalization group method and the Monte Carlo method. A few results from Monte Carlo simulations will also be discussed.

123

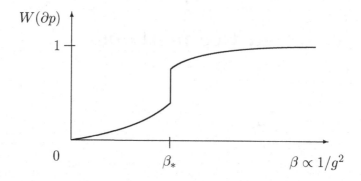

Fig. 7.1. Typical β-dependence of the plaquette average for a first-order phase transition which occurs at $\beta = \beta_*$.

7.1 Phase transitions

As was pointed out in Sect. 6.7, analytic calculations of the string tension are available only in the strong-coupling regime $g^2 \rightarrow \infty$, while one needs $g^2 \rightarrow 0$ for the continuum limit. A question arises as to what happens with lattice systems when g^2 is decreased. In particular, does an actual picture of the dependence of the string tension on g^2 look like that shown in Fig. 6.10?

We know from statistical mechanics that lattice systems can undergo phase transitions with a change of parameters, say the temperature, which completely alters the macroscopic properties. The simplest example is that of the first-order phase transition which occurs in a teapot.

First-order phase transitions very often happen in lattice gauge theories. They are usually seen as a discontinuity in the β- (or $1/g^2$-) dependence of the plaquette average (6.65) as is depicted in Fig. 7.1. The form of $W(\partial p)$ at small β is given to the leading order of the strong-coupling expansion by Eq. (6.72), while that at large β is prescribed by the lattice perturbation theory* to be

$$ W(\partial p) \;=\; 1 - \frac{d_G}{\beta d} + \mathcal{O}(\beta^{-2}), \tag{7.1} $$

where d_G is the dimensionality of the gauge group G ($d_G = N^2 - 1$ for $SU(N)$, $d_G = N^2$ for $U(N)$) and d is the dimensionality of the lattice as before.

This behavior of the plaquette average is quite analogous to the dependence of the internal energy per unit volume (called the specific energy) in statistical systems. In order to see the analogy between the specific en-

---

* It is often called, for obvious reasons, the weak-coupling expansion.

ergy and $(1 - W(\partial p))$, let us remember that β is analogous to the inverse temperature and rewrite Eq. (6.65) as

$$W(\partial p) = 1 + \frac{1}{N_p} \frac{\partial}{\partial \beta} \ln Z(\beta), \qquad (7.2)$$

where the partition function is given by Eq. (6.31) and the number of plaquettes N_p is analogous to the volume of a statistical system.

Problem 7.1 Derive Eq. (7.1) for the $SU(N)$ gauge group.

Solution The partition function (6.31) can be calculated at large β (weak coupling) using the saddle-point method. The saddle-point configurations are given by solutions of the classical equation (6.24). The appropriate solution reads as

$$U_\mu^{\mathrm{sp}}(x) = Z_\mu, \qquad (7.3)$$

where Z_μ is an element of the $Z(N)$ group, the center of $SU(N)$,

$$Z_\mu = \mathbb{I} \cdot \mathrm{e}^{2\pi \mathrm{i} n_\mu / N}, \qquad n_\mu = 1, \ldots, N. \qquad (7.4)$$

It is evident that this is a solution because elements of the center commute so that Z_μ and $Z_{-\mu}$ cancel each other in $U_{\mu,\nu}(x) \equiv U(\partial p)$.

In order to take into account fluctuations around the saddle-point solution (7.3), let us expand

$$U_\mu(x) = U_\mu^{\mathrm{sp}}(x) \, \mathrm{e}^{\mathrm{i} t^a \epsilon_\mu^a(x)}, \qquad (7.5)$$

where the order of multiplication is not essential since Z_μ commute with the generators t^a. The expansion of $\mathrm{tr}\, U(\partial p)$ to the quadratic order in ϵ^a is given by

$$\frac{1}{N} \mathrm{tr}\, U_{\mu,\nu}(x) = 1 - \frac{1}{2N} \mathcal{E}_{\mu,\nu}^2(x), \qquad (7.6)$$

where

$$\mathcal{E}_{\mu,\nu}^a(x) = \epsilon_\mu^a(x) + \epsilon_\nu^a(x + a\hat{\mu}) - \epsilon_\mu^a(x + a\hat{\nu}) - \epsilon_\nu^a(x). \qquad (7.7)$$

Owing to the local gauge invariance, we can always choose, say, $\epsilon_d(x) = 0$ so that there are only $N_l - N_s$ independent ϵs.

Substituting into Eq. (6.31) and expanding the Haar measure, we obtain

$$Z(\beta) \propto \prod_{\nu=1}^{d} \sum_{n_\nu=1}^{N} \prod_{a,x,\mu<d} \int_{-\infty}^{+\infty} \mathrm{d}\epsilon_\mu^a(x) \, \mathrm{e}^{-\beta \mathcal{E}_{\mu,\nu}^2(x)/2N}. \qquad (7.8)$$

The sum over n_ν, which arises from the degenerate saddle points, is just an irrelevant constant.

We see from Eq. (7.8) that only

$$\epsilon_\mu^a(x) \sim \frac{1}{\sqrt{\beta}} \qquad (7.9)$$

are essential which justifies the expansion in ϵ. Rescaling the integration variables in Eq. (7.8), we therefore find

$$Z(\beta) \propto \beta^{-(N_l - N_s)\, d_G/2}. \tag{7.10}$$

Substituting into Eq. (7.2) and remembering that $(N_l - N_s)/N_p = 2/d$ (see Eq. (6.5)), we obtain Eq. (7.1).

Problem 7.2 Repeat the derivation of the previous Problem for the adjoint action (6.29).

Solution The only difference with respect to the Wilson action (6.16) is that the saddle-point solution (7.3) is now modified as

$$U_\mu^{\mathrm{sp}}(x) = Z_\mu(x), \tag{7.11}$$

i.e. may take on different values at different links. It is evident that this is a minimum of the action (6.29).

The only modification of Eq. (7.8) is

$$\prod_{\nu=1}^{d} \sum_{n_\nu=1}^{N} \implies \prod_x \prod_{\nu=1}^{d} \sum_{n_\nu(x)=1}^{N}, \tag{7.12}$$

which only changes an irrelevant overall constant. Therefore, Eq. (7.1) remains unchanged providing the plaquette average is also taken in the adjoint representation. This supports the expectation that the continuum limits for both actions coincide.

The first-order phase transitions of the type given in Fig. 7.1 are usually harmless and are not associated with deconfinement. They are related with dynamics of some lattice degrees of freedom (say, with large fluctuations of the link variable $U_\mu(x)$ which occur independently at adjacent links) which do not affect the continuum limit and are called *lattice artifacts*. Moreover, these lattice degrees of freedom become frozen for $\beta > \beta_*$, which is necessary for the continuum limit to exist.

Another possibility for a lattice system is to undergo a second-order phase transition in analogy with spin systems. In this case $W(\partial p)$ is continuous but the derivative $\partial W(\partial p)/\partial \beta$ becomes infinite at the critical point $\beta = \beta_*$ as depicted in Fig. 7.2. Given Eq. (7.2), this derivative is to be considered as an analog of the specific heat of statistical systems. Its behavior at small and large β is governed by Eqs. (6.72) and (7.1), respectively.

Differentiating Eq. (6.65) with respect to β, the derivative $\partial W(\partial p)/\partial \beta$ can be expressed via the sum of the connected correlators:

$$\frac{\partial W(\partial p)}{\partial \beta} = \frac{1}{2} \sum_{\text{orient } p'} \left\langle \frac{1}{N} \operatorname{tr} U(\partial p) \frac{1}{N} \operatorname{tr} U(\partial p') \right\rangle_{\text{conn}}. \tag{7.13}$$

This formula also shows that $\partial W(\partial p)/\partial \beta$ is positive definite, since the

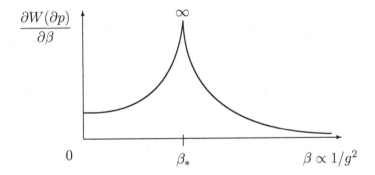

Fig. 7.2. Typical β-dependence of $\partial W(\partial p)/\partial\beta$ for a second-order phase transition which occurs at $\beta = \beta_*$.

RHS can be rewritten using translational invariance as

$$\frac{1}{2} \sum_{\text{orient } p'} \left\langle \frac{1}{N}\text{tr}\, U(\partial p)\, \frac{1}{N}\text{tr}\, U(\partial p') \right\rangle_{\text{conn}}$$

$$= \frac{1}{4N_p} \left\langle \left[\sum_{\text{orient } p} \frac{1}{N}\text{tr}\, U(\partial p) \right]^2 \right\rangle - \frac{1}{4N_p} \left[\left\langle \sum_{\text{orient } p} \frac{1}{N}\text{tr}\, U(\partial p) \right\rangle \right]^2 \geq 0\,,$$

$$(7.14)$$

where the equality is possible only for a Gaussian averaging, i.e. for a free theory. This repeats the standard proof of the positivity of specific heat in statistical mechanics.

Since each term of the sum in Eq. (7.13) is finite (remember that the trace of a unitary matrix takes on values between $-N$ and N), the only possibility for the RHS to diverge is for the sum over plaquettes p' to diverge. This is possible only when long-range (in the units of the lattice spacing) correlations are essential or, in other words, the correlation length is infinite. Thus, once again we have reproduced the argument that the continuum limit of lattice theories is reached at the points of second-order phase transitions.

Such a second-order phase transition seems to occur in compact QED (i.e. the $U(1)$ lattice gauge theory with fermions) at $e_*^2 \sim 1$. It is associated there with deconfinement of electrons. Electrons are confined for $e^2 > e_*^2$, similarly to quarks in lattice QCD, and are liberated for $e^2 < e_*^2$. The interaction potential looks like that of Fig. 6.9b for $e^2 < e_*^2$ and like that of Fig. 6.9a in the confinement region $e^2 > e_*^2$.[*] In order to reach the continuum limit with deconfined electrons, the bare charge

---

[*] The latter statement is not quite correct for reasons which are discussed in Sect. 9.5.

e^2 should be chosen to be slightly below the critical value. Then the renormalized physical charge can be made as small as the experimental value ($\alpha \approx 1/137$) according to the renormalization group arguments which are presented in the Remarks in Sect. 6.7.

The nature of the phase transition in a four-dimensional compact $U(1)$ lattice gauge theory without fermions was investigated using numerical methods. While the very first paper [LN80] indicated that the phase transition is of second order, some more advanced later investigations noted [EJN85] that it may be weakly first order. Anyway, we need fermions which usually weaken a phase transition that happens in a pure lattice gauge theory.

There are no indications that a second-order phase transition occurs in non-Abelian pure lattice gauge theories at intermediate values of β. This supports very strongly the behavior of the string tension being of the type depicted in Fig. 6.10. The second-order phase transition occurs in four dimensions at $\beta = \infty$ (or $g^2 = 0$) according to the general arguments of Sect. 6.7, which is necessary for the continuum limit to exist.

Remark on confinement in $4 + \epsilon$ dimensions

In $4+\epsilon$ dimensions ($\epsilon > 0$), a second-order deconfining phase transition always occurs in non-Abelian pure lattice gauge theories at some finite value of $\beta < \infty$ (or $g^2 > 0$). The case of $\epsilon \ll 1$ can be considered to be analogous to the ϵ-expansion in statistical mechanics [WK74]. An ultraviolet-stable fixed point exists at $g_*^2 \sim \epsilon$ since the theory is asymptotically free in $d = 4$. This phase transition is associated with deconfinement quite analogously to compact QED in $d = 4$. The deconfining phase is realized when the bare coupling $g < g_*$, while the confining phase is realized when $g > g_*$.

7.2 Mean-field method

The idea of applying the mean-field method, which is widely used in statistical systems, to study phase transitions in the lattice gauge theories was proposed by Wilson [Wil74] and first implemented for Abelian theories by Balian, Drouffe and Itzykson [BDI74]. A mean field usually works well when there are many neighboring degrees of freedom, interacting with a given one.

In the simplest version of the mean-field method, the link variable $U_\mu(x)$ is replaced by the mean-field value $m \cdot \mathbb{I}$ everywhere but at a given link (see Fig. 7.3) at which the self-consistency condition

$$\left\langle [U_\mu(x)]^{ij} \right\rangle_0 = m\,\delta^{ij} \qquad (7.15)$$

is imposed.

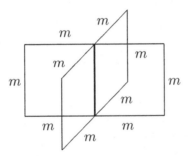

Fig. 7.3. Graphical representation of the self-consistency condition (7.17). The link variables are replaced by $m \cdot \mathbb{I}$ at all links except for a given one denoted by the bold line.

The average on the LHS of Eq. (7.15) is calculated with the action which is obtained from (6.16) by the substitution of $m \cdot \mathbb{I}$ for all the link variables (or their Hermitian conjugates) except at the given link. Since the given link enters $2(d-1)$ plaquettes, the average on the LHS of Eq. (7.15) is to be calculated with the action

$$S_0[U] = 2(d-1)m^3 \operatorname{Re} \operatorname{tr} U_\mu(x) + \text{const}. \tag{7.16}$$

Therefore, the self-consistency condition (7.15) can be written using the substitution of the mean-field ansatz into the lattice partition function (6.31) as

$$\frac{\displaystyle\int dU \, e^{\bar{\beta}N \operatorname{Re} \operatorname{tr} U} \frac{1}{N} \operatorname{tr} U}{\displaystyle\int dU \, e^{\bar{\beta}N \operatorname{Re} \operatorname{tr} U}} = m \tag{7.17}$$

with

$$\bar{\beta} = 2(d-1)m^3 \frac{\beta}{N^2}. \tag{7.18}$$

The meaning of Eq. (7.17) is very simple: the average of the normalized trace of the link variable at the given link should coincide with m, which is substituted for all other links of the lattice.

In order to verify whether the self-consistency condition (7.17) admits nontrivial solutions, one should first calculate the group integral on the LHS and then solve the self-consistency equation for m versus β. Typical behavior of the solution is depicted in Fig. 7.4. For all values of β, there exists a trivial solution $m = 0$ that is associated with no mean field. At

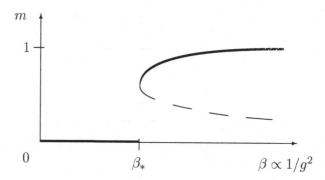

Fig. 7.4. Typical behavior of the mean-field solutions of the self-consistency equation (7.17). The only solution with $m = 0$ exists for $\beta < \beta_*$. Two more solutions appear for $\beta > \beta_*$. The solution depicted by the dashed line is unstable. The actual value of m versus β is depicted by the bold lines. A first-order phase transition is associated with $\beta = \beta_*$.

some value β_*, two more solutions of the self-consistency equation appear. The upper one is associated with positive specific heat, while the lower one corresponds to negative specific heat. This can be seen by noting that

$$W(\partial p) = m^4 \qquad (7.19)$$

in the mean-field approximation which follows from the substitution of the link variables in the definition (6.65) by the mean-field values. This nontrivial solution is preferred for $\beta > \beta_*$, since the partition function for it is larger (or the free energy is smaller) than for the $m = 0$ solution. The value of β_* is often associated with the point of a first-order phase transition.

The mean-field method in such a simple form was first applied to non-Abelian lattice gauge theories in [GL81, CGL81]. For the cases when a first-order phase transition occurs (say, for the $SU(N)$ groups with $N > 3$ or for the $SO(3)$ group), agreement with numerically calculated positions of the phase transitions is remarkable.

Problem 7.3 Calculate β_* for the $SU(\infty)$ lattice gauge theory, when the group integral on the LHS of Eq. (7.17) equals $\bar{\beta}/2$ for $\bar{\beta} \leq 1$ (a strong-coupling phase) and $1 - 1/2\bar{\beta}$ for $\bar{\beta} \geq 1$ (a weak-coupling phase).

Solution For the strong-coupling phase, the self-consistency equation

$$(d-1)\, m^3 \frac{\beta}{N^2} = m \qquad (7.20)$$

has the only solution $m = 0$. The other solutions are unacceptable owing to the stability criterion.

The nontrivial solutions of the self-consistency equation appear in the weak-coupling phase when $dm/d\beta = \infty$ or $d\beta/dm = 0$. Differentiating, we obtain then in $d = 4$

$$\frac{\partial \beta^{-1} N^2}{\partial m} = 12 \left(3m^2 - 4m^3 \right), \tag{7.21}$$

which yields

$$m_* = \frac{3}{4}, \qquad \frac{\beta_*}{N^2} = \frac{4^3}{3^4} \approx 0.79. \tag{7.22}$$

It is still left to verify that the proper $\bar{\beta}$ is indeed associated with the weak-coupling phase. From Eq. (7.18), we find $\bar{\beta}_* = 2$ and this is the case.

How one can calculate the one-matrix integral on the LHS of Eq. (7.17) at large N is explained in Sect. 12.9.

7.3 Mean-field method (variational)

There are some puzzles with the simplest mean-field ansatz described above. First of all, the average value of the link variable $U_\mu(x)$ in a lattice gauge theory must vanish owing to the gauge invariance (remember that $U_\mu(x)$ changes under the gauge transformation according to Eq. (6.13), while the action and the measure are gauge invariant). This is in accordance with Elitzur's theorem [Eli75], which says that a local gauge symmetry cannot be broken spontaneously, so that any order parameter for phase transitions in lattice gauge theories must be gauge invariant.

A way out of this is to reformulate the mean-field method in lattice gauge theories as a variational method [BDI74] which is similar to that proposed by R. Peierls in the 1930s. It is based on Jensen's inequality*

$$\left\langle e^F \right\rangle_0 \geq e^{\langle F \rangle_0} \tag{7.23}$$

which arises from the convexity of the exponential function, where $\langle \cdots \rangle_0$ denotes averaging with respect to a trial action.

Let us choose the trial partition function

$$Z_0 = \int \prod_{x,\mu} dU_\mu(x) \, e^{\bar{\beta} N \sum_{x,\mu} \operatorname{Re} \operatorname{tr} U_\mu(x)} \tag{7.24}$$

as a product of one-link integrals. Adding and subtracting the trial action, we write down the following bound on the partition function (6.31):

$$Z \geq Z_0 \exp \left\langle \frac{\beta}{N} \sum_p \operatorname{Re} \operatorname{tr} U(\partial p) - \bar{\beta} N \sum_{x,\mu} \operatorname{Re} \operatorname{tr} U_\mu(x) \right\rangle_0, \tag{7.25}$$

---

* More detail can be found, for example, in the books [Fey72, Sak85].

where $\langle \cdots \rangle_0$ denotes averaging with respect to the same action as in Eq. (7.24).

Since the expression that is averaged in the exponent in Eq. (7.25) is linear in each of the link variables, it can be calculated via the one-matrix integral given by the LHS of Eq. (7.17). Therefore, we find

$$
\left\langle \frac{\beta}{N} \sum_p \operatorname{Re} \operatorname{tr} U(\partial p) - \bar{\beta} N \sum_{x,\mu} \operatorname{Re} \operatorname{tr} U_\mu(x) \right\rangle_0 = \beta N_p m^4 - \bar{\beta} N^2 N_l m \,,
$$

(7.26)

where Eq. (7.19) has been used.

The idea of the variational mean-field method is to fix $\bar{\beta}$ from the condition for the trial ansatz (7.24) to give the best approximation to Z in the given class. Calculating the derivative of the RHS of Eq. (7.25) with respect to $\bar{\beta}$ and taking into account the fact that m depends on $\bar{\beta}$ according to Eq. (7.17), we find the maximum at $\bar{\beta}$ given by Eq. (7.18), which reproduces the simplest version of the mean-field method described above.

To restore Elitzur's theorem, a more sophisticated trial ansatz [Dro81] can be considered:

$$
Z_0 = \int \prod_{x,\mu} dU_\mu(x) \, e^{N \sum_{x,\mu} \operatorname{Re} \operatorname{tr} B_\mu^\dagger(x) U_\mu(x)} \,,
$$

(7.27)

where we choose $B_\mu(x)$ to be an arbitrary complex $N \times N$ matrix. Now the best approximation is reached for

$$
B_\mu(x) = \bar{\beta} \, \Omega(x) \, \Omega^\dagger(x + a\hat{\mu}) \,,
$$

(7.28)

where $\bar{\beta}$ is given by exactly the same equation as before, while $\Omega(x) \in SU(N)$ but is arbitrary otherwise. Now $\langle U_\mu^{ij}(x) \rangle_0$ vanishes after summing over equivalent maxima which results in integrations over $d\Omega(x)$.

Problem 7.4 Perform the variational mean-field calculation with the ansatz (7.27).

Solution Let us denote

$$
M_\mu^{ij}(x) = \frac{\int \prod_{x,\mu} dU_\mu(x) \, e^{N \sum_{x,\mu} \operatorname{Re} \operatorname{tr} B_\mu^\dagger(x) U_\mu(x)} \, U_\mu^{ij}(x)}{\int \prod_{x,\mu} dU_\mu(x) \, e^{N \sum_{x,\mu} \operatorname{Re} \operatorname{tr} B_\mu^\dagger(x) U_\mu(x)}} \,.
$$

(7.29)

Then the analog of Eq. (7.26) is

$$
\left\langle \frac{\beta}{N} \sum_p \operatorname{Re} \operatorname{tr} U(\partial p) - N \sum_{x,\mu} \operatorname{Re} \operatorname{tr} B_\mu^\dagger(x) U_\mu(x) \right\rangle_0
$$

$$
= \frac{\beta}{N} \sum_p \operatorname{Re} \operatorname{tr} M(\partial p) - N \sum_{x,\mu} \operatorname{Re} \operatorname{tr} B_\mu^\dagger(x) M_\mu(x)
$$

(7.30)

so that the inequality (7.25) takes the form

$$Z \geq Z_0 \exp\left[\frac{\beta}{N}\sum_p \operatorname{Re}\operatorname{tr} M(\partial p) - N\sum_{x,\mu} \operatorname{Re}\operatorname{tr} B_\mu^\dagger(x)M_\mu(x)\right]. \qquad (7.31)$$

$B_\mu(x)$ can now be determined by maximizing with respect to $B_\mu(x)$ and taking into account Eq. (7.29).

It is easy to see that if $B_\mu(x) = \bar{\beta}\cdot\mathbb{I}$ is a solution as before, then (7.28) is also a solution. Therefore, we find

$$\langle U_\mu^{ij}(x)\rangle_0 = m\int d\Omega(x+a\hat{\mu})\,d\Omega(x)\,\Omega(x+a\hat{\mu})\,\Omega^\dagger(x) = 0\,, \qquad (7.32)$$

where the integration over Ω takes into account different equivalent maxima. Thus, all gauge-invariant quantities for the ansatz (7.27) are the same as for the ansatz (7.24), while gauge-noninvariant quantities now vanish in agreement with Elitzur's theorem.

Remark on the criterion for phase transition

Another puzzle with the simplest mean-field method is why the point of the first-order phase transition is chosen as explained in Fig. 7.4 but not when the free energy of both phases coincide (the standard Maxwell rule in statistical physics). Perhaps, the criterion of Fig. 7.4 should be chosen if a barrier between two phases is impenetrable, which happens at large N or if quantum fluctuations are not taken into account such as for the simplest mean field. The mean-field calculations of [FLZ82], which take into account fluctuations around the mean-field solution (7.28), agree for the Maxwell-rule criterion with numerical data. These results are reviewed in [DZ83].

7.4 Lattice renormalization group

While the mean-field method is useful for studying the first-order phase transitions, the second-order phase transitions in lattice statistical systems are better described by the renormalization group method (see, for example, the review by Wilson and Kogut [WK74]). The idea of applying a similar method to lattice gauge theories is due to Migdal [Mig75].

A simple renormalization group transformation in lattice gauge theories is associated with doubling of the lattice spacing a. Originally one has a lattice as depicted in Fig. 7.5a. The lattice renormalization group (r.g.) transformation consists in integrating over the link variables $U_\mu(x)$ on the links shown by the thin lines which results in a lattice with spacing $2a$,

$$a \overset{\text{r.g.}}{\Longrightarrow} 2a\,, \qquad (7.33)$$

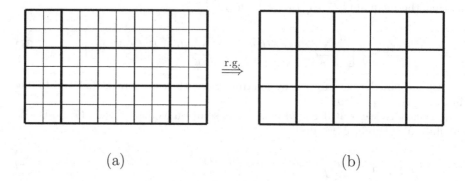

(a)   (b)

Fig. 7.5. Lattice renormalization group transformation (7.33). The thin lines of the old lattice (a) represent links on which integration is performed. The new lattice (b) has a lattice spacing of $2a$ but the same spatial extent La.

which is depicted in Fig. 7.5b. The space size of the lattice is L before the transformation and becomes $L/2$ after the transformation,

$$L \overset{\text{r.g.}}{\Longrightarrow} \frac{L}{2}, \tag{7.34}$$

so that the lattice extent is $L \cdot a$ in both cases, which is expected to reduce the influence of finite-size effects on the transformation.

The Wilson action on the lattice of Fig. 7.5a becomes a more general one under the renormalization group transformation:

$$S[U] = \sum_p \beta \frac{1}{N} \operatorname{tr} U(\partial p)$$

$$\overset{\text{r.g.}}{\Longrightarrow} \quad S'[U] = \sum_p \beta_1' \frac{1}{N} \operatorname{tr} U(\partial p) + \sum_{p_2} \beta_2' \frac{1}{N} \operatorname{tr} U(\partial p_2)$$

$$+ \sum_{p_3} \beta_3' \frac{1}{N} \operatorname{tr} U(\partial p_3) + \cdots . \tag{7.35}$$

The new action $S'[U]$ is not necessarily a single-plaquette action and can involve traces of the Wilson loops for boundaries of double plaquettes, triple plaquettes and so on.

The new action would be the same as the old one only at a fixed point. This usually happens after the renormalization group transformation is applied several times when the lattice theory does have a fixed point. The resulting action is then associated with an action of the continuum theory.

The great success of non-Abelian lattice gauge theories with the Wilson action in describing the continuum limit even at a relatively small spatial extent or, which is the same, at relatively large g^2 and a, is because it is not far away from the fixed-point action of the renormalization group. The proper numerical results will be presented in a moment (Fig. 7.6).

If both actions $S[U]$ and $S'[U]$ are the single-plaquette Wilson actions, then

$$\beta \stackrel{\text{r.g.}}{\Longrightarrow} \beta' = \beta - \Delta\beta \tag{7.36}$$

under the renormalization group transformation on the lattice.

Since the Gell-Mann–Low function $\mathcal{B}(g^2)$ in the continuum is known, $\Delta\beta$ versus β is determined by the equation

$$\int_{\beta-\Delta\beta}^{\beta} \frac{dx}{x^2 \mathcal{B}(3/x)} = -\frac{\ln 2}{3}. \tag{7.37}$$

Here $\ln 2$ on the RHS arises from Eq. (7.33) and the relation (6.32) between β and g^2 is used with $N = 3$.

For the pure $SU(3)$ gauge theory, we obtain from Eq. (7.37)

$$\Delta\beta = 0.579 + \frac{0.204}{\beta} + \mathcal{O}(\beta^{-2}) \tag{7.38}$$

at asymptotically large β.

One can integrate over the thin links in Fig. 7.5a either approximately or numerically. The following procedure for an approximate integration is known as the Migdal–Kadanoff recursion relations.

Let us expand the exponential of the old action in the characters

$$e^{-S[U]} = \sum_r f_r d_r \chi_r(U), \tag{7.39}$$

where

$$d_r = \chi_r(\mathbb{I}) \tag{7.40}$$

is the dimension of a given representation r and f_r are the coefficients which depend on the form of $S[U]$.

Migdal [Mig75] proposed to approximate the new action, which appears after

$$a \implies \rho\, a, \tag{7.41}$$

by the formula

$$e^{-S'[U']} = \left[\sum_r (f_r)^{\rho^2} d_r \chi_r(U') \right]^{\rho^{d-2}}, \tag{7.42}$$

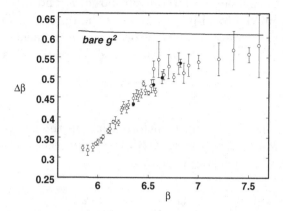

Fig. 7.6. Monte Carlo data from Akemi *et al.* [Ake93] for $\Delta\beta$. The error bars represent statistical errors. The solid line represents the asymptote (7.38).

which is exact in $d = 2$ dimensions. Kadanoff [Kad76] modified slightly the recursion relation (7.42).

The study of the Migdal–Kadanoff recursion relations was historically the first argument that second-order phase transitions do not occur in the non-Abelian lattice gauge theory when g^2 is decreased. Moreover, these relations in $d = 4$ are the same as for spin systems (with the same symmetry group) in $d = 2$ where this phenomenon is known. A disadvantage of the method is that it is difficult to estimate its accuracy.

A final answer to the question of whether or not a second-order phase transition occurs in the non-Abelian lattice gauge theory was given by the numerical integration. This is known as the Wilson Monte Carlo renormalization group. Some typical results [Ake93] for $\Delta\beta$, which is defined by Eq. (7.36), versus β are depicted in Fig. 7.6. The solid line represents the asymptote (7.38). The agreement confirms that the continuum limit is reached already at these values of β, while the deviation of the Monte Carlo data from the asymptotic behavior for smaller values of β is owing to lattice nonperturbative effects.

7.5 Monte Carlo method

The idea of the Monte Carlo method is to calculate the partition function (6.31) and the averages (6.39) for arbitrary values of β numerically, using the fact that the multiplicity of the integral is large. For an $L \times L \times L \times L$ lattice in 4 dimensions, a typical multiplicity of the integral is as large as $4 \cdot (N^2 - 1) \cdot L^4$ ($\sim 10^7$ for $L = 24$). It is hopeless to calculate such an integral exactly. In contrast, the larger the multiplicity the better the Monte Carlo method works.

As usual, the Monte Carlo method is applied not to sequential integrals over $U_\mu(x)$ at each link but rather to the multiple integral as a whole, which can be viewed as the sum over states of a statistical system.

A state is identified with a gauge field configuration which is described by the values of the link variables at all the links of the lattice:

$$C = \{U_\mu^{ij}(x), \ldots, \ldots, \ldots, \ldots, \ldots, \ldots, \ldots\}. \tag{7.43}$$

There are as many positions in this row as the multiplicity of the integral.

Then the sequential integral can be represented as

$$\int \prod_{x,\mu} dU_\mu(x) \cdots = \sum_C \cdots. \tag{7.44}$$

The averages (6.39) can be rewritten as

$$\langle F(C) \rangle = \frac{\sum_C e^{-\beta S(C)} F(C)}{\sum_C e^{-\beta S(C)}}, \tag{7.45}$$

where $S(C)$ and $F(C)$ are the values of S and F for the given configuration C.

The task of Monte Carlo calculations is not to sum over all possible configurations, the number of which is infinite, but rather to construct an ensemble, say, of n configurations

$$E = \{C_1, \ldots, C_n\} \tag{7.46}$$

such that a given configuration C_k is encountered with the Boltzmann probability

$$P_{\text{Bol}}(C_k) = Z^{-1}(\beta) e^{-\beta S(C_k)}. \tag{7.47}$$

Such a sample of configurations is called the *equilibrium ensemble*.

Given an equilibrium ensemble, the averages (7.45) take the form of the arithmetic mean

$$\langle F[U] \rangle = \frac{1}{n} \sum_{k=1}^{n} F(C_k) \tag{7.48}$$

because each configuration "weights" already as much as is required. In particular, the Wilson loop average for a rectangular contour is given by

$$W(R \times T) = \frac{1}{n} \sum_{k=1}^{n} \frac{1}{N} \operatorname{tr} U(R \times T; C_k). \tag{7.49}$$

If all configurations in the equilibrium ensemble are independent, then the RHS of Eq. (7.49) will approximate the exact value of $W(R \times T)$ with an accuracy of $\sim \sqrt{n}$.

The analogy between this method of calculating averages and statistical physics is obvious. The equilibrium ensemble simulates actual states of a statistical system, while the index k describes the time evolution.

A crucial point in the Monte Carlo method is to construct the equilibrium ensemble. It is not simple to do that because the Boltzmann probability is not known at the outset. A way around this problem is to establish a random process for which each new configuration in the sequence (7.46) is obtained from the previous one by a definite algorithm but stochastically. In other words, the random process is completely determined by the probability $P(C_{k-1} \to C_k)$ for a transition from a state C_{k-1} to a state C_k and does not depend on the history of the system, i.e.

$$P(C_{k-1} \to C_k) \;=\; P(C_{k-1}, C_k) \,. \tag{7.50}$$

Such a random process is known as the *Markov process*.

The transition probability $P(C, C')$ should be chosen in such a way as to provide the Boltzmann distribution (7.47). This is ensured if $P(C, C')$ satisfies the detailed balance condition

$$e^{-\beta S(C)} P(C, C') \;=\; e^{-\beta S(C')} P(C', C) \,. \tag{7.51}$$

Then

(1) an equilibrium sequence of states will transform into another equilibrium sequence,

(2) a nonequilibrium sequence will approach an equilibrium one when moving through the Markov chain.

Problem 7.5 Prove statements (1) and (2) listed in the previous paragraph using the detailed balance condition (7.51).

Solution Let a state C be encountered in ensembles E and E' with probability densities $P(C)$ and $P'(C)$, respectively. Then the distance between the two ensembles can be defined as

$$\|E - E'\| \;=\; \sum_C |P(C) - P'(C)| \,. \tag{7.52}$$

For a Markov process when Eq. (7.50) holds, we have

$$P'(C) \;=\; \sum_{C'} P(C, C') \, P(C') \tag{7.53}$$

if E' is obtained from E by a Monte Carlo algorithm. The transition probability

$P(C, C')$ is nonnegative and obeys

$$\sum_C P(C, C') = \sum_{C'} P(C, C') = 1 \tag{7.54}$$

since each new state is obtained from an old one and vice versa.

It is now easy to prove statement (1). Summing the detailed balance condition (7.51) over C', we obtain

$$P_{\mathrm{Bol}}(C) = \sum_{C'} P(C, C') P_{\mathrm{Bol}}(C'), \tag{7.55}$$

i.e. the Boltzmann distribution is an eigenvector of $P(C, C')$. Comparing with Eq. (7.53), we see that the new distribution is again the Boltzmann one, which proves statement (1).

To prove statement (2), let us compare the distances from E and E' to some equilibrium ensemble E_{Bol} associated with the Boltzmann distribution (7.47). We have the inequality

$$
\begin{aligned}
\left\| E' - E_{\mathrm{Bol}} \right\| &= \sum_C \left| P'(C) - P_{\mathrm{Bol}}(C) \right| \\
&= \sum_C \left| \sum_{C'} P(C, C') \left[P(C') - P_{\mathrm{Bol}}(C') \right] \right| \\
&\leq \sum_{CC'} P(C, C') \left| P(C') - P_{\mathrm{Bol}}(C') \right| \\
&= \sum_{C'} \left| P(C') - P_{\mathrm{Bol}}(C') \right| \\
&= \left\| E - E_{\mathrm{Bol}} \right\|, \tag{7.56}
\end{aligned}
$$

where Eqs. (7.53), (7.55) and (7.54) are used. Thus, statement (2) is proven.

Specific Monte Carlo algorithms differ in the choice of the transition probability $P(C, C')$, while the detailed balance condition (7.51) is always satisfied. The two most popular algorithms, which act at one link, are as follows.

Heat bath algorithm

A new link variable $U'_\mu(x)$ is selected randomly from the group manifold with a probability given by the Boltzmann factor

$$P\big(U'_\mu(x)\big) \propto e^{-\beta S(C')}. \tag{7.57}$$

Then this procedure is repeated for the next link and so on until the whole lattice is passed. This can be imagined as if a reservoir at temperature $1/\beta$ touches each link of the lattice in succession. It is clear from physical intuition that the system will be brought to thermodynamic equilibrium sooner or later.

Metropolis algorithm

This algorithm is used in statistical physics since the 1950s and consists of several steps.

(1) A trial new link variable $U'_\mu(x)$ is selected (suppose randomly on the group manifold).

(2) The difference between the action for this trial configuration and that for the old one is calculated:

$$\Delta S = S(C') - S(C). \tag{7.58}$$

(3) A random number $r \in [0, 1]$ is generated.

(4) If

$$e^{-\beta \Delta S} > r, \tag{7.59}$$

then $U'_\mu(x)$ is accepted. Otherwise, $U'_\mu(x)$ is rejected and the old value $U_\mu(x)$ is kept.

(5) All of this is repeated for the next links.

An advantage of the Metropolis algorithm is that it is usually more easy implemented in practical calculations.

A new configuration C', which is obtained by applying once either Monte Carlo algorithm to each link of the lattice (this procedure is often called the Monte Carlo sweep), will be strongly correlated with the old one, C. This is because the lattice action depends not only on the variable at the given link but also on those at the neighboring links which form plaquettes with the given one. In order for C' to become independent of C, this procedure should be repeated many times or special tricks should be used to reduce the correlations. Then this new configuration can be added to the equilibrium ensemble (7.46) as C_k.

More details concerning the Monte Carlo algorithms as well as their practical implementation in lattice gauge theories can be found in the review [CJR83] and the books [Cre83, MM94].

7.6 Some Monte Carlo results

The first Monte Carlo calculation in non-Abelian lattice gauge theories, which is relevant for the continuum limit, was performed by Creutz [Cre79] who evaluated the string tension for the $SU(2)$ gauge group. His result is reproduced in Fig. 7.7 and looks very much like what is expected in Fig. 6.10 on p. 120. This calculation was the first demonstration that the continuum limit sets in for relatively large $g^2 \approx 0.91$

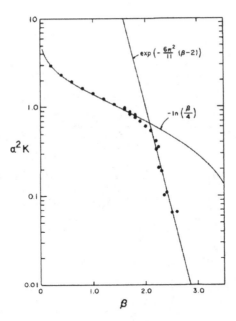

Fig. 7.7. Monte Carlo data from Creutz [Cre79] for the string tension in the $SU(2)$ pure lattice gauge theory.

($\beta \approx 2.2$) and that results for the continuum can therefore be extracted from relatively small lattices.

The restoration of rotational symmetry for these values of g^2 was demonstrated explicitly by Land and Rebbi [LR82]. They calculated equipotential surfaces for the interaction between static quarks. In the strong-coupling region $g^2 \rightarrow \infty$, they appear as in Fig. 7.8a since the interaction potential is given by

$$E(x,y,z) \;=\; K\,(\,|x| + |y| + |z|\,) \tag{7.60}$$

because the distance between the quarks is measured along the lattice. This is associated with the cubic symmetry on the lattice (i.e. rotations through an angle which is a multiple of $\pi/2$ around each axis and translations by a multiple of the lattice spacing along each axis) rather than with the Poincaré group. The rotational symmetry must be restored in the continuum limit.

The Monte Carlo data of Land and Rebbi [LR82] are shown in Figs. 7.8b and c. They demonstrate the restoration of rotational symmetry when passing from $\beta = 2$ (Fig. 7.8b) to $\beta = 2.25$ (Fig. 7.8c).

The early Monte Carlo calculations played a very important role in the development of the method. Their main result is that the Monte Carlo

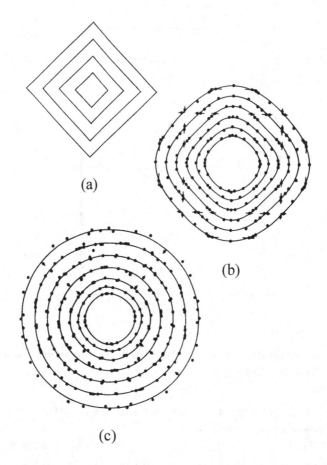

(a)

(b)

(c)

Fig. 7.8.   Behavior of equipotential lines at different values of β: (a) the strong-coupling limit $\beta = 0$; (b) $\beta = 2$; (c) $\beta = 2.25$. (b) and (c), taken from the paper by Lang and Rebbi [LR82], show how the rotational symmetry is restored as β is increased.

calculation of physical quantities in QCD is possible on relatively small lattices.

A dramatic improvement of the Monte Carlo technology in lattice gauge theories has taken place over the last 20 years. New calculations are performed on larger lattices and with better statistics. The best way to follow current developments of the subject is via plenary talks published in the proceedings of the annual Lattice Conference (currently [Lat00]).

8

Fermions on a lattice

It turned out to be most difficult in the lattice approach to QCD to deal with fermions. Putting fermions on a lattice is an ambiguous procedure since the cubic symmetry of a lattice is less restrictive than the continuous Lorentz group.

The simplest chiral-invariant formulations of lattice fermions lead to a doubling of fermionic degrees of freedom, as was first noted by Wilson [Wil75], and describe from 16 to four relativistic continuum fermions, depending on the formulation. One-half of them have a positive axial charge and the other half have a negative one, so that the chiral anomaly cancels. There is a no-go theorem which says that the fermionic doubling is always present under natural assumptions concerning a lattice gauge theory.

A practical way out of this problem is to choose the fermionic lattice action to be explicitly noninvariant under the chiral transformation and to have, by tuning the mass of the lattice fermion, one relativistic fermion in the continuum and the masses of the doublers to be of the order of the inverse lattice spacing. The chiral anomaly is recovered in this way.

In this chapter we consider various formulations of lattice fermions and the doubling problem. We discuss briefly the results on spontaneous breaking of the chiral symmetry in QCD.

8.1 Chiral fermions

The quark fields are generically matter fields, the gauge transformation of which in the continuum is given by Eqs. (5.1) and (5.3), and can be put on a lattice according to Eq. (6.7). Then the lattice gauge transformation is

$$\psi_x \xrightarrow{\text{g.t.}} \psi'_x = \Omega(x)\psi_x \,, \qquad \bar{\psi}_x \xrightarrow{\text{g.t.}} \bar{\psi}'_x = \bar{\psi}_x \,\Omega^\dagger(x) \,. \qquad (8.1)$$

143

The lattice analog of the QCD action (5.13) is given as[*]

$$S[U, \bar{\psi}, \psi] = \beta S_{\text{lat}}[U] + M \sum_x \bar{\psi}_x \psi_x$$

$$+ \frac{1}{2} \sum_{x,\mu>0} \left[\bar{\psi}_x \gamma_\mu U_\mu^\dagger(x) \psi_{x+a\hat{\mu}} - \bar{\psi}_{x+a\hat{\mu}} \gamma_\mu U_\mu(x) \psi_x \right].$$

(8.2)

The first term on the RHS is the pure gauge lattice action (6.16). The second term is a quark mass term on a lattice. The sum in the third term is over all lattice links (i.e. over all sites x and positive directions μ). This action is Hermitian and invariant under the lattice gauge transformation (6.13) and (8.1) with finite lattice spacing.

The partition function of lattice QCD with fermions is defined by

$$Z(\beta, M) = \int \prod_{x,\mu} dU_\mu(x) \prod_x d\bar{\psi}_x \, d\psi_x \, e^{-S[U, \bar{\psi}, \psi]}, \quad (8.3)$$

where the action is given by Eq. (8.2). The integration over $U_\mu(x)$ is as in Eq. (6.31), and the integral over the quark field is the Grassmann one. The averages are defined by

$$\langle F[U, \bar{\psi}, \psi] \rangle$$

$$= Z^{-1}(\beta, M) \int \prod_{x,\mu} dU_\mu(x) \prod_x d\bar{\psi}_x \, d\psi_x \, e^{-S[U, \bar{\psi}, \psi]} \, F[U, \bar{\psi}, \psi],$$

(8.4)

which extends Eq. (6.39) to the case of fermions. Since both the action and the measure in Eq. (8.4) are gauge invariant at finite lattice spacing, a nonvanishing result only occurs when the integrand, $F[U, \bar{\psi}, \psi]$, is gauge invariant as well.

In order to show how the lattice action (8.2) reproduces (5.13) in the naive continuum limit $a \to 0$, let us assume that the lattice quark field ψ_x varies slowly from site to site and substitute

$$\left. \begin{array}{rcl} \psi_x & \to & a^{3/2} \psi(x), \\[2mm] \psi_{x+a\hat{\mu}} & \to & a^{3/2} [\psi(x) + a\partial_\mu \psi(x)] \end{array} \right\} \quad (8.5)$$

in $d = 4$. Here $\psi(x)$ is a continuum quark field and the power of a arises from the dimensional consideration (remember that ψ_x is dimensionless).

---

[*] The standard formula differs from this one by an interchange of U and $U^\dagger$ owing to the inverse ordering of matrices in the phase factors (see the footnote on p. 88). It does not matter how one defines $U_\mu(x)$ since the Haar measure is invariant under Hermitian conjugation.

Equation (8.5) together with Eq. (6.10) yields

$$\bar{\psi}_x \gamma_\mu U_\mu^\dagger(x) \psi_{x+a\hat{\mu}} \quad \rightarrow \quad a^3 \bar{\psi} \gamma_\mu \psi + a^4 \bar{\psi} \nabla_\mu^{\text{fun}} \gamma_\mu \psi + \mathcal{O}(a^5), \qquad (8.6)$$

where there is no summation over μ in the second term on the RHS as earlier in this part. The first term cancels when substituted into Eq. (8.2), while the second one reproduces the fermionic part of the continuum action. The mass term is also reproduced if $M = am$.

The fermionic lattice action (8.2) was proposed in [Wil74]. For $M = 0$ it is invariant under the global chiral transformation

$$\psi_x \xrightarrow{\text{c.t.}} e^{i\alpha\gamma_5} \psi_x, \qquad \bar{\psi}_x \xrightarrow{\text{c.t.}} \bar{\psi}_x e^{i\alpha\gamma_5}. \qquad (8.7)$$

For this reason, these lattice fermions are called *chiral fermions*. Since the lattice action is both gauge and chiral invariant, there is no Adler–Bell–Jackiw anomaly according to the general arguments of Chapter 3.

Problem 8.1 Show that the lattice action (8.2) is invariant under

$$\psi_x \quad \rightarrow \quad i\gamma_4\gamma_5 (-1)^{t/a} \psi_x. \qquad (8.8)$$

Find 15 further similar transformations.

Solution Let us define the generators T_A by

$$\psi_x \rightarrow T_A \psi_x, \qquad \bar{\psi}_x \rightarrow \bar{\psi}_x T_A^\dagger. \qquad (8.9)$$

The transformation (8.8) can be performed for each of the $d = 4$ axes which gives

$$T_A = i\gamma_\mu\gamma_5 (-1)^{x_\mu/a}. \qquad (8.10)$$

The other generators are given by products of (8.10). Their explicit form is [KS81a]

$$T_A = \mathbb{I}, \; i\gamma_\mu\gamma_5 (-1)^{x_\mu/a}, \; i\gamma_\mu\gamma_\nu (-1)^{(x_\mu+x_\nu)/a} \; (\mu > \nu),$$
$$\gamma_4 (-1)^{(x+y+z)/a}, \; \ldots, \; \gamma_1 (-1)^{(y+z+t)/a}, \; \gamma_5 (-1)^{(x+y+z+t)/a}. \qquad (8.11)$$

All together there are $1 + 4 + 6 + 4 + 1 = 16$ independent transformations which form a discrete subgroup of the $U(4)$ group.

8.2 Fermion doubling

As was pointed out at the end of the previous section, the lattice fermionic action (8.2) is both gauge and chiral invariant (for $M = 0$) so that there is no chiral anomaly in the continuum. Since the anomaly is present for one continuum fermion, this suggests that the action (8.2) is associated with more than one species of continuum fermions.

In order to verify this explicitly, let us calculate the poles of the lattice fermionic propagator.

As usual, it is easier to work with the Fourier image of ψ_x:

$$\psi_k = a^{5/2} \sum_x \psi_x e^{-ikx} . \tag{8.12}$$

The free fermionic action then reads as

$$S_0[\bar\psi, \psi] = \int_{-\pi/a}^{\pi/a} \frac{d^4 k}{(2\pi)^4} \, \bar\psi_k \, G^{-1}(k) \, \psi_k \tag{8.13}$$

with

$$G^{-1}(k) = \frac{1}{a} \sum_{\mu=1}^{4} i\gamma_\mu \sin k_\mu a \tag{8.14}$$

for $M = 0$.

In the naive continuum limit, the sin function in Eq. (8.14) can be expanded as a power series in a, which results in the free (inverse) continuum propagator

$$G^{-1}(k) \;\rightarrow\; i \sum_{\mu=1}^{4} \gamma_\mu k_\mu = i\hat k . \tag{8.15}$$

Lorentz invariance has been restored after summing over μ.

When passing from the lattice expression (8.14) to the continuum one (8.15), it was implicitly assumed that the momentum k_μ is not of the order of $1/a$ because otherwise the sin function cannot be expanded in a. The doubling of relativistic continuum fermionic states occurs exactly for this reason.

To find the poles of the propagator, let us return to Minkowski space by substituting $k_4 = iE$, where E is the energy. The poles are then determined from the dispersion law

$$\sinh^2 Ea = \sum_{\mu=1}^{3} \sin^2 p_\mu a . \tag{8.16}$$

Let us look for solutions of Eq. (8.16) with positive energy $E > 0$ (solutions with negative energy are associated as usual with antiparticles). Suppose that a particle moves along the z-axis so that components of the four-momentum

$$p^{(1)} = (E, 0, 0, p_z) \tag{8.17}$$

are related by

$$\sinh Ea = \sin p_z a, \tag{8.18}$$

which follows from the substitution of (8.17) into the dispersion law.

Since sin is a periodic function, the four-vector

$$p^{(2)} = \left(E, 0, 0, \frac{\pi}{a} - p_z\right) \tag{8.19}$$

is also a solution of Eq. (8.18) if (8.17) is. Quite analogously, the four-vectors

$$\left. \begin{aligned} p^{(3)} &= \left(E, \frac{\pi}{a}, 0, p_z\right), \\ &\vdots \quad , \\ p^{(8)} &= \left(E, \frac{\pi}{a}, \frac{\pi}{a}, \frac{\pi}{a} - p_z\right), \end{aligned} \right\} \tag{8.20}$$

which are obtained from $p^{(1)}$ and $p^{(2)}$ by changing zeros for π/a, also satisfy Eq. (8.18). Therefore, a quark state with energy E is eightfold degenerate.

The quark states with four-momenta $p^{(1)}, \dots, p^{(8)}$ are different states. Their wave functions equal

$$\Psi^{(j)}(t, x, y, z) \propto \exp\left[iEt - ip_x^{(j)}x - ip_y^{(j)}y - ip_z^{(j)}z\right]. \tag{8.21}$$

The wave function in the state with momentum $p^{(3)}$ differs, say, from the wave function in the state $p^{(1)}$ by an extra factor of $(-1)^{x/a}$. In other words, it changes strongly as $a \to 0$ with one step along the lattice in the x-direction. One more step returns the wave function to the initial value.

For such functions, the naive continuum limit of the lattice action (8.2) is as good as for the slowly varying functions when Eq. (8.5) holds. In order to see that, let us rewrite the action (8.2) as

$$S[U, \bar\psi, \psi] = \beta S_{\mathrm{lat}}[U] + M \sum_x \bar\psi_x \psi_x$$

$$+ \frac{1}{2} \sum_{x,\mu>0} \left\{ \bar\psi_x \gamma_\mu \left[U_\mu^\dagger(x)\psi_{x+a\hat\mu} - U_{-\mu}^\dagger(x)\psi_{x-a\hat\mu} \right] \right\}. \tag{8.22}$$

Even if ψ_x has opposite signs at neighboring lattice sites along the μ-axis,

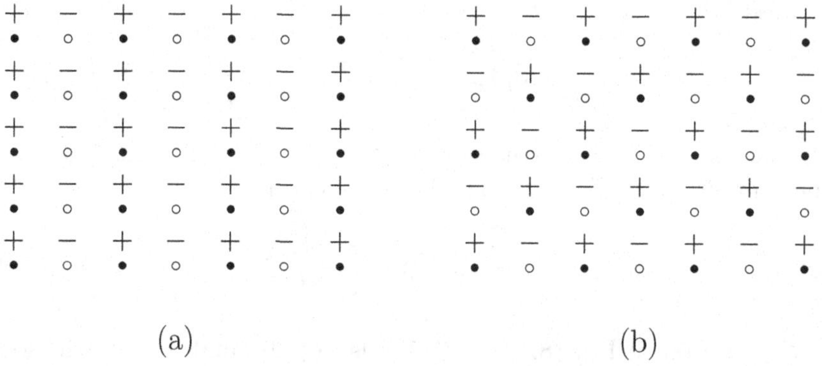

Fig. 8.1.   Altering signs of ψ_x on a lattice along (a) one axis and (b) two axes.

as illustrated by Fig. 8.1a, i.e.

$$\psi_{x+a\hat{\mu}} \quad \rightarrow \quad -\psi_x\,, \tag{8.23}$$

then the difference $\psi_{x+a\hat{\mu}} - \psi_{x-a\hat{\mu}}$ on the RHS of Eq. (8.22) is still of the correct order in a:

$$\left.\begin{array}{rcl} \psi_x & \rightarrow & a^{3/2}\,\psi(x)\,, \\[2mm] \psi_{x+a\hat{\mu}} - \psi_{x-a\hat{\mu}} & \rightarrow & -2a^{5/2}\partial_\mu\psi(x)\,, \end{array}\right\} \tag{8.24}$$

so that the continuum fermionic action is reproduced except for the sign of the γ_μ-matrix which is opposite to that in Eq. (5.13).

This extra minus sign can be absorbed in the redefinition of the continuum fermionic field $\psi(x) \rightarrow i\gamma_\mu\gamma_5\psi(x)$, which changes its chirality. Therefore, the axial charge of the doublers is opposite. Analogously, four of the eight doublers have a positive axial charge and the four others have a negative one dependent on whether the sign of ψ_x alters at neighboring sites along an even or odd number of axes (see Fig. 8.1). In Euclidean space the doubling also occurs along the temporal axis, so the number of doublers is equal to $2^d = 16$: eight of them with positive and eight with negative axial charge. This explains why the chiral anomaly cancels.

Problem 8.2 Calculate the vector and axial charges of the doublers deriving the vector and axial currents on a lattice.

Solution   The vector and axial currents on a lattice can be derived using a lattice analog of the Noether theorem. The invariance of the lattice fermionic action under

$$\psi_x \rightarrow e^{i\alpha(x)}\,\psi_x\,, \qquad \bar{\psi}_x \rightarrow \bar{\psi}_x\,e^{-i\alpha(x)} \tag{8.25}$$

results in the lattice vector current

$$J_\mu^{\rm V}(x) \;=\; \frac{1}{2}\left[\bar{\psi}_x\gamma_\mu U_\mu^\dagger(x)\psi_{x+a\hat{\mu}} + \bar{\psi}_{x+a\hat{\mu}}\gamma_\mu U_\mu(x)\psi_x\right]\,, \tag{8.26}$$

which is conserved in the sense that

$$\sum_{\mu>0} \left[J^{\mathrm{V}}_\mu(x) - J^{\mathrm{V}}_\mu(x - a\hat{\mu}) \right] = 0. \tag{8.27}$$

This can be proven using the lattice (quantum) Dirac equation

$$\frac{1}{2}\sum_{\mu>0} \gamma_\mu \left[U^\dagger_\mu(x)\psi_{x+a\hat{\mu}} - U^\dagger_{-\mu}(x)\psi_{x-a\hat{\mu}} \right] + M\psi_x \overset{\mathrm{w.s.}}{=} \frac{\delta}{\delta\bar{\psi}_x}, \tag{8.28}$$

which is the lattice analog of Eq. (3.19).

Analogously, the lattice chiral transformation

$$\psi_x \rightarrow \mathrm{e}^{\mathrm{i}\alpha(x)\gamma_5} \psi_x, \qquad \bar{\psi}_x \rightarrow \bar{\psi}_x\, \mathrm{e}^{\mathrm{i}\alpha(x)\gamma_5} \tag{8.29}$$

results in the lattice axial current

$$J^{\mathrm{A}}_\mu(x) = \frac{\mathrm{i}}{2} \left[\bar{\psi}_x \gamma_\mu \gamma_5 U^\dagger_\mu(x)\psi_{x+a\hat{\mu}} + \bar{\psi}_{x+a\hat{\mu}}\gamma_\mu\gamma_5 U_\mu(x)\psi_x \right], \tag{8.30}$$

which reproduces (3.10) as $a \to 0$. The current (8.30) is conserved for $M = 0$.

It is now easy to verify that 16 generators (8.11) commute with the lattice $U(1)$ transformation (8.25) so that the lattice vector current (8.26) is left invariant. Analogously, the lattice axial $U(1)$ transformation (8.29) commutes only with $1 + 6 + 1 = 8$ of 16 generators (8.11) which are constructed from the products of an even number of the generators (8.10) and does not commute with the $4 + 4 = 8$ other generators. Therefore, the axial current (8.30) is invariant under the $1+6+1 = 8$ transformations, which are the products of an even number of the generators (8.10), and alters its sign under the other $4 + 4 = 8$ transformations, which are the products of an odd one. Thus, the vector charge of all the doublers is the same, while the axial charge is positive for eight and negative for the other eight doublers.

It is worth noting that the mass term in Eq. (8.2) is not γ_5 invariant, but does not remove the fermion doubling.

One might think of removing the doubling problem by modifying the expression for the inverse lattice propagator $G^{-1}(k)$ in the free fermionic lattice action (8.13), for instance, by adding next-to-neighbor terms. It is easy to see that this does not help if the function $G^{-1}(k)$ is periodic as it should be on a lattice. A typical form of $G^{-1}(k)$ as a function of, say, k_4 is depicted in Fig. 8.2. The behavior around $k_4 = 0$ is prescribed by Eq. (8.15) and is just a straight line with a positive slope. Therefore, $G^{-1}(k)$ will inevitably have another zero at $k_4 = \pi/a$ owing to periodicity.

This is the difference between the fermionic and bosonic cases. For bosons $G^{-1}(k)$ is quadratic in k_4 near $k_4 = 0$ rather than linear as for fermions. The typical behavior of $G^{-1}(k)$ for bosons is shown in Fig. 8.3. There is no doubling of states in the bosonic case.

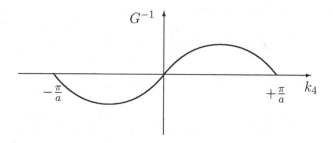

Fig. 8.2. Momentum dependence of G^{-1} for the chiral lattice fermions. The periodicity leads to an extra zero at $k_4 = \pi/a$.

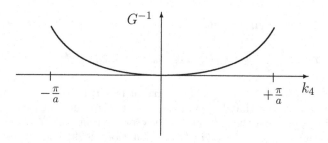

Fig. 8.3. Momentum dependence of G^{-1} for the lattice bosons. No doubling of states is associated with this behavior.

Remark on the Nielsen–Ninomiya theorem

A general proof of the theorem which states that there is no way to avoid fermion doubling under natural assumptions concerning the structure of a lattice gauge theory was given by Nielsen and Ninomiya [NN81]. It is sometimes formulated as an absence of neutrinos on the lattice. In other words, this is a no-go theorem for putting theories with an unequal number of left- and right-handed massless Weyl particles on a lattice, such as in the standard electroweak theory.

A naive way to bypass the Nielsen–Ninomiya theorem is, say, to choose a fermionic lattice action which is highly nonlocal. Then it is possible to replace $\sin k_\mu a$ in Eq. (8.14) by $k_\mu a$ itself to obtain an expression which is similar to the continuum propagator (8.15). However, such a nonlocal modification is useless in practice.

Some recent progress [Neu98, Lus98] in formulating chiral gauge theories on the lattice has been based on the idea of modifying the lattice chiral transformation in the spirit of Ginsparg and Wilson [GW82] and using a sophisticated lattice approximation of the Dirac operator which has no doublers but is manifestly invariant under such a modified lattice chiral transformation. Ordinary chiral symmetry is then reproduced in the continuum limit.

8.3 Kogut–Susskind fermions

The number of continuum fermion species is not necessarily equal to 16. It can be reduced down to four by a trick which was proposed for the Hamiltonian formulation in [KS75, Sus77] and elaborated for the Euclidean formulation in [STW81, KS81b].

Let us substitute

$$\psi_x = \gamma_1^{x/a} \gamma_2^{y/a} \gamma_3^{z/a} \gamma_4^{t/a} \phi_x \qquad (8.31)$$

into the free fermionic action. Then it takes the form

$$S_0[\bar{\psi}, \psi] = \frac{1}{2} \sum_x \sum_i \sum_{\mu>0} \eta_\mu(x) \left([\phi_x^\dagger]^i [\phi_{x+a\hat{\mu}}]^i - [\phi_{x+a\hat{\mu}}^\dagger]^i [\phi_x]^i \right), \qquad (8.32)$$

which is diagonal with respect to the spinor indices, since

$$\eta_\mu(x) = (-1)^{(x_1 + \cdots + x_{\mu-1})/a}, \qquad (8.33)$$

or explicitly

$$\left. \begin{aligned} \eta_1(x) &= 1, \\ \eta_2(x) &= (-1)^{x_1/a}, \\ &\vdots \\ \eta_d(x) &= (-1)^{(x_1 + \cdots + x_{d-1})/a}, \end{aligned} \right\} \qquad (8.34)$$

does not depend on spinor indices.

The idea is to leave only one component of ϕ_x^i in order to reduce the degeneracy:

$$\phi_x^i = \begin{pmatrix} \chi_x \\ 0 \\ 0 \\ 0 \end{pmatrix}, \qquad \bar{\phi}_x^i = \begin{pmatrix} \bar{\chi}_x \\ 0 \\ 0 \\ 0 \end{pmatrix}. \qquad (8.35)$$

These lattice fermions are known as the *staggered fermions*, since $\eta_\mu(x)$ is staggering from one lattice site to another. They are also often called the *Kogut–Susskind fermions* because of their relation to those of [KS75, Sus77].

The action of the Kogut–Susskind fermions is

$$
\begin{aligned}
S[U, \bar{\psi}, \psi] \;=\; & \beta S_{\text{lat}}[U] + M \sum_{x} \bar{\chi}_x \chi_x \\
& + \frac{1}{2} \sum_{x,\mu>0} \eta_\mu(x) \left[\bar{\chi}_x \, U_\mu^\dagger(x)\, \chi_{x+a\hat{\mu}} - \bar{\chi}_{x+a\hat{\mu}} \, U_\mu(x)\, \chi_x \right].
\end{aligned}
$$

$$(8.36)$$

It describes $2^{d/2} = 4$ species for complex χ_x or $2^{d/2-1} = 2$ species for Majorana χ_x. Components of a continuum bispinor are distributed in this approach over four lattice sites.

There is no chiral anomaly for the Kogut–Susskind fermions as with the chiral fermions.

Remark on four generations

It might seem plausible to identify four species of Kogut–Susskind fermions with four generations of quarks and leptons (see, for example, [KMN83]). Remember that one of the motivations for adding the fourth generation to the standard model is to cancel the anomaly. However, there are problems with this idea concerning the splitting of fermion masses for the four generations.

8.4 Wilson fermions

The chiral lattice fermions were proposed by Wilson [Wil74]. Soon after that he recognized [Wil75] the problem of fermion doubling and proposed a lattice fermionic action that describes only one relativistic fermion in the continuum. The latter fermions are called *Wilson fermions*.

The lattice action for the Wilson fermions reads

$$
\begin{aligned}
S[U, \bar{\psi}, \psi] \;=\; & \beta S_{\text{lat}}[U] + M \sum_{x} \bar{\psi}_x \psi_x \\
& - \frac{1}{2} \sum_{x,\mu>0} \left[\bar{\psi}_x \, (1 - \gamma_\mu)\, U_\mu^\dagger(x)\, \psi_{x+a\hat{\mu}} + \bar{\psi}_{x+a\hat{\mu}} \, (1 + \gamma_\mu)\, U_\mu(x)\, \psi_x \right].
\end{aligned}
$$

$$(8.37)$$

The difference between this action and the action (8.2) for chiral fermions arises from the projectors $(1 \pm \gamma_\mu)$ which pick only one fermionic state.

Substituting the expansion (8.5) in the action (8.37), we obtain, in the naive continuum limit, the continuum fermionic action (5.13) with the mass being

$$
m = \frac{M - 4}{a}.
$$

$$(8.38)$$

Therefore, the Wilson lattice fermions describe a relativistic fermion of the mass m in the continuum when

$$M \rightarrow 4 + ma. \tag{8.39}$$

In order to see that there are no other relativistic fermion states in the limit (8.39), let us consider the fermionic propagator which is given by

$$G^{-1}(k) = M - \frac{1}{2} \sum_{\mu=1}^{4} \left[(1 - \gamma_\mu) e^{ik_\mu a} + (1 + \gamma_\mu) e^{-ik_\mu a} \right]. \tag{8.40}$$

Introducing the Minkowski-space energy $E = -ik_4$, we obtain the following dispersion law:

$$\cosh Ea = \frac{1 + \left(M - \sum_{\mu=1}^{3} \cos p_\mu a \right)^2 + \sum_{\mu=1}^{3} \sin^2 p_\mu a}{2 \left(M - \sum_{\mu=1}^{3} \cos p_\mu a \right)}. \tag{8.41}$$

Let a particle be at rest, i.e. $p_1 = p_2 = p_3 = 0$ and $E = m > 0$. Then Eq. (8.41) reduces for $ma \ll 1$ to the relation (8.39). It is easy to show that a particle at rest is the only solution to Eq. (8.41) with finite energy as $a \rightarrow 0$.

The difference between the dispersion laws for the chiral and Wilson fermions is because the function on the RHS of Eq. (8.41) is no longer periodic. It reduces for $a \rightarrow 0$ and $M \rightarrow 4$ to a usual relation

$$E^2 = \vec{p}^2 + m^2 \tag{8.42}$$

between the energy and momentum of a relativistic particle.

Problem 8.3 Show that the solution to the dispersion law (8.41) is unique for $M \approx 4$.

Solution For $M \approx 4$, we can replace the LHS of Eq. (8.41) by 1 and substitute $M = 4$ on the RHS. Then Eq. (8.41) reduces to the equation for spatial components of the four-momentum:

$$\left(3 - \sum_{\mu=1}^{3} \cos p_\mu a \right)^2 + \left(3 - \sum_{\mu=1}^{3} \cos^2 p_\mu a \right) = 0, \tag{8.43}$$

for which the only solution is $p_1 = p_2 = p_3 = 0$, since both terms on the LHS are nonnegative.

It is instructive to discuss what happens with the fermion doublers under the change of $\pm \gamma_\mu$ by $(1 \pm \gamma_\mu)$ in the lattice fermionic action. Let us consider one such state, such as that with $p_1 = \pi/a$, $p_2 = p_3 = 0$. Its energy is determined by Eq. (8.41) to be $\sim 1/a$ so that this state is not essential as $a \to 0$.

The chiral anomaly is correctly recovered using the Wilson fermions. The 15 states of the mass $\sim 1/a$ play the role of regulators, which results in an anomaly as $a \to 0$.

Problem 8.4 Calculate the masses of all 16 fermionic states.

Solution Substituting Eqs. (8.23), (8.24) and so on into the action (8.37), we obtain

$$m = \frac{M - \sum_{\mu=1}^{4} s_\mu}{a}, \tag{8.44}$$

where

$$s_\mu = e^{ip_\mu a} \begin{cases} +1 & p_\mu = 0 \\ -1 & p_\mu = \dfrac{\pi}{a} \end{cases}. \tag{8.45}$$

Therefore, one state is relativistic as $M \to 4$, while 15 others have masses $\sim 1/a$.

Remark on backtrackings for Wilson fermions

Another way to understand why the doubling problem is removed for the Wilson fermions is to consider how they propagate on a lattice. The projectors

$$P_\mu^\pm = \frac{1 \pm \gamma_\mu}{2} \qquad \boxed{\text{Wilson fermions}} \tag{8.46}$$

restrict the propagation of the Wilson fermions. One-half of the states propagate only in positive directions and the other half propagate only in negative directions. In particular, there are no backtrackings in the (lattice) sum over paths, since

$$P_\mu^+ P_\mu^- = 0. \tag{8.47}$$

This removes the doubling.

Problem 8.5 Represent the fermion propagator in an external Yang–Mills field as a sum over paths on a lattice, performing an expansion in $1/M$.

Solution Let us rescale the fermion field, absorbing the parameter M in front of the mass term. The fermionic part of the new action is given by

$$S_\psi = \sum_x \bar{\psi}_x \psi_x - \kappa \sum_{x,\mu>0} \left[\bar{\psi}_x P_\mu^- U_\mu^\dagger(x) \psi_{x+a\hat\mu} + \bar{\psi}_{x+a\hat\mu} P_\mu^+ U_\mu(x) \psi_x \right], \tag{8.48}$$

where $\kappa = 1/M$ is usually called the *hopping parameter*. The large-mass expansion in $1/M$ is now represented as the hopping parameter expansion in κ.

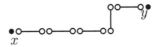

Fig. 8.4. A path Γ_{yx} made out of the string bits, which leads to a nonvanishing term of the hopping parameter expansion for the quark propagator (8.49) on a lattice. Each site involves at least two quark fields (depicted by the circles). Otherwise the Grassmann integral at a given site vanishes.

It is convenient to depict each of the two terms in square brackets in Eq. (8.48) by a string bit as in Fig. 6.2 on p. 101 with the quark fields at the ends and the gauge variable at the link. The first term corresponds to the negative direction of the link, and the second term corresponds to the positive direction. Substituting Eq. (8.48) into definition (8.4) and expanding the exponential in κ, we obtain a combination of terms constructed from the string bits. A nonvanishing contribution to the quark propagator

$$G^{ij}_{mn}(x,y;U) = \langle \psi^i_m(x) \bar{\psi}^j_n(y) \rangle_\psi \,, \tag{8.49}$$

where i, j and m, n represent, respectively, color and spinor indices, emerges when the links form a path Γ_{yx} that connects x and y on the lattice as depicted in Fig. 8.4. Otherwise, the average over ψ and $\bar{\psi}$ vanishes owing to the rules of integration over Grassmann variables described in Problem 2.2 on p. 37.

Therefore, we obtain

$$G^{ij}_{mn}(x,y;U) = \sum_{\Gamma_{yx}} \frac{1}{M^{L(\Gamma)+1}} U^{ij}[\Gamma_{yx}] \left[\prod_{\Gamma_{yx}} P^\pm_\mu \right]_{mn} \,, \tag{8.50}$$

where P^+_μ or P^-_μ are associated with the positive or negative direction of a given link $\in \Gamma_{yx}$. For the Wilson fermions, they are given by Eq. (8.46), while

$$P^\pm_\mu = \pm \frac{\gamma_\mu}{2} \qquad \boxed{\text{chiral fermions}} \tag{8.51}$$

for chiral fermions. The sum in Eq. (8.50) runs over all the paths between x and y on the lattice, while $L(\Gamma)$ denotes the length of the path Γ_{yx} in the lattice units. A continuum counterpart of Eq. (8.50) is derived in Problem 12.1.

Problem 8.6 Represent the integral over fermions in Eq. (8.3) as a sum over closed paths on a lattice, performing an expansion in $1/M$.

Solution The calculation is analogous to that of the previous Problem. The result can be written as

$$\int \prod_x d\bar{\psi}_x \, d\psi_x \, e^{-S_\psi} = e^{-S_{\text{ind}}[U]} \tag{8.52}$$

with

$$S_{\text{ind}}[U] = -\sum_\Gamma \frac{\text{tr}\, U\,[\Gamma]}{L(\Gamma)\, M^{L(\Gamma)}} \, \text{sp} \prod_\Gamma P^\pm_\mu, \tag{8.53}$$

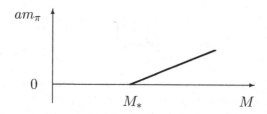

Fig. 8.5. Dependence of the π-meson mass on the lattice quark mass M. At $M = M_*$ the π-meson becomes massless and the chiral symmetry is restored.

where the combinatoric factor $1/L(\Gamma)$ arises from the identity of L links forming the closed contour Γ, and the minus sign is because of fermions.

Equation (8.53) defines an effective (or induced) action of a pure lattice gauge theory, which is nonlocal since it involves arbitrarily large loops. However, it can be made the single-plaquette lattice action (6.16) by introducing many flavors of lattice fermions [Ban83, Ham83].

8.5 Quark condensate

The lattice action (8.37) is not invariant under the chiral transformation. Therefore, the chiral symmetry is broken explicitly for the Wilson fermions.

Nevertheless, one expects a restoration of chiral symmetry as $a \to 0$ when the relativistic fermion is massless (say, for $M = 4$ in the free case), while heavy states with $m \sim 1/a$ play the role of regulators. For the interaction theory, this restoration happens at some value $M = M_*$, which is no longer equal to 4. A signal of this restoration is the vanishing of the mass of the π-meson (as illustrated by Fig. 8.5). $m_\pi = 0$ is usually associated with the fact that the chiral symmetry is realized in a spontaneously broken phase and the π-meson is the corresponding Goldstone boson.

For the chiral or Kogut–Susskind fermions with $M = 0$, the lattice action is invariant under the global chiral transformation (8.7). The order parameter for breaking the chiral symmetry is

$$\bar{\psi}\psi \xrightarrow{\text{c.t.}} \bar{\psi}\, e^{2i\alpha\gamma_5}\psi \,, \tag{8.54}$$

which is not invariant under the chiral transformation. Therefore, the average of $\bar{\psi}\psi$ must vanish if the symmetry is not broken spontaneously.[*]

---

[*] Spontaneous symmetry breaking usually occurs when the vacuum state is not invariant under the symmetry transformation.

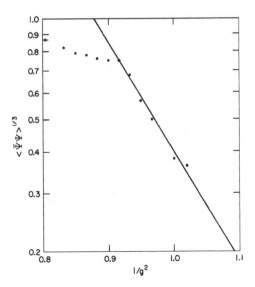

Fig. 8.6. Monte Carlo data from Hamber and Parisi [HP81] for the quark condensate in the quenched approximation.

Such spontaneous breaking results in

$$\langle \bar{\psi}\psi \rangle \ \neq \ 0 \,. \tag{8.55}$$

This nonvanishing value of the average of $\bar{\psi}_x \psi_x$ does not depend on x owing to translational invariance and is called the *quark condensate*.

The spontaneous breaking of the chiral symmetry in QCD was demonstrated using Monte Carlo calculations of the quark condensate. This quantity has a dimension of $[\text{mass}]^3$ and should depend on g^2 at small g^2 as prescribed by the asymptotic scaling. The Monte Carlo data for the quark condensate from the pioneering paper by Hamber and Parisi [HP81] are shown in Fig. 8.6. Its agreement with asymptotic scaling demonstrates that the chiral symmetry is spontaneously broken in the continuum QCD.

Remark on Monte Carlo simulations with fermions

Monte Carlo simulations with quarks are much more difficult than in a pure gauge theory. Integrating over the quark fields using Eq. (2.15), one is left with the determinant, say for the Kogut–Susskind fermions, of the matrix

$$D[U] \ = \ M\delta_{xy} + \frac{1}{2} \sum_{\mu>0} \left[\eta_\mu(x)\, U_\mu^\dagger(x)\, \delta_{x(x+a\hat{\mu})} - \eta_\mu(x)\, U_\mu(x)\, \delta_{x(x-a\hat{\mu})} \right]$$

$$\tag{8.56}$$

for a given configuration of the gluon field $U_\mu(x)$. This results in a pure gauge-field problem with the effective action given by

$$e^{-\beta S_{\text{eff}}[U]} \;=\; \det D[U]\, e^{-\beta S[U]} . \tag{8.57}$$

The matrix that appears in this determinant has at least $NL^4 \times NL^4$ elements, and is to be calculated at each Monte Carlo upgrading of $U_\mu(x)$.

Several methods are proposed to manage the quark determinant exactly or approximately. The simplest one is not to take it into account at all. This approximation is known as the *quenched* approximation when only valence quarks are considered, while the effects of virtual quark loops are disregarded. Recently, progress in the full theory has been achieved using some tricks to evaluate the quark determinants (see, for example, [Aok00] for a review of the subject).

9

Finite temperatures

Finite-temperature quantum field theories at thermodynamic equilibrium are naturally described by Euclidean path integrals. The time-variable in this approach is compactified and varies between 0 and the inverse temperature $1/T$. Periodic boundary conditions are imposed on Bose fields, while antiperiodic ones are imposed on Fermi fields in order to reproduce the standard Bose or Fermi statistics, respectively.

The lattice formulation of QCD at finite temperature is especially simple, since the Euclidean lattice has a finite extent in the temporal direction. The Wilson criterion of confinement is not applicable at finite temperatures and is replaced by another one based on the thermal Wilson lines passing through the lattice in the temporal direction. They are closed owing to the periodic boundary condition for the gauge field.

When the temperature increases, QCD undergoes [Pol78, Sus79] a deconfining phase transition which is associated with a liberation of quarks. At low temperatures below the phase transition, thermodynamical properties of the hadron matter are well described by a gas of noninteracting hadrons while at high temperatures above the phase transition these are well described by an ideal gas of quarks and gluons.

The situation with the deconfining phase transition becomes less definite when the effects of virtual quarks are taken into account. The deconfining phase transition makes strict sense only for large values of the quark mass. For light quarks, a phase transition associated with the chiral symmetry restoration at high temperatures occurs with increasing temperature. It makes strict sense only for massless quarks.

In this chapter we first derive a path-integral representation of finite-temperature quantum field theories starting from the Boltzmann distribution. Then we apply this technique to QCD and discuss the confinement criterion at finite temperatures as well as the deconfining and chiral symmetry restoration phase transitions.

159

9.1 Feynman–Kac formula

Thermodynamic properties of an equilibrium system in $3+1$ dimensions are determined by the thermal partition function

$$Z(T,V) \;=\; \sum_n e^{-E_n/T} \;\equiv\; \mathrm{Tr}\, e^{-\boldsymbol{H}/T} \qquad (9.1)$$

which is associated with the Boltzmann distribution at the temperature T. Here $\boldsymbol{H}$ is a Hamiltonian of the system and Tr is calculated over any complete set of states, say, over eigenstates of the Hamiltonian, eigenvalues of which are characterized by the energy levels E_n.

For a quantum theory of a single scalar field $\varphi(\vec{x},t)$, the (Schrödinger) states are described by the bra- and ket-vectors $\langle g|$ and $|f\rangle$:

$$\langle g|\vec{x}\rangle \;=\; g(\vec{x})\,, \qquad \langle \vec{x}|f\rangle \;=\; f(\vec{x})\,, \qquad (9.2)$$

as is explained in Sect. 1.1. A matrix element of the evolution operator $\exp(-\boldsymbol{H}/T)$ is given by the formula

$$\langle g|\, e^{-\boldsymbol{H}/T}\,|f\rangle \;=\; \int_{\substack{\varphi(\vec{x},0)=f(\vec{x})\\ \varphi(\vec{x},1/T)=g(\vec{x})}} \mathcal{D}\varphi(\vec{x},t)\, e^{-\int_0^{1/T} dt\, \mathcal{L}[\varphi]}\,, \qquad (9.3)$$

where $\mathcal{L}$ is a proper Lagrangian, say for example,

$$\mathcal{L}[\varphi] \;=\; \int_V d^3\vec{x} \left[\frac{1}{2}\left(\partial_\mu \varphi\right)^2 + \frac{1}{2}m^2\varphi^2 + \frac{\lambda}{3!}\varphi^3 \right] \qquad (9.4)$$

for the cubic self-interaction of φ. The derivation is quite analogous to that of Problem 1.9 on p. 22.

In order to calculate the trace over states, one should put $g(\vec{x}) = f(\vec{x})$ and perform the additional integration over $f(\vec{x})$. This yields the Feynman–Kac formula[*]

$$\mathrm{Tr}\, e^{-\boldsymbol{H}/T} \;=\; \int \mathcal{D}f(\vec{x})\, \langle f|\, e^{-\boldsymbol{H}/T}\,|f\rangle$$

$$=\; \int_{\varphi(\vec{x},1/T)=\varphi(\vec{x},0)} \mathcal{D}\varphi(\vec{x},t)\, e^{-\int_0^{1/T} dt\, \mathcal{L}[\varphi]}\,. \qquad (9.5)$$

Note that the path integral in Eq. (9.5) is taken with periodic boundary conditions for the field φ:

$$\varphi(\vec{x},1/T) \;=\; \varphi(\vec{x},0)\,. \qquad (9.6)$$

---

[*] Its derivation in the modern context of non-Abelian gauge theories, which extends the Feynman derivation [Fey53] for statistical mechanics, is due to Bernard [Ber74].

As $T \to 0$ it reproduces the standard Euclidean formulation of quantum field theory which is discussed in Chapter 2. The point is that nothing depends on real time for a system at thermodynamic equilibrium. The variable t in Eq. (9.5) is just the proper time of the disentangling procedure. This analogy between the partition functions of statistical systems and the Euclidean formulation of quantum field theory has already been mentioned in the Remark on p. 33.

Remark on thermal density matrix

A statistical-mechanical counterpart of the propagator in Euclidean quantum field theory is the (unnormalized) thermal density matrix

$$\left\langle y \left| e^{-\boldsymbol{H}/T} \right| x \right\rangle = \sum_n e^{-E_n/T} \, \Psi_n^*(y) \Psi_n(x) , \qquad (9.7)$$

where $\Psi_n(x)$ denotes the wave function of the nth eigenstate. This equality can be derived by inserting a complete set of states. The thermal partition function (9.1) is then given by the space integral of the diagonal element:

$$Z(T,V) = \int_V d^d x \left\langle x \left| e^{-\boldsymbol{H}/T} \right| x \right\rangle . \qquad (9.8)$$

For a quantum particle with the nonrelativistic Hamiltonian (1.107), the path-integral representation of the thermal density matrix (9.7) is given by Eq. (1.118) with $\tau = 1/T$. This pursues the analogy between Euclidean quantum field theory and statistical mechanics.

More concerning the thermal density matrix (9.7) can be found in the book [Fey72].

Problem 9.1 Derive the Feynman–Kac formula for a quantum particle with the nonrelativistic Hamiltonian (1.107).

Solution The matrix element $\langle x \, | \exp\left(-\boldsymbol{H}/T\right)| \, x \rangle$ is determined by Eq. (1.118) to be

$$\left\langle x \left| e^{-\boldsymbol{H}/T} \right| x \right\rangle = \int_{\substack{z_\mu(0)=x_\mu \\ z_\mu(1/T)=x_\mu}} \mathcal{D}z_\mu(t) \, e^{-\int_0^{1/T} dt \, \mathcal{L}(t)}, \qquad (9.9)$$

where the Lagrangian $\mathcal{L}(t)$ is given by Eq. (1.119). Using Eq. (9.8), we obtain [Fey53]

$$\mathrm{Tr}\, e^{-\boldsymbol{H}/T} = \int_V d^d x \left\langle x \left| e^{-\boldsymbol{H}/T} \right| x \right\rangle$$

$$= \int_{z_\mu(0)=z_\mu(1/T)} \mathcal{D}z_\mu(t) \, e^{-\int_0^{1/T} dt \, \mathcal{L}(t)}. \qquad (9.10)$$

This integral is over the trajectories with periodic boundary conditions

$$z_\mu(0) = z_\mu(1/T). \tag{9.11}$$

Problem 9.2 Calculate the partition function (9.10) for the free case.

Solution The Gaussian path integral with the boundary conditions

$$z_\mu(0) = z_\mu(1/T) = x_\mu \tag{9.12}$$

is calculated in Sect. 1.5 with the result given by Eq. (1.90). In order to calculate the partition function (9.10), we need to integrate this expression over x_μ which yields [Fey53]

$$Z(T, V) = \int_V d^d x \, \mathcal{F}(1/mT) = V \left(\frac{mT}{2\pi} \right)^{d/2}. \tag{9.13}$$

The formula (9.13) is to be compared with that given by the Boltzmann distribution in classical statistics. Since the energy of a free nonrelativistic particle is

$$E(\vec{p}) = \frac{\vec{p}^2}{2m}, \tag{9.14}$$

the Boltzmann distribution is given by the sum over positions of the particle in a box of volume V and the integration over its momentum $\vec{p}$:

$$Z(T, V) = V \int \frac{d^d \vec{p}}{(2\pi)^d} e^{-E(\vec{p})/T} = V \left(\frac{mT}{2\pi} \right)^{d/2}, \tag{9.15}$$

which coincides with Eq. (9.13) derived from the path integral.

Problem 9.3 Calculate the thermal density matrix (9.7) for the free case.

Solution The calculation is the same as in Sect. 1.5 for $\tau = 1/mT$. The result is

$$\left\langle y \left| e^{-H/T} \right| x \right\rangle = \left(\frac{mT}{2\pi} \right)^{d/2} e^{-mT(x-y)^2/2}. \tag{9.16}$$

This formula can alternatively be derived using Eq. (9.7) for the wave functions associated with the plane waves

$$\Psi_{\vec{p}}(x) = \frac{1}{\sqrt{V}} e^{-i\vec{p}\vec{x}}. \tag{9.17}$$

Then we obtain

$$\sum_n e^{-E_n/T} \Psi_n^*(y) \Psi_n(x) = \int \frac{d^d p}{(2\pi)^d} e^{i\vec{p}(\vec{y}-\vec{x}) - p^2/2mT}$$

$$= \left(\frac{mT}{2\pi} \right)^{d/2} e^{-mT(x-y)^2/2} \tag{9.18}$$

which reproduces Eq. (9.16).

Problem 9.4 Calculate the partition function (9.10) for a harmonic oscillator with $V(x) = m\omega^2 x^2/2$.

Solution The path integral in Eq. (9.10) can be calculated using the mode expansion

$$z(t) = a_0 + \sqrt{2} \sum_{n=1}^{\infty} \left[a_n \cos\left(2\pi n t T\right) + b_n \sin\left(2\pi n t T\right) \right], \qquad (9.19)$$

where the sin and cos functions form a set of orthogonal basis functions on the interval $[0, 1/T]$ and satisfy the boundary condition (9.11). The expansion (9.19) is of the same type as Eq. (1.82).

Substituting (9.19) into the action, we have

$$\frac{m}{2} \int_0^{1/T} dt \left(\dot{z}^2 + \omega^2 z^2 \right) = \frac{m\omega^2}{2T} a_0^2 + \frac{m}{2T} \sum_{n=1}^{\infty} \left[(2\pi n T)^2 + \omega^2 \right] \left(a_n^2 + b_n^2 \right). \qquad (9.20)$$

Representing the measure as

$$\mathcal{D}z(t) = \frac{d^d a_0}{(2\pi)^{d/2}} \prod_{n=1}^{\infty} \frac{d^d a_n}{(2\pi)^{d/2}} \frac{d^d b_n}{(2\pi)^{d/2}}, \qquad (9.21)$$

which is of the same type as Eq. (1.83), and performing the Gaussian integral over the a_n and b_n, we obtain for the partition function (9.10)

$$Z(T) = \left[\frac{\sqrt{T}}{\sqrt{m\omega}} \prod_{n=1}^{\infty} \frac{T/m}{(2\pi n T)^2 + \omega^2} \right]^d. \qquad (9.22)$$

The infinite product can be calculated by virtue of the formula

$$\prod_{n=1}^{\infty} \left(A + \frac{n^2}{B} \right) = \frac{2}{\sqrt{A}} \sinh(\pi \sqrt{AB}) \qquad (9.23)$$

which implies a zeta-function regularization. Finally, we obtain

$$Z(T) = \left[\frac{1}{2 \sinh\left(\omega/2T\right)} \right]^d \qquad (9.24)$$

for the thermal partition function of a nonrelativistic harmonic oscillator with frequency ω. Equation (9.24) can be derived alternatively by simply substituting the oscillator spectrum $E_n = \omega \left(n + \frac{1}{2} \right)$ into the Boltzmann formula (9.1).

In contrast with Eq. (9.13), there is no volume-dependence in Eq. (9.24), which comes usually from the translational zero mode, since now the particle oscillates near the origin. It is clear from the integral over a_0 that the volume factor is reproduced as $V \sim (T/m\omega^2)^{d/2}$ when $\omega \to 0$. Then Eq. (9.13) is reproduced as $\omega \to 0$.

Problem 9.5 Calculate the thermal density matrix (9.7) of the harmonic oscillator.

Solution It is convenient to use the mode expansion

$$z(t) = z_{cl}(t) + \sqrt{2} \sum_{n=1}^{\infty} c_n \sin(\pi n t T), \tag{9.25}$$

where

$$z_{cl}(t) = x \frac{\sinh[\omega(1/T - t)]}{\sinh(\omega/T)} + y \frac{\sinh(\omega t)}{\sinh(\omega/T)} \tag{9.26}$$

obeys the classical equation of motion

$$\ddot{z}_{cl} - \omega^2 z_{cl} = 0 \tag{9.27}$$

with the boundary condition $z(0) = x$, $z(1/T) = y$. This reproduces Eq. (1.84) with $\tau = 1/T$ as $\omega \to 0$. The sin functions form an appropriate set of orthogonal basis functions for the interval $[0, 1/T]$.

Inserting the mode expansion (9.25) into the action, we obtain

$$\frac{m}{2} \int_0^{1/T} dt \left(\dot{z}^2 + \omega^2 z^2\right) = S_{cl}(x,y) + \frac{m}{2T} \sum_{n=1}^{\infty} \left[(\pi n T)^2 + \omega^2\right] c_n^2, \tag{9.28}$$

where

$$S_{cl}(x,y) = \frac{m\omega}{2} \left[(x^2 + y^2)\coth(\omega/T) - 2xy \frac{1}{\sinh(\omega/T)}\right]. \tag{9.29}$$

Substituting the measure as in Eq. (9.21) and performing the Gaussian integration over c_n, we have

$$\left\langle y \left| e^{-H/T} \right| x \right\rangle \propto \prod_{n=1}^{\infty} \left[\frac{T/m}{(\pi n T)^2 + \omega^2}\right]^{d/2} e^{-S_{cl}(x,y)}. \tag{9.30}$$

Finally, using Eq. (9.23), we obtain

$$\left\langle y \left| e^{-H/T} \right| x \right\rangle = \left[\frac{m\omega}{2\pi \sinh(\omega/T)}\right]^{d/2} e^{-S_{cl}(x,y)} \tag{9.31}$$

for the thermal density matrix of a nonrelativistic harmonic oscillator with frequency ω. The formulas of Sect. 1.5 are reproduced as $\omega \to 0$ which fixes an ω-independent normalization factor in Eq. (9.30). The partition function (9.24) is reproduced when we set $y = x$ in Eq. (9.31) and integrate over x.

Problem 9.6 Calculate the partition function (9.5) for the free case.

Solution Since the path integral over $\varphi(\vec{x}, t)$ is Gaussian, it can be represented as

$$
\ln Z(T, V) = -\frac{1}{2} \ln \det \left(-\partial_\mu^2 + m^2 \right) = -\frac{1}{2} \operatorname{Tr} \ln \left(-\partial_\mu^2 + m^2 \right)
$$
$$
= -\frac{1}{2} V \int \frac{d^d \vec{p}}{(2\pi)^d} \operatorname{Tr}_t \ln \left(-D^2 + \omega^2 \right), \tag{9.32}
$$

where

$$
\omega = \sqrt{\vec{p}^2 + m^2} . \tag{9.33}
$$

We have used the fact that the $\vec{x}$ variable is not restricted, while the remaining trace of the one-dimensional operator is to be calculated with periodic boundary conditions.

We shall perform the calculation by expressing the trace via the diagonal resolvent of the same operator as has already been done in Problem 4.4 on p. 73. The Green function $G_\omega(t - t')$ is no longer given by Eq. (1.38) because of the periodic boundary conditions. Instead, we obtain the sum over even Matsubara frequencies:

$$
G_\omega(t - t') = T \sum_{n=-\infty}^{+\infty} \frac{e^{2\pi i n T(t' - t)}}{(2\pi n T)^2 + \omega^2}, \tag{9.34}
$$

which satisfies $G_\omega(1/T) = G_\omega(0)$, as it should for periodic boundary conditions, and reproduces Eq. (1.38) as $T \to 0$. The diagonal resolvent is given by

$$
G_\omega(0) = T \sum_{n=-\infty}^{+\infty} \frac{1}{(2\pi n T)^2 + \omega^2}
$$
$$
= \frac{1}{2\omega} \coth \frac{\omega}{2T} . \tag{9.35}
$$

Therefore,

$$
\operatorname{Tr}_t \ln \left(-D^2 + \omega^2 \right) = \int^{\omega^2} d\omega^2 \int_0^{1/T} dt \, G_\omega(0)
$$
$$
= \int^{\omega} d\omega \frac{1}{T} \coth \frac{\omega}{2T}
$$
$$
= \frac{\omega}{T} + 2 \ln \left(1 - e^{-\omega/T} \right) \tag{9.36}
$$

modulo an ω-independent constant. Substituting into Eq. (9.32), we obtain

$$
\ln Z(T, V) = -V \int \frac{d^d \vec{p}}{(2\pi)^d} \left[\frac{\omega}{2T} + \ln \left(1 - e^{-\omega/T} \right) \right], \tag{9.37}
$$

which is the standard result for an ideal Bose gas in quantum statistics modulo the first term on the RHS associated with the zero-point energy of the vacuum.

9.2 QCD at finite temperature

QCD at finite temperatures is described by the partition function

$$Z(T,V) = \int \mathcal{D}A_\mu\, \mathcal{D}\bar{\psi}\, \mathcal{D}\psi\, e^{-\int_0^{1/T} dt \int_V d^3\vec{x}\, \mathcal{L}[A_\mu, \psi, \bar{\psi}]}, \qquad (9.38)$$

which is the proper analog of Eq. (9.5). The path integral is taken with the boundary conditions

$$A_\mu(\vec{x}, 1/T) = A_\mu(\vec{x}, 0), \qquad (9.39)$$
$$\psi(\vec{x}, 1/T) = -\psi(\vec{x}, 0), \qquad (9.40)$$
$$\bar{\psi}(\vec{x}, 1/T) = -\bar{\psi}(\vec{x}, 0), \qquad (9.41)$$

which are periodic for the gauge field (gluon) and antiperiodic for the Fermi fields (quarks). The antiperiodicity of the Fermi fields is related, roughly speaking, with the famous extra minus sign of fermionic loops in the vacuum energy.

Problem 9.7 Calculate the partition function for free massive one-dimensional fermions with antiperiodic boundary conditions

$$\psi(1/T) = -\psi(0), \qquad \bar{\psi}(1/T) = -\bar{\psi}(0). \qquad (9.42)$$

Solution The calculation is analogous to that of Problem 9.6. We obtain

$$\ln Z(T,V) = \ln \det(D+m) = \operatorname{Tr} \ln(D+m). \qquad (9.43)$$

The fermion Green function $G_m(t-t')$ is given by the sum over odd Matsubara frequencies:

$$G_m(t-t') = T \sum_{n=-\infty}^{+\infty} \frac{e^{\pi i(2n+1)T(t'-t)}}{i\pi(2n+1)T + m}, \qquad (9.44)$$

which satisfies $G_m(1/T) = -G_m(0)$, as it should for antiperiodic boundary conditions.

As $T \to 0$, we obtain

$$G_m(t-t') = \int_{-\infty}^{+\infty} \frac{d\epsilon}{2\pi} \frac{e^{i\epsilon(t'-t)}}{i\epsilon + m} = \theta(t-t') \qquad (9.45)$$

since the contour of integration over ϵ can be closed for $t > t'$ ($t < t'$) in the lower (upper) half-plane. We have thus reproduced the fermionic Green function (5.34) from Problem 5.3 on p. 90.

The diagonal resolvent is given by

$$G_m(0) = T \sum_{n=-\infty}^{+\infty} \frac{1}{i\pi(2n+1)T + m}$$
$$= \frac{1}{2} \tanh \frac{m}{2T}, \qquad (9.46)$$

which differs from Eq. (9.35) by the change of the coth for tanh. Therefore,

$$\ln Z(T, V) = \int^m dm \frac{1}{T} \tanh \frac{m}{2T}$$

$$= \frac{m}{2T} + \ln\left(1 + e^{-m/T}\right) \tag{9.47}$$

modulo an m-independent constant. The second term on the RHS involves a plus sign, which characterizes Fermi statistics (remember that $\omega = m$ if there are no spatial dimensions). If we were choose periodic boundary conditions instead of antiperiodic ones, we would have a minus sign as in Eq. (9.37) which is wrong for fermions. The first term on the RHS is again associated with the zero-point energy of the vacuum.

An extension of Eq. (9.47) to d dimensions can be obtained on substituting m by ω, given by Eq. (9.33), and integrating over the phase space, which results in a formula of the type of Eq. (9.37) but with the plus sign in the second term on the RHS.

The discussion of the previous section concerning the relation between the finite-temperature and Euclidean formulations explains why the latter allows one to calculate only static quantities in QCD, say hadron masses or interaction potentials, which do not depend on time. It is also worth noting that we did not add a gauge-fixing term in Eq. (9.38), having in mind a lattice quantization as before.

The lattice formulation of finite-temperature QCD is especially simple. One should take an asymmetric lattice whose size along the temporal axis is much smaller than that along the spatial ones:

$$L_t = \frac{1}{Ta} \ll L. \tag{9.48}$$

This guarantees that the system is in the thermodynamic limit. Then the temperature is given by

$$T = \frac{1}{aL_t}, \tag{9.49}$$

i.e. it coincides with the inverse extent of the lattice along the temporal axis. The periodic boundary conditions are usually imposed on the lattice by construction.

Since the lattice spacing a and the bare coupling constant g^2 are related by Eq. (6.85), the temperature (9.49) can be rewritten as

$$T = \frac{1}{L_t} \Lambda_{\text{QCD}} \exp\left[\int \frac{dg^2}{\mathcal{B}(g^2)}\right]. \tag{9.50}$$

Therefore, one can change the temperature on the lattice by varying either the size along the temporal axis, L_t, or g^2.

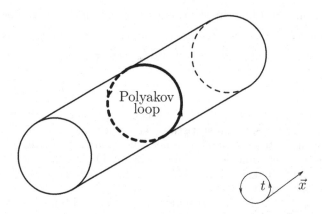

Fig. 9.1. Polyakov loop which winds around compactified temporal direction.

9.3 Confinement criterion at finite temperature

Wilson's confinement criterion, which is discussed in Sect. 6.6, is not applicable at finite temperatures. A proper criterion for confinement at finite temperatures was proposed by Polyakov [Pol78].

The Polyakov criterion of confinement at finite temperature uses the thermal Wilson loop which goes along the temporal direction:

$$L(\vec{x}) \;=\; \mathrm{tr}\, \boldsymbol{P}\, \mathrm{e}^{\mathrm{i}\int_0^{1/T}\mathrm{d}t\, \mathcal{A}_d(\vec{x},t)}. \tag{9.51}$$

It is gauge invariant because of the periodic boundary conditions for the gauge field and is called the *Polyakov loop* or the thermal Wilson line. One can imagine that the time-variable $t \equiv x_d$ is compactified so that the Polyakov loop winds around the temporal direction as shown in Fig. 9.1.

The lattice Polyakov loop

$$L_{\vec{x}} \;=\; \mathrm{tr}\, \prod_{x_d} U_d(x) \tag{9.52}$$

is just the trace of the product of the link variables along a line which goes in the temporal direction through the lattice with imposed periodic boundary conditions.

Using the lattice gauge transformation (6.13), almost all link variables, associated with links pointing in the temporal direction, can be set equal 1 except for one time slice since the gauge transformation is periodic:

$$\Omega(\vec{x},0) \;=\; \Omega(\vec{x},1/T)\,. \tag{9.53}$$

The average of the Polyakov loop is related to the free energy $F_0(\vec{x})$ of a single quark (minus that of the vacuum) located at the point $\vec{x}$ of a three-dimensional space by

$$\langle L(\vec{x}) \rangle = e^{-F_0/T}. \tag{9.54}$$

If F_0 is infinite, which is associated with confinement, then

$$\langle L(\vec{x}) \rangle = 0 \qquad \boxed{\text{confinement}}. \tag{9.55}$$

In contrast,

$$\langle L(\vec{x}) \rangle \neq 0 \qquad \boxed{\text{deconfinement}} \tag{9.56}$$

is associated with deconfinement. This is the Polyakov criterion of confinement at finite temperature.

This criterion establishes a connection on a lattice between confinement and the $Z(3)$ symmetry – the center of $SU(3)$. The $Z(3)$ transformation of the link variables

$$U_d(x) \rightarrow Z_{x_d} U_d(x) \qquad (Z_{x_d} \in Z(3)) \tag{9.57}$$

leaves the lattice action invariant. This transformation is not of the same type as the local gauge transformation (6.13) since only the temporal link variables are transformed. The parameter Z_{x_d} of the transformation (9.57) depends on x_d, but is independent of the spatial coordinates $\vec{x}$ so the symmetry is a global one.

While the lattice action is invariant under the transformation (9.57), the Polyakov loop transforms as

$$L_{\vec{x}} \rightarrow Z L_{\vec{x}} \qquad (Z \in Z(3)), \tag{9.58}$$

where

$$Z = \prod_{x_d} Z_{x_d}. \tag{9.59}$$

Therefore, Eq. (9.55) holds if the symmetry is unbroken, while Eq. (9.56) signals spontaneous breaking of the symmetry. Thus, confinement or deconfinement are associated with the unbroken or broken global $Z(3)$ symmetry, respectively.

On a lattice of finite volume, the number of degrees of freedom is finite and spontaneous breaking of the $Z(3)$ symmetry is impossible. Then, it is more convenient to use a criterion which is based on the correlator of two Polyakov loops separated by a distance R along a spatial direction. This correlator determines the interaction energy $E(R)$ between a quark and an antiquark by the formula

$$\left\langle L(\vec{x}) L^{\dagger}(\vec{y}) \right\rangle_{\text{conn}} = e^{-E(R)/T}. \tag{9.60}$$

A finite correlation length is now associated with confinement, while an infinite one corresponds to deconfined quarks.

More details concerning the $Z(3)$ symmetry in finite-temperature lattice gauge theories can be found in the review by Svetitsky [Sve86].

Problem 9.8 Calculate the correlator (9.60) to the leading order of the strong-coupling expansion.

Solution The calculation is analogous to that of Sect. 6.5. The group integral is nonvanishing when the plaquettes completely fill a cylinder, spanned by two Polyakov loops, with area equal to R/T. This is analogous to the filling shown in Fig. 6.8. Contracting the indices, we find

$$\left\langle L_{\vec{x}} L_{\vec{y}}^{\dagger} \right\rangle_{\text{conn}} = [W(\partial p)]^{R/T}, \tag{9.61}$$

where $W(\partial p)$ is given by Eq. (6.72). This yields the same interaction potential $E(R)$ as before (see Eqs. (6.76) and (6.77)).

Remark on high temperatures

At high temperatures $T \to \infty$, the temporal direction shrinks and the partition function (9.38) reduces to a three-dimensional one with the coupling constant

$$g_{3\text{D}}^2 = g^2 T, \tag{9.62}$$

which has the dimension of [mass] in three dimensions. Three-dimensional QCD and QED always confine. If we take a Wilson loop in the form of a rectangle along spatial directions in four-dimensional QCD at high temperature, its average coincides with that in three dimensions and obeys the area law. This does not mean, however, that we are in a confining phase since the confinement criterion at finite temperature is different [Pol78].

9.4 Deconfining transition

The effects of finite temperatures are negligible under normal circumstances in QCD where the typical energy scale is of the order of hundreds of MeV, while a temperature of, say, $T \approx 300$ K is associated with the energy* $kT \approx 3 \times 10^{-8}$ MeV. However, for times of the order of 10^{-4} seconds after the big bang in the very early universe, the energies of thermal fluctuations were ~ 100 MeV, i.e. of the order of the mass of the π-meson. Therefore, π-mesons can be created out of the vacuum at those times, while their density in a unit volume is described by the thermodynamics of an ideal gas. Heavier hadrons are suppressed at these energies by the Boltzmann factor.

---

* Here $k = 8.6 \times 10^{-11}$ MeV K^{-1} is the Boltzmann constant.

The energy density $\mathcal{E}(T)$ of the hadron matter is given by the standard thermodynamical relation

$$\mathcal{E}(T) = \frac{1}{V}\frac{\partial}{\partial(1/T)} \ln Z(T, V)\bigg|_{V}, \qquad (9.63)$$

with $Z(T, V)$ being given by Eq. (9.38).

When the density of hadrons is small, $\mathcal{E}(T)$ is given by the formula

$$\mathcal{E}_{\text{h}}(T) = \frac{T}{2\pi^2} \sum_{i=\pi,\rho,\omega,\ldots} g_i \left[m_i^3 \text{K}_1(m_i/T) + 3m_i^2 \text{K}_2(m_i/T) \right], \qquad (9.64)$$

where $g_\pi = 3$, $g_\rho = 9$, $g_\omega = 3$, ... are the statistical weights of the π, ρ, ω, ... mesons, while K_1 and K_2 are the modified Bessel functions.

Problem 9.9 Derive Eq. (9.64) starting from the partition function (9.37).

Solution For a dilute gas, the logarithm in Eq. (9.37) can be expanded in $\exp(-E/T)$. Therefore, we find

$$\begin{aligned}
\ln Z(T, V) &= \frac{\text{const}}{T} + V \int \frac{\text{d}^3\vec{p}}{(2\pi)^3} \, \text{e}^{-\sqrt{\vec{p}^2 + m^2}/T} \\
&= \frac{\text{const}}{T} + \frac{VTm^2}{2\pi^2} \text{K}_2(m/T).
\end{aligned} \qquad (9.65)$$

The second term on the RHS describes the classical statistics of an ideal gas of relativistic particles. Equation (9.64) can now be derived by differentiating this formula with respect to $1/T$ according to Eq. (9.63) and taking into account the statistical weights of the hadron states. The zero-point energy term gives a T-independent contribution to $\mathcal{E}_{\text{h}}$, which only changes the energy reference level.

At low temperatures, the hadron matter is in the confinement phase. However, when the temperature is increased, a phase transition associated with deconfinement occurs at some temperature $T = T_{\text{c}}$ as was first pointed out by Polyakov [Pol78] and Susskind [Sus79]. For $T < T_{\text{c}}$ the interaction potential between static quarks is linear, as is shown in Fig. 6.9a on p. 117, while for $T > T_{\text{c}}$ the potential is deconfining, as is shown in Fig. 6.9b. The state of the hadron matter with deconfined quarks and gluons is often called the *quark–gluon plasma*.

There exists a very simple physical argument as to why the deconfining phase transition must occur in QCD when the temperature is increased. It is based on the string picture of confinement which was considered in Sect. 6.6. The string is made of the gluon field between static quarks in the confining phase, which are associated with the string end points. With increasing temperature, condensation of strings of infinite length will inevitably occur owing to the large entropy of such states, which corresponds to a deconfining phase transition.

Problem 9.10 Derive the temperature of a phase transition for an elastic string by analyzing the temperature dependence of its free energy.

Solution Let us consider the thermodynamics of an elastic string with fixed end points. For low temperatures, thermal fluctuations of the length of the string are suppressed by the Boltzmann factor since the energy is proportional to the length. Therefore, the string is tightened along the shortest distance between the quarks which leads to a linear potential.

When the temperature is increased, entropy effects associated with fluctuations of the shape of the string become essential. An increment of the string length l by Δl increases energy by

$$\Delta E = \frac{\partial E}{\partial l} \Delta l = K \Delta l, \tag{9.66}$$

where K is the string tension as before, but causes a gain of the entropy

$$\Delta S = \frac{\partial S}{\partial l} \Delta l. \tag{9.67}$$

The change of free energy is given by

$$\Delta F = \Delta E - T \Delta S = \left(K - T \frac{\partial S}{\partial l} \right) \Delta l. \tag{9.68}$$

A phase transition occurs at the temperature

$$T_c = K \left(\frac{\partial S}{\partial l} \right)^{-1}, \tag{9.69}$$

when the changes of energy and entropy compensate each other, so that the free energy ceases to depend on Δl. Therefore, the phase transition is associated with a condensation of arbitrarily long strings.

The energy density $\mathcal{E}(T)$ is described by a free gas of hadrons for low temperatures, as has already been mentioned, and by a free gas of quarks and gluons at high temperatures. The latter statement is a result of asymptotic freedom, which says that the effective coupling constant describing a strong interaction at temperature T is given by

$$g^2(T) = \frac{1}{b \ln \left(\frac{\Lambda_{\text{QCD}}}{T} \right)} \tag{9.70}$$

with

$$b = \frac{1}{4\pi^2} \left(-11 + \frac{2}{3} N_f \right) \tag{9.71}$$

and N_f being the number of fermion species (or flavors) with mass much less than T. This formula has the same structure as the running constant $g^2(Q)$, which describes the strong interaction at a momentum of Q. Since

$Q \sim T$ for thermal fluctuations, these two coupling constants coincide with logarithmic accuracy.*

The energy density $\mathcal{E}(T)$ of the quark–gluon plasma is given by Boltzmann's law

$$\mathcal{E}_{\mathrm{p}}(T) = g_{\mathrm{p}}\frac{\pi^2}{30}T^4 + B, \qquad (9.72)$$

where

$$g_{\mathrm{p}} = 2 \cdot 8 + \frac{7}{8} \cdot 2 \cdot 2 \cdot 3 \cdot N_{\mathrm{f}} \qquad (9.73)$$

is the statistical weight, i.e. the number of independent internal degrees of freedom of the particles of the ideal gas. There are two spin and eight color states for gluons, and two spin, two particle–antiparticle, three color and N_{f} flavor states for quarks ($N_{\mathrm{f}} = 2$ for the u- and d-quarks). The factor of $7/8$ is the usual one for fermions.

The T-independent constant $B > 0$ in Eq. (9.72) is associated with the fact that the vacuum energy in the plasma phase is higher than in the hadron phase. In other words, the energy density of the perturbative vacuum is larger by B than that of a nonperturbative one. It is because of this energy difference that hadrons are stable at low temperatures. Such a shift of energy densities between perturbative and nonperturbative vacua is typical for bag models of hadrons.

Numerical Monte Carlo simulations of lattice gauge theory at finite temperature indicate that the deconfining phase transition is of first order and occurs at $T_{\mathrm{c}} \approx 200$ MeV. The actual dependence of the energy density on T, calculated by the Monte Carlo method, is well described by Eq. (9.64) for $T < T_{\mathrm{c}}$ and Eq. (9.72) for $T > T_{\mathrm{c}}$. This behavior is illustrated in Fig. 9.2.

Problem 9.11 Calculate T_{c} and the latent heat $\Delta\mathcal{E}$, approximating $\mathcal{E}_{\mathrm{h}}$ by an ideal gas of massless π-mesons.

Solution It is reasonable to disregard the mass of the π-mesons for $T \gtrsim 200$ MeV. Then,

$$\mathcal{E}_{\mathrm{h}}(T) = g_{\mathrm{h}}\frac{\pi^2}{30}T^4, \qquad (9.74)$$

where $g_{\mathrm{h}} = 3$ as a result of the three isotopic states (π^+, π^-, and π°). $\mathcal{E}_{\mathrm{h}}(T)$ for the plasma state is given by Eq. (9.72).

---

* The perturbative calculations in QCD at finite temperature are described in the book by Kapusta [Kap89] and in the more recent review by Smilga [Smi97] and the book by Le Bellac [Bel00].

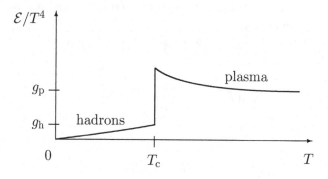

Fig. 9.2.   Temperature dependence of the energy density for hadron matter. $\mathcal{E}(T)$ for the hadron and the plasma phases are given by Eqs. (9.64) and (9.72). The difference $\mathcal{E}_\mathrm{p} - \mathcal{E}_\mathrm{h}$ at the temperature T_c of the deconfining phase transition is equal to the latent heat $\Delta\mathcal{E}$.

The pressure for the relativistic gases with the energy densities (9.74) and (9.72) is given, respectively, by

$$\mathcal{P}_\mathrm{h}(T) \;=\; g_\mathrm{h}\frac{\pi^2}{90}T^4 \tag{9.75}$$

and

$$\mathcal{P}_\mathrm{p}(T) \;=\; g_\mathrm{p}\frac{\pi^2}{90}T^4 - B\,. \tag{9.76}$$

The positive constant B in the energy density (9.72) leads to a negative pressure in the plasma state at low temperatures. Therefore, the hadron phase is preferable at low temperatures. This is in the spirit of the bag model of hadrons. At high energies the pressure is higher for the plasma phase, since

$$g_\mathrm{p} = 37 \;>\; g_\mathrm{h} = 3\,, \tag{9.77}$$

so that the plasma phase is realized. The behavior of the pressure versus T^4 is shown in Fig. 9.3 for both phases of hadron matter.

The deconfining phase transition occurs when the pressures in both phases coincide. Therefore, we obtain

$$T_\mathrm{c}^4 \;=\; \frac{B}{\frac{\pi^2}{90}\left(g_\mathrm{p} - g_\mathrm{h}\right)} \tag{9.78}$$

and

$$\Delta\mathcal{E} \;\equiv\; \mathcal{E}_\mathrm{p}(T) - \mathcal{E}_\mathrm{h}(T) \;=\; 4B\,. \tag{9.79}$$

If we were set $g_\mathrm{h} = 0$ in Eq. (9.78), this would change the value of T_c by a few per cent. This justifies the approximation of massless π-mesons.

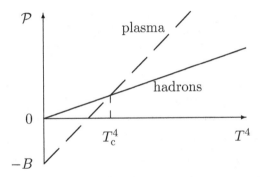

Fig. 9.3. Pressure versus T^4 for the two phases of hadron matter. The solid and dashed lines represent Eqs. (9.75) and (9.76), respectively. The hadron phase is stable for $T < T_c$, while the plasma phase is stable for $T > T_c$.

Remark on the deconfining phase transition in the early universe

The confining phase transition from a quark–gluon plasma to hadrons happened in the early universe when its age was $\approx 10^{-5}$ seconds. The equation of state of the hadron matter is described by Eqs. (9.72) and (9.76) before that time and by Eqs. (9.64) and (9.75) after that time. There are presumably no cosmological consequences of this phase transition, which survive to our time, since it happened too long ago. For instance, fluctuations of the hadron matter density which might have occurred just after the phase transition were washed out by further expansion. The confining phase transition in the early universe is considered in the review [CGS86], Section 6.

9.5 Restoration of chiral symmetry

The chiral symmetry is broken spontaneously in QCD at $T = 0$, as is discussed in Sect. 8.5. With increasing temperature, the chiral symmetry should be restored at some temperature T_{ch} (which does not necessarily coincide with T_c) since perturbation theory is applicable at high T. This restoration occurs as a phase transition with $\langle \bar{\psi}\psi \rangle$ being the proper order parameter. Therefore, the quark condensate is destroyed at $T = T_{ch}$. Monte Carlo simulations indicate that this chiral phase transition is of first order.

However, there is a subtlety in the above string picture of quark confinement when virtual quarks are taken into account. The effects of virtual quarks are suppressed when their mass m is infinitely large and the picture of confinement is the same as in pure gluodynamics: quarks are permanently confined by strings constructed from the flux tubes of the

Fig. 9.4. Breaking of the flux tube by creating a quark–antiquark pair (depicted by the open circles) out of the vacuum.

gluon field. This is associated with a linear interaction potential.

For light virtual quarks, the flux tube can break creating a quark–antiquark pair out of the vacuum, as is shown in Fig. 9.4. This happens when the energy saved in the flux tube is large enough to compensate the kinetic energy of the particles produced. Hence, the linear growth of the potential will stop at such distances.

The average of the Polyakov loop (9.51) is no longer a criterion for quark confinement in the presence of virtual quarks. The test static quark can always be screened by an antiquark created out of the vacuum (a quark created at the same time will go to infinity). Therefore, the free energy F_0 in Eq. (9.54) is always finite so that $\langle L(\vec{x}) \rangle \neq 0$ in both phases.

The effects of virtual quarks usually weaken a phase transition in a pure gauge theory. If the deconfining phase transition in the $SU(3)$ pure gauge theory was of second order rather than first order, it would presumably disappear for an arbitrarily large but finite value of m. Such a phenomenon happens in the Ising model where an arbitrarily small external magnetic field (which is an analog of the quark mass) destroys the second-order phase transition. A discontinuity of $\langle L(\vec{x}) \rangle$ at the first-order deconfining transition continues in the (T, m)-plane as illustrated by Fig. 9.5. It seems to terminate at some value m_c of the quark mass.

This situation with the order parameter for the deconfining phase transition is somewhat similar to that for the chiral phase transition. $\langle \bar{\psi}\psi \rangle$ vanishes in the unbroken phase only for $m = 0$. If $m \neq 0$ but is small, there is a small explicit breaking of chiral symmetry owing to the quark mass. Since the chiral phase transition is of first order for $m = 0$, it is natural to expect that a discontinuity of $\langle \bar{\psi}\psi \rangle$ continues in the (T, m)-plane up to some value m_{ch} of the quark mass.

If $m_{ch} < m_c$, the phase diagram in the (T, m)-plane may look like that shown in Fig. 9.5. In the intermediate region $m_{ch} < m < m_c$, the behavior of neither $\langle L(\vec{x}) \rangle$ nor $\langle \bar{\psi}\psi \rangle$ can answer the question of whether a phase transition (or two separate transitions) occurs. A proper parameter, which signals a phase transition is this region, could be the temperature-dependence of the energy density $\mathcal{E}(T)$ that undergoes discontinuities at the points of first-order phase transitions.

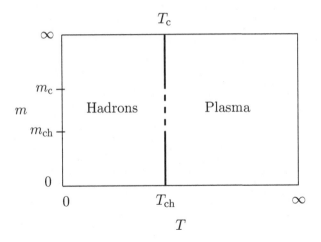

Fig. 9.5. Expected phase diagram of the hadron matter in the (T, m)-plane. The deconfining phase transition starts at $T = T_\mathrm{c}$ for $m = \infty$. $\langle L(\vec{x}) \rangle$ is its order parameter for $m > m_\mathrm{c}$. The chiral phase transition starts at $T = T_\mathrm{ch}$ for $m = 0$. $\bar{\psi}\psi$ is its order parameter for $m < m_\mathrm{ch}$.

It is worth noting that an alternative behavior of the phase diagram in the (T, m)-plane, when $m_\mathrm{ch} > m_\mathrm{c}$, is not confirmed by Monte Carlo simulations.

Bibliography to Part 2

Reference guide

There are many very good introductory lectures/reviews on lattice gauge theory. For a perfect description of motivations and the lattice formulation, I would recommend the original paper [Wil74] and lectures [Wil75] by Wilson. Original papers on lattice gauge theories are collected in the book edited by Rebbi [Reb83]. Various topics within lattice gauge theory are covered in the well-written book by Creutz [Cre83]. The book by Seiler [Sei82] contains some mathematically rigorous results. The more recently published book by Montvay and Münster [MM94] contains a comprehensive look at lattice gauge theory.

I shall also list some of the old reviews on lattice gauge theory which might be useful for deeper studies of the lattice methods. The strong-coupling expansion and the mean-field method are discussed in the review by Drouffe and Zuber [DZ83]. The Monte Carlo method and some results of numerical simulations are considered in [CJR83, Mak84]. The fermion doubling problem and the Wilson fermions are discussed in the lectures [Wil75].

An introduction to quantum field theory at finite temperature is given in the book by Kapusta [Kap89], which contains, in particular, a discussion of perturbation theory in QCD at finite temperature. Lattice gauge theory aspects of the deconfining phase transition at finite temperature are considered in the review by Svetitsky [Sve86]. A description of the thermal phases of hadron matter, a comparison with results of Monte Carlo simulations and a discussion of the deconfining phase transition in the early universe are contained in the review [CGS86]. Various physical aspects of thermal QCD are considered in the recent review by Smilga [Smi97] and book by Le Bellac [Bel00].

Most of the reviews mentioned above were written in the 1980s and contain a description of lattice methods as well as early Monte Carlo results. The best way to follow current developments of the subject is via plenary talks at the annual Lattice Conference ([Lat00] and those for the preceding years).

References

[Ake93] AKEMI K. *et al.* 'Scaling study of pure gauge lattice QCD by Monte Carlo renormalization group method'. *Phys. Rev. Lett.* **71** (1993) 3063.

[Aok00] AOKI S. 'Unquenched QCD simulation results'. *Nucl. Phys. B (Proc. Suppl.)* **94** (2001) 3.

[Are80] AREF'EVA I.YA. 'Non-Abelian Stokes theorem'. *Theor. Math. Phys.* **43** (1980) 353.

[Ban83] BANDER M. 'Equivalence of lattice gauge and spin theories'. *Phys. Lett.* **B126** (1983) 463.

[BC81] BHANOT G. AND CREUTZ M. 'Variant actions and phase structure in lattice gauge theory'. *Phys. Rev.* **D24** (1981) 3212.

[BCL77] BARDUCCI A., CASALBUONI R., AND LUSANNA L. 'Classical scalar and spinning particles interacting with external Yang–Mills field'. *Nucl. Phys.* **B124** (1977) 93.

[BDI74] BALIAN R., DROUFFE J.M., AND ITZYKSON C. 'Gauge fields on a lattice. 1. General outlook'. *Phys. Rev.* **D10** (1974) 3376.

[Bel00] LE BELLAC M. *Thermal field theory* (Cambridge Univ. Press, 2000).

[Ber74] BERNARD C.W. 'Feynman rules for gauge theories at finite temperature'. *Phys. Rev.* **D9** (1974) 3312.

[Bra80] BRALIĆ N. 'Exact computation of loop averages in two-dimensional Yang–Mills theory'. *Phys. Rev.* **D22** (1980) 3090.

[BSS77] BALACHANDRAN A.P., SALOMONSON P., SKAGERSTAM B., AND WINNBERG J. 'Classical description of a particle interacting with a non-Abelian gauge field'. *Phys. Rev.* **D15** (1977) 2308.

[CGL81] CVITANOVICH P., GREENSITE J., AND LAUTRUP B. 'The crossover point in lattice gauge theories with continuous gauge groups'. *Phys. Lett.* **B105** (1981) 197.

[CGS86] CLEYMANS J., GAVAI R.V., AND SUHONEN E. 'Quarks and gluons at high temperatures and densities'. *Phys. Rep.* **130** (1986) 217.

[CJR79] CREUTZ M., JACOBS L., AND REBBI C. 'Experiments with a gauge-invariant Ising system'. *Phys. Rev. Lett.* **42** (1979) 1390; 'Monte Carlo study of Abelian lattice gauge theories'. *Phys. Rev.* **D20** (1979) 1915.

[CJR83] CREUTZ M., JACOBS L., AND REBBI C. 'Monte Carlo computations in lattice gauge theories'. *Phys. Rep.* **95** (1983) 201.

[Cre79] CREUTZ M. 'Confinement and critical dimensionality of space-time'. *Phys. Rev. Lett.* **43** (1979) 553; 'Monte Carlo study of quantized $SU(2)$ gauge theory'. *Phys. Rev.* **D21** (1980) 2308.

[Cre80] CREUTZ M. 'Asymptotic-freedom scales'. *Phys. Rev. Lett.* **45** (1980) 313.

[Cre83] CREUTZ M. *Quarks gluons and lattices* (Cambridge Univ. Press, 1983).

[Dro81] DROUFFE J.M. 'The mean-field approximation in the SU(2) pure lattice gauge theory'. *Phys. Lett.* **B105** (1981) 46.

[DZ83] DROUFFE J.-M. AND ZUBER J.-B. 'Strong coupling and mean field methods in lattice gauge theories'. *Phys. Rep.* **102** (1983) 1.

[EJN85] EVERTZ H.G., JERSAK J., NEUHAUS T., AND ZERWAS P.M. 'Tricritical point in lattice QED'. *Nucl. Phys.* **B251** (1985) 279.

[Eli75] ELITZUR S. 'Impossibility of spontaneously breaking local symmetries'. *Phys. Rev.* **D12** (1975) 3978.

[Fey53] FEYNMAN R.P. 'Atomic theory of the λ transition in helium'. *Phys. Rev.* **91** (1953) 1291.

[Fey72] FEYNMAN R.P. *Statistical mechanics. A set of lectures* (Benjamin, Reading, 1972).

[FLZ82] FLYVBJERG H., LAUTRUP B., AND ZUBER J.B. 'Mean field with corrections in lattice gauge theory'. *Phys. Lett.* **B110** (1982) 279.

[GL81] GREENSITE J. AND LAUTRUP B. 'Phase transition and mean-field methods in lattice gauge theories'. *Phys. Lett.* **B104** (1981) 41.

[GN80] GERVAIS J.L. AND NEVEU A. 'The slope of the leading Regge trajectory in quantum chromodynamics'. *Nucl. Phys.* **B163** (1980) 189.

[Gri78] GRIBOV V.N. 'Quantization of non-Abelian gauge theories'. *Nucl. Phys.* **B139** (1978) 1.

[GW82] GINSPARG P. AND WILSON K. 'A remnant of chiral symmetry on the lattice'. *Phys. Rev.* **D25** (1982) 2649.

[Ham83] HAMBER H.W. 'Lattice gauge theories at large N_f'. *Phys. Lett.* **B126** (1983) 471.

[HJS77] HALPERN M.B., JEVICKI A., AND SENJANOVIĆ P. 'Field theories in terms of particle-string variables: spin, internal symmetries, and arbitrary dimension'. *Phys. Rev.* **D16** (1977) 2476.

[HP81] HAMBER H. AND PARISI G. 'Numerical estimates of hadron masses in a pure SU(3) gauge theory'. *Phys. Rev. Lett.* **47** (1981) 1792.

[Kad76] KADANOFF L.P. 'Recursion relations in statistical physics and field theory'. *Ann. Phys.* **100** (1976) 359.

[Kap89] KAPUSTA J.I. *Finite-temperature field theory* (Cambridge Univ. Press, 1989).

[Kob79] KOBE D.H. 'Aharonov–Bohm effect revisited'. *Ann. Phys.* **123** (1979) 381.

[KM81] KHOKHLACHEV S.B. AND MAKEENKO YU.M. 'Phase transition over the gauge group center and quark confinement in QCD'. *Phys. Lett.* **B101** (1981) 403.

[KMN83] KLUBERG-STERN H., MOREL A., NAPOLY O., AND PETERSSON B. 'Flavors of Lagrangian Susskind fermions'. *Nucl. Phys.* **B220** (1983) 447.

[KS75] KOGUT J. AND SUSSKIND L. 'Hamiltonian formulation of Wilson's lattice gauge theories'. *Phys. Rev.* **D11** (1975) 395.

[KS81a] KARSTEN L.H. AND SMIT J. 'Lattice fermions: species doubling, chiral invariance and the triangle anomaly'. *Nucl. Phys.* **B183** (1981) 103.

[KS81b] KAWAMOTO N. AND SMIT J. 'Effective Lagrangian and dynamical symmetry breaking in strongly coupled lattice QCD'. *Nucl. Phys.* **B192** (1981) 100.

[Lat00] *Lattice 2000*, Proc. of the XVIIIth Int. Symp. on lattice field theory, ed. T. Bhattacharya *et al. Nucl. Phys. B (Proc. Suppl.)* **94**, 2001.

[LN80] LAUTRUP B. AND NAUENBERG M. 'Phase transition in four-dimensional compact QED'. *Phys. Lett.* **B95** (1980) 63.

[Lon27] LONDON F. 'Quantenmechanische Deutung der Theorie von Weyl'. *Z. Phys.* **42** (1927) 375.

[LR82] LANG C.B. AND REBBI C. 'Potential and restoration of rotational symmetry in SU(2) lattice gauge theory'. *Phys. Lett.* **B115** (1982) 137.

[Lus98] LÜSCHER M. 'Exact chiral symmetry on the lattice and the Ginsparg–Wilson relation'. *Phys. Lett.* **B428**, 342 (1998)

[Mak84] MAKEENKO YU.M. 'The Monte Carlo method in lattice gauge theories'. *Sov. Phys. Usp.* **27** (1984) 401.

[Mig75] MIGDAL A.A. 'Recursion equations in gauge theories'. *Sov. Phys. JETP* **42** (1975) 413.

[MM94] MONTVAY I. AND MÜNSTER G. *Quantum fields on a lattice* (Cambridge Univ. Press, 1994).

[Neu98] NEUBERGER H. 'Exactly massless quarks on the lattice'. *Phys. Lett.* **B417** (1998) 141.

[NN81] NIELSEN H.B. AND NINOMIYA M. 'Absence of neutrinos on a lattice. 1. Proof by homotopy theory'. *Nucl. Phys.* **B185** (1981) 20; '2. Intuitive topological proof'. *Nucl. Phys.* **B193** (1981) 173.

[Pol78] POLYAKOV A.M. 'Thermal properties of gauge fields and quark liberation'. *Phys. Lett.* **B72** (1978) 477.

[Reb83]   REBBI C. *Lattice gauge theories and Monte Carlo simulations* (World Scientific, Singapore, 1983).

[Sak85]   SAKITA B. *Quantum theory of many-variable systems and fields* (World Scientific, Singapore, 1985).

[Sei82]   SEILER E. *Gauge theories as a problem of constructive quantum field theory and statistical mechanics*, Lect. Notes in Physics **159** (Springer-Verlag, Berlin, 1982).

[Smi97]   SMILGA A.V. 'Physics of thermal QCD'. *Phys. Rep.* **291** (1997) 1.

[STW81]   SHARATCHANDRA H.S., THUN H.J., AND WEISZ P. 'Susskind fermions on a Euclidean lattice'. *Nucl. Phys.* **B192** (1981) 205.

[Sus77]   SUSSKIND L. 'Lattice fermions'. *Phys. Rev.* **D16** (1977) 3031.

[Sus79]   SUSSKIND L. 'Lattice models of quark confinement at high temperature'. *Phys. Rev.* **D20** (1979) 2610.

[Sve86]   SVETITSKY B. 'Symmetry aspects of finite-temperature confinement transitions'. *Phys. Rep.* **132** (1986) 1.

[Weg71]   WEGNER F.J. 'Duality in generalized Ising models and phase transitions without local order parameter'. *J. Math. Phys.* **12** (1971) 2259.

[Wey19]   WEYL H. 'Eine neue Erweiterung der Relativitätstheorie'. *Ann. der Phys.* **59** (1919) v. 10, p. 101.

[Wil74]   WILSON K.G. 'Confinement of quarks'. *Phys. Rev.* **D10** (1974) 2445.

[Wil75]   WILSON K.G. 'Quarks and strings on a lattice', in *New phenomena in a subnuclear physics* (Erice 1975), ed. A. Zichichi (Plenum, New York, 1977), p. 69.

[Wil77]   WILSON K.G. 'Quantum chromodynamics on a lattice', in *New developments in quantum field theory and statistical mechanics*, eds. M. Levy and P. Mitter (Plenum, New York, 1977), p. 143.

[WK74]   WILSON K.G. AND KOGUT J. 'The renormalization group and the ϵ-expansion'. *Phys. Rep.* **12C** (1974) 75.

[Won70]   WONG S.K. 'Field and particle equations for the classical Yang–Mills field and particles with isotopic spin'. *Nuovo Cim.* **65A** (1970) 689.

[YM54]   YANG C.N. AND MILLS R.L. 'Conservation of isotopic spin and isotopic gauge invariance'. *Phys. Rev.* **96** (1954) 191.

Part 3

$1/N$ Expansion

In many physical problems, especially when fluctuations of scales of different orders of magnitude are essential, there is no small parameter which
could simplify a study. A typical example is QCD where the effective
coupling, describing strong interaction at a given distance, becomes large
at large distances so that the interaction really becomes strong.

't Hooft [Hoo74a] proposed in 1974 to use the dimensionality of the
gauge group $SU(N)$ as such a parameter, considering the number of colors, N, as a large number and performing an expansion in $1/N$. The
motivation was an expansion in the inverse number of field components
N in statistical mechanics where it is known as the $1/N$-expansion, and
is a standard method for nonperturbative investigations.

The expansion of QCD in the inverse number of colors rearranges diagrams of perturbation theory in a way which is consistent with a string
picture of strong interaction, the phenomenological consequences of which
agree with experiment. The accuracy of the leading-order term, which is
often called multicolor QCD or large-N QCD, is expected to be of the
order of the ratios of meson widths to their masses, i.e. about 10–15%.

While QCD is simplified in the large-N limit, it is still not yet solved.
Generically, it is a problem of infinite matrices, rather than of infinite
vectors as in the theory of second-order phase transitions in statistical
mechanics.

We shall start this part by showing how the $1/N$-expansion works for
the $O(N)$-vector models, and describing some applications to the four-
Fermi interaction, the φ^4 theory and the nonlinear sigma model. Then
we shall concentrate on multicolor QCD.

10

$O(N)$ vector models

The simplest models, which become solvable in the limit of a large number of field components, deal with a field which has N components forming an $O(N)$ vector in an internal symmetry space. A model of this kind was first considered by Stanley [Sta68] in statistical mechanics and is known as the spherical model. The extension to quantum field theory was made by Wilson [Wil73] both for the four-Fermi and φ^4 theories.

Within the framework of perturbation theory, the four-Fermi interaction is renormalizable only in $d = 2$ dimensions and is nonrenormalizable for $d > 2$. The $1/N$-expansion resums perturbation-theory diagrams after which the four-Fermi interaction becomes renormalizable to each order in $1/N$ for $2 \leq d < 4$. An analogous expansion exists for the nonlinear $O(N)$ sigma model. The φ^4 theory remains "trivial" in $d = 4$ to each order of the $1/N$-expansion and has a nontrivial infrared-stable fixed point for $2 < d < 4$.

The $1/N$-expansion of the vector models is associated with a resummation of Feynman diagrams. A very simple class of diagrams – the bubble graphs – survives to the leading order in $1/N$. This is why the large-N limit of the vector models is solvable. Alternatively, the large-N solution is nothing but a saddle-point solution in the path-integral approach. The existence of the saddle point is a result of the fact that N is large. This is to be distinguished from a perturbation-theory saddle point which arises from the fact that the coupling constant is small. Taking into account fluctuations around the saddle-point results in the $1/N$-expansion of the vector models.

We begin this chapter with a description of the $1/N$-expansion of the N-component four-Fermi theory analyzing the bubble graphs. Then we introduce functional methods and construct the $1/N$-expansion of the $O(N)$-symmetric φ^4 theory and nonlinear sigma model. Finally, we discuss the factorization in the $O(N)$ vector models at large N.

10.1 Four-Fermi theory

The action of the $O(N)$-symmetric four-Fermi theory in a d-dimensional Euclidean space* is defined by

$$S[\bar{\psi}, \psi] = \int d^d x \left[\bar{\psi}\, \hat{\partial}\, \psi + m\, \bar{\psi}\psi - \frac{G}{2} \left(\bar{\psi}\psi \right)^2 \right]. \qquad (10.1)$$

Here $\hat{\partial} = \gamma_\mu \partial_\mu$ and $\psi = (\psi_1, \dots, \psi_N)$ is a spinor field which forms an N-component vector in an internal-symmetry space so that

$$\bar{\psi}\psi = \sum_{i=1}^{N} \bar{\psi}_i \psi_i. \qquad (10.2)$$

The dimension of the four-Fermi coupling constant G is

$$\dim [G] = m^{2-d}. \qquad (10.3)$$

For this reason, the perturbation theory for the four-Fermi interaction is renormalizable in $d = 2$ but is nonrenormalizable for $d > 2$ (and, in particular, in $d = 4$). This is why the old Fermi theory of weak interactions was replaced by the modern electroweak theory, where the interaction is mediated by the $W^\pm$ and Z bosons.

The action (10.1) can be rewritten equivalently as

$$S[\bar{\psi}, \psi, \chi] = \int d^d x \left(\bar{\psi}\, \hat{\partial}\, \psi + m\, \bar{\psi}\psi - \chi\, \bar{\psi}\psi + \frac{\chi^2}{2G} \right), \qquad (10.4)$$

where χ is an auxiliary field. The two forms of the action, (10.1) and (10.4), are equivalent owing to the equation of motion which reads in the operator notation as

$$\chi = G : \bar{\psi}\psi :, \qquad (10.5)$$

where $: \cdots :$ denotes the normal ordering of operators. Equation (10.5) can be derived by varying the action (10.4) with respect to χ.

In the path-integral quantization, where the partition function is defined by

$$Z = \int \mathcal{D}\chi\, \mathcal{D}\bar{\psi}\, \mathcal{D}\psi\, e^{-S[\bar{\psi}, \psi, \chi]} \qquad (10.6)$$

with $S[\bar{\psi}, \psi, \chi]$ given by Eq. (10.4), the action (10.1) appears after performing the Gaussian integral over χ. Therefore, alternatively one obtains

$$Z = \int \mathcal{D}\bar{\psi}\, \mathcal{D}\psi\, e^{-S[\bar{\psi}, \psi]} \qquad (10.7)$$

with $S[\bar{\psi}, \psi]$ given by Eq. (10.1).

---

* In $d = 2$ this model was studied in the large-N limit in [GN74] and is often called the Gross–Neveu model.

The perturbative expansion of the $O(N)$-symmetric four-Fermi theory can be represented conveniently using the formulation (10.4) via the auxiliary field χ. Then the diagrams are of the same type as those in Yukawa theory, and resemble those for QED with $\bar{\psi}$ and ψ being an analog of the electron–positron field and χ being an analog of the photon field.

However, the auxiliary field $\chi(x)$ does not propagate, since it follows from the action (10.4) that

$$D_0(x-y) \equiv \langle \chi(x)\chi(y) \rangle_{\text{Gauss}} = G\,\delta^{(d)}(x-y) \qquad (10.8)$$

or

$$D_0(p) \equiv \langle \chi(-p)\chi(p) \rangle_{\text{Gauss}} = G \qquad (10.9)$$

in momentum space.

It is convenient to represent the four-Fermi vertex

$$\Gamma_{ij}^{kl} = G\left(\delta_i^k \delta_j^l - \delta_i^l \delta_j^k\right) \qquad (10.10)$$

as the sum of two terms

$$\qquad (10.11)$$

where the empty space inside the vertex is associated with the propagator (10.8) (or (10.9) in momentum space). The relative minus sign makes the vertex antisymmetric in both incoming and outgoing fermions as is prescribed by the Fermi statistics.

The diagrams that contribute to second order in G for the four-Fermi vertex are depicted, in this notation, in Fig. 10.1. The $O(N)$ indices propagate through the solid lines so that the closed line in the diagram in Fig. 10.1b corresponds to the sum over the $O(N)$ indices which results in a factor of N. Analogous one-loop diagrams for the propagator of the ψ-field are depicted in Fig. 10.2.

Problem 10.1 Calculate the one-loop Gell-Mann–Low function of the four-Fermi theory in $d = 2$.

Solution Evaluating the diagrams in Fig. 10.1 that are logarithmically divergent in $d = 2$, and noting that the diagrams in Fig. 10.2 do not contribute to the wave-function renormalization of the ψ-field, which emerges to the next order in G, one obtains

$$\mathcal{B}(G) = -\frac{(N-1)\,G^2}{2\pi}. \qquad (10.12)$$

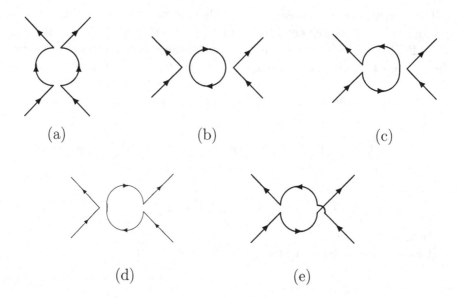

Fig. 10.1. Diagrams of second-order perturbation theory for the four-Fermi vertex. Diagram (b) involves the sum over the $O(N)$ indices.

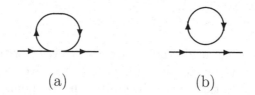

Fig. 10.2. One-loop diagrams for the propagator of the ψ-field. Diagram (b) involves the sum over the $O(N)$ indices.

The four-Fermi theory in two dimensions is asymptotically free as was first noted by Anselm [Ans59] and rediscovered by Gross and Neveu [GN74].

The vanishing of the one-loop Gell-Mann–Low function in the Gross–Neveu model for $N = 1$ is related to the same phenomenon in the Thirring model. The latter model is associated with the vector-like interaction $(\bar\psi\gamma_\mu\psi)^2$ of one species of fermions, where γ_μ are the γ-matrices in two dimensions. Since in $d = 2$ a bispinor has only two components ψ_1 and ψ_2, both the vector-like and the scalar-like interaction (10.1) for $N = 1$ reduce to $\bar\psi_1\psi_1\bar\psi_2\psi_2$, since the square of a Grassmann variable vanishes. Therefore, these two models coincide. For the Thirring model, the vanishing of the Gell-Mann–Low function for any G was shown by Johnson [Joh61] to all loops.

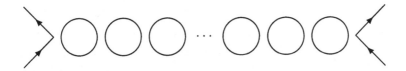

Fig. 10.3. Bubble diagram which survives the large-N limit of the $O(N)$ vector models.

Remark on auxiliary fields

The introduction of the auxiliary field is often called the Hubbard–Stratonovich transformation in statistical mechanics. The proper term used in quantum field theory is just "auxiliary field".

10.2 Bubble graphs as the zeroth order in 1/N

The perturbation-theory expansion of the $O(N)$-symmetric four-Fermi theory contains, in particular, diagrams of the type depicted in Fig. 10.3, which are called *bubble graphs*. Since each bubble has a factor of N, the contribution of the n-bubble graph is $\propto G^{n+1} N^n$, which is of the order of

$$G^{n+1} N^n \sim G \qquad (10.13)$$

as $N \to \infty$, since

$$G \sim \frac{1}{N}. \qquad (10.14)$$

Therefore, all the bubble graphs are essential to the leading order in $1/N$.

Let us denote

$$\text{~~~~} = G + \cdots + G^2 \bigcirc + G^{n+1} \underbrace{\bigcirc \cdots \bigcirc}_{n \text{ loops}} + \cdots . \qquad (10.15)$$

In fact, the wavy line is nothing but the propagator D of the χ field with the bubble corrections included. The first term G on the RHS of Eq. (10.15) is nothing but the free propagator (10.9).

Summing the geometric series of the fermion-loop chains on the RHS of Eq. (10.15), one obtains analytically[*]

$$D^{-1}(p) = \frac{1}{G} - N \int \frac{d^d k}{(2\pi)^d} \frac{\text{sp}\left[(\hat{k} + im)(\hat{k} + \hat{p} + im)\right]}{(k^2 + m^2)\left[(k + p)^2 + m^2\right]}. \qquad (10.16)$$

---

[*] Recall that the free Euclidean fermionic propagator is given by $S_0(p) = (i\hat{p} + m)^{-1}$ from Eqs. (10.4) and (10.6), and the additional minus sign is associated with the fermion loop.

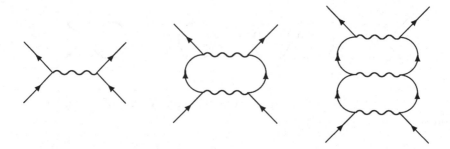

Fig. 10.4.   Some diagrams of the $1/N$-expansion for the $O(N)$ four-Fermi theory. The wavy line represents the (infinite) sum of the bubble graphs (10.15).

This determines the exact propagator of the χ field at large N. It is $\mathcal{O}(N^{-1})$ since the coupling G is included in the definition of the propagator.

The idea is now to change the order of summation of diagrams of perturbation theory using $1/N$ rather than G as the expansion parameter. Therefore, the zeroth-order propagator of the expansion in $1/N$ is defined as the sum over the bubble graphs (10.15), which is given by Eq. (10.16).

Some of the diagrams of the new expansion for the four-Fermi vertex are depicted in Fig. 10.4. The first diagram is proportional to G, while the second and third ones are proportional to G^2 or G^3, respectively, and therefore are of order $\mathcal{O}(N^{-1})$ or $\mathcal{O}(N^{-2})$ with respect to the first diagram. The perturbation theory is thus rearranged as a $1/N$-expansion.

The general structure of the $1/N$-expansion is the same for all vector models, say, for the N-component φ^4 theory which is considered in the next section.

The main advantage of the expansion in $1/N$ for the four-Fermi interaction, over the perturbation theory, is that it is renormalizable in $d < 4$, while the perturbation-theory expansion in G is renormalizable only in $d = 2$. Moreover, the $1/N$-expansion of the four-Fermi theory in $2 < d < 4$ demonstrates [Wil73] the existence of an ultraviolet-stable fixed point, i.e. a nontrivial zero of the Gell-Mann–Low function.

Problem 10.2 Show that the $1/N$-expansion of the four-Fermi theory is renormalizable in $2 \leq d < 4$ (but not in $d = 4$).

Solution   In order to demonstrate renormalizability, let us analyze indices of the diagrams of the $1/N$-expansion.

First of all, we shall remove an ultraviolet divergence of the integral over the d-momentum k in Eq. (10.16). The divergent part of the integral is proportional to Λ^{d-2} (logarithmically divergent in $d = 2$), where Λ is an ultraviolet cutoff. It can be canceled by choosing

$$G = \frac{g^2}{N} \Lambda^{2-d}, \tag{10.17}$$

where g^2 is a proper dimensionless constant which is not necessarily positive since the four-Fermi theory is stable with either sign of G. The power of Λ in Eq. (10.17) is consistent with the dimension of G. This prescription works for $2 < d < 4$ where there is only one divergent term, while another divergence $\propto p^2 \ln \Lambda$ emerges additionally in $d = 4$. This is why the consideration is not applicable in $d = 4$.

The propagator $D(p)$ is therefore finite, and behaves at large momenta $|p| \gg m$ as

$$D(p) \propto |p|^{2-d}. \tag{10.18}$$

The standard power-counting arguments then show that the only divergent diagrams appear in the propagators of the ψ and χ fields, and in the $\bar{\psi}$–χ–ψ three-vertex. These divergences can be removed by a renormalization of the coupling g, mass, and wave functions of ψ and χ.

This completes a demonstration of renormalizability of the $1/N$-expansion for the four-Fermi interaction in $2 \leq d < 4$. For more details, see [Par75].

Problem 10.3 Calculate in $d = 3$ the value of g in Eq. (10.17).

Solution To calculate the divergent part of the integral in Eq. (10.16), we set $p = 0$ and $m = 0$. Remembering that the γ-matrices are 2×2 matrices in $d = 3$, we obtain

$$\int^{\Lambda} \frac{d^3k}{(2\pi)^3} \frac{\mathrm{sp}\,\hat{k}\hat{k}}{k^2 k^2} = 2 \int^{\Lambda} \frac{d^3k}{(2\pi)^3} \frac{1}{k^2} = \frac{1}{\pi^2} \int_0^{\Lambda} d|k| = \frac{\Lambda}{\pi^2}. \tag{10.19}$$

Note that the integral is linearly divergent in $d = 3$ and Λ is the cutoff for the integration over $|k|$. This divergence can be canceled by choosing G according to Eq. (10.17) with g equal to

$$g_* = \pi. \tag{10.20}$$

Problem 10.4 Calculate in $d = 3$ the coefficient of proportionality in Eq. (10.18).

Solution Let us choose $G = \pi^2/N\Lambda$, as prescribed by Eqs. (10.17) and (10.20), and in Eq. (10.16) set $m = 0$, since we are interested in the asymptotic behavior at $|p| \gg m$. Then the RHS of Eq. (10.16) can be rearranged as

$$\begin{aligned}
D^{-1}(p) &= -2N \int \frac{d^3k}{(2\pi)^3} \left[\frac{k^2 + kp}{k^2(k+p)^2} - \frac{1}{k^2} \right] \\
&= 2N \int \frac{d^3k}{(2\pi)^3} \frac{p^2 + kp}{k^2(k+p)^2}. \tag{10.21}
\end{aligned}$$

This integral is obviously convergent.

To calculate it, we apply the standard technique of α-parametrization, which is based on the formula

$$\frac{1}{k^2} = \int_0^\infty d\alpha \, e^{-\alpha k^2} . \qquad (10.22)$$

We have

$$\int \frac{d^3k}{(2\pi)^3} \frac{p^2 + kp}{k^2(k+p)^2} = \int_0^\infty d\alpha_1 \int_0^\infty d\alpha_2 \int \frac{d^3k}{(2\pi)^3} (p^2 + kp) \, e^{-\alpha_1 k^2 - \alpha_2 (k+p)^2} \qquad (10.23)$$

after which the Gaussian integral over d^3k can easily be performed. We then obtain

$$D^{-1}(p) = \frac{N}{4\pi^{3/2}} p^2 \int_0^\infty d\alpha_1 \int_0^\infty d\alpha_2 \frac{\alpha_1}{(\alpha_1 + \alpha_2)^{5/2}} \exp\left(-\frac{\alpha_1 \alpha_2}{\alpha_1 + \alpha_2} p^2\right). \qquad (10.24)$$

The remaining integration over α_1 and α_2 can easily be performed by introducing the new variables $\alpha \in [0, \infty]$ and $x \in [0, 1]$ so that

$$\alpha_1 = \alpha x, \qquad \alpha_2 = \alpha(1 - x), \qquad \frac{\partial(\alpha_1, \alpha_2)}{\partial(x, \alpha)} = \alpha . \qquad (10.25)$$

Finally, this gives

$$D(p) = \frac{8}{N |p|} . \qquad (10.26)$$

Equation (10.26) (or (10.18) in d dimensions) is remarkable since it shows that the scale dimension of the field χ, which is defined in Sect. 3.5 by Eq. (3.65), changes its value from $l_\chi = d/2$ in perturbation theory to $l_\chi = 1$ in the zeroth order of the $1/N$ expansion (remember that the momentum-space propagator of a field with the scale dimension l is proportional to $|p|^{2l-d}$). This appearance of scale invariance in the $1/N$-expansion of the four-Fermi theory at $2 < d < 4$ was first pointed out by Wilson [Wil73] and implies that the Gell-Mann–Low function $\mathcal{B}(g)$ has a zero at $g = g_*$, which is given in $d = 3$ by Eq. (10.20).

Problem 10.5 Find the (logarithmic) anomalous dimensions of the fields ψ and χ, and of the $\bar{\psi}$–χ–ψ three-vertex in $d = 3$ to order $1/N$.

Solution The $1/N$-correction to the propagator of the ψ-field is given by the diagram depicted in Fig. 10.5a. Since we are interested in the ultraviolet behavior, we can again set $m = 0$. Analytically, we have

$$S^{-1}(p) = i\hat{p} + \frac{8i}{N} \int^\Lambda \frac{d^3k}{(2\pi)^3} \frac{\hat{k} + \hat{p}}{|k|(k+p)^2} . \qquad (10.27)$$

The (logarithmically) divergent contribution emerges from the domain of integration $|k| \gg |p|$ so we can expand the integrand in p. The p-independent term

(a)   (b)

Fig. 10.5.   Diagrams for the $1/N$-correction to the ψ-field propagator (a) and the three-vertex (b).

vanishes after integration over the directions of k so that we obtain

$$S^{-1}(p) = i\hat{p}\left(1 + \frac{8}{N}\int^{\Lambda}\frac{d^3k}{(2\pi)^3}\frac{1}{|k|^3}\right) = i\hat{p}\left(1 + \frac{2}{3\pi^2 N}\ln\frac{\Lambda^2}{p^2} + \frac{\text{finite}}{N}\right).$$

(10.28)

The diagram, which gives a nonvanishing contribution to the three-vertex at order $1/N$, is depicted in Fig. 10.5b. It gives analytically

$$\Gamma(p_1, p_2) = 1 + \frac{8}{N}\int^{\Lambda}\frac{d^3k}{(2\pi)^3}\frac{(\hat{k}+\hat{p}_1)(\hat{k}+\hat{p}_2)}{|k|(k+p_1)^2(k+p_2)^2},$$

(10.29)

where p_1 and p_2 are the incoming and outgoing fermion momenta, respectively. The logarithmic domain is $|k| \gg |p|_{\max}$, with $|p|_{\max}$ being the largest of $|p_1|$ and $|p_2|$. This gives

$$\Gamma(p_1, p_2) = 1 - \frac{2}{\pi^2 N}\ln\frac{\Lambda^2}{p_{\max}^2} + \frac{\text{finite}}{N}.$$

(10.30)

The analogous calculation of the $1/N$ correction for the field χ is slightly more complicated since it involves three two-loop diagrams (see, for example, [CMS93]). The resulting expression for $D^{-1}(p)$ is given by

$$[ND(p)]^{-1} = \frac{\Lambda}{g^2} + \left(-\frac{\Lambda}{\pi^2} + \frac{|p|}{8}\right) + \frac{1}{\pi^2 N}\left[2\Lambda - |p|\left(\frac{2}{3}\ln\frac{\Lambda^2}{p^2} + \text{finite}\right)\right].$$

(10.31)

The linear divergence is canceled to order $1/N$, providing g is equal to

$$g_* = \pi\left(1 + \frac{1}{N}\right),$$

(10.32)

which determines g_* to order $1/N$. After this $D^{-1}(p)$ takes the form

$$D^{-1}(p) = \frac{N|p|}{8}\left(1 - \frac{16}{3\pi^2 N}\ln\frac{\Lambda^2}{p^2}\right).$$

(10.33)

To make all three expressions (10.28), (10.30), and (10.33) finite, we need logarithmic renormalizations of the wave functions of ψ- and χ-fields and of the vertex Γ. This can be achieved by multiplying them by the renormalization constants

$$Z_i(\Lambda) = 1 - \gamma_i \ln \frac{\Lambda^2}{\mu^2}, \tag{10.34}$$

where μ denotes a reference mass scale and γ_i are anomalous dimensions. The index i denotes ψ, χ, or v for the ψ- and χ-propagators or the three-vertex Γ, respectively. We have, therefore, calculated

$$\gamma_\psi = \frac{2}{3\pi^2 N}, \qquad \gamma_v = -\frac{2}{\pi^2 N}, \qquad \gamma_\chi = -\frac{16}{3\pi^2 N} \tag{10.35}$$

to order $1/N$. Owing to Eq. (10.5) γ_χ coincides with the anomalous dimension of the composite field $\bar\psi\psi$: $\gamma_{\bar\psi\psi} = \gamma_\chi$.

Note that

$$Z_\psi^2 Z_v^{-2} Z_\chi = 1. \tag{10.36}$$

This implies that the effective charge is not renormalized and is given by Eq. (10.32). Thus, the nontrivial zero of the Gell-Mann–Low function persists to order $1/N$ (and, in fact, to all orders of the $1/N$-expansion).

Remark on scale invariance at the fixed point

The renormalization group says that

$$\mu = \Lambda \exp\left[-\int \frac{\mathrm{d}g^2}{\mathcal{B}(g^2)}\right], \tag{10.37}$$

which is essentially the same as Eq. (6.85). If $\mathcal{B}$ has a nontrivial fixed point g_*^2 near which

$$\mathcal{B}(g^2) = b\left(g^2 - g_*^2\right) \tag{10.38}$$

with $b < 0$, then the substitution into Eq. (10.37) gives

$$g^2 = g_*^2 + \left(\frac{\mu}{\Lambda}\right)^{-b}. \tag{10.39}$$

Therefore, the approach to the critical point is power-like rather than logarithmic as for the case of $g_*^2 = 0$ when

$$\mathcal{B}(g^2) = bg^4. \tag{10.40}$$

The latter behavior of $\mathcal{B}$ results, after the substitution into Eq. (10.37), in the logarithmic dependence

$$g^2 = \frac{1}{b \ln(\mu/\Lambda)} \tag{10.41}$$

when $b < 0$, which is associated with asymptotic freedom.

If g is chosen exactly at the critical point g_*, then the renormalization-group equations

$$\mu \frac{d \ln \Gamma_i}{d\mu} = \gamma_i(g^2), \tag{10.42}$$

where Γ_i denotes generically either vertices or inverse propagators, possess the scale-invariant solutions

$$\Gamma_i \propto \mu^{\gamma_i(g_*^2)}. \tag{10.43}$$

This complements the heuristic consideration of Sect. 3.5 on the relation between scale invariance and the vanishing of the Gell-Mann–Low function.

For the four-Fermi theory in $d = 3$, Eq. (10.43) yields

$$S(p) = \frac{1}{i\hat{p}} \left(\frac{p^2}{\mu^2} \right)^{\gamma_\psi}, \tag{10.44}$$

$$D(p) = \frac{8}{N\,|p|} \left(\frac{p^2}{\mu^2} \right)^{\gamma_\chi}, \tag{10.45}$$

$$\Gamma(p_1, p_2) = \left(\frac{\mu^2}{p_1^2} \right)^{\gamma_v} f\left(\frac{p_2^2}{p_1^2}, \frac{p_1 p_2}{p_1^2} \right), \tag{10.46}$$

where f is an arbitrary function of the dimensionless ratios which is not determined by scale invariance. Here the indices obey the relation

$$\gamma_v = \gamma_\psi + \frac{1}{2}\gamma_\chi \tag{10.47}$$

which guarantees that Eq. (10.36), implied by scale invariance, is satisfied.

The indices γ_i are given to order $1/N$ by Eqs. (10.35). When expanded in $1/N$, Eqs. (10.44) and (10.45) obviously reproduce Eqs. (10.28) and (10.33). Therefore, one obtains the exponentiation of the logarithms which emerge in the $1/N$-expansion. The calculation of the next terms of the $1/N$-expansion for the indices γ_i is given in [Gra91, DKS93, Gra93].

Remark on conformal invariance at fixed point

Scale invariance implies, in a renormalizable quantum field theory, more general conformal invariance as was first pointed out in [MS69, GW70]. The conformal group in a d-dimensional space-time has $(d + 1)(d + 2)/2$ parameters as illustrated by Table 10.1. More details concerning the conformal group can be found in the lecture by Jackiw [Jac72].

A heuristic proof [MS69] of the fact that scale invariance implies conformal invariance is based on the explicit form of the conformal current

Table 10.1.   Contents and the number of parameters of groups of space-time symmetry.

Group	Transformations		Parameters
Lorentz	$\frac{d(d-1)}{2}$ rotations	$x'_\mu = \Omega_{\mu\nu} x_\nu$	$\frac{d(d-1)}{2}$
Poincaré	$+ \, d$ translations	$x'_\mu = x_\mu + a_\mu$	$\frac{d(d+1)}{2}$
Weyl	$+ \, 1$ dilatation	$x'_\mu = \rho \, x_\mu$	$\frac{d^2+d+2}{2}$
Conformal	$+ \, d$ special conformal	$\frac{x'_\mu}{(x')^2} = \frac{x_\mu}{x^2} + \alpha_\mu$	$\frac{(d+1)(d+2)}{2}$

K^α_μ, which is associated with the special conformal transformation, via the energy–momentum tensor:

$$K^\alpha_\mu = \left(2x_\nu x^\alpha - x^2 \delta^\alpha_\nu\right) \theta_{\mu\nu}. \qquad (10.48)$$

Differentiating, we obtain

$$\partial_\mu K^\alpha_\mu = 2x^\alpha \theta_{\mu\mu}, \qquad (10.49)$$

which is analogous to Eqs. (3.66) and (3.67) for the dilatation current. Therefore, both the dilatation and conformal currents vanish simultaneously when $\theta_{\mu\nu}$ is traceless which is provided, in turn, by the vanishing of the Gell-Mann–Low function.

Conformal invariance completely fixes three-vertices as was first shown by Polyakov [Pol70] for scalar theories. The proper formula for the four-Fermi theory (the same as for Yukawa theory [Mig71]) is given by

$$\Gamma(p_1, p_2) = \mu^{2\gamma_v} \frac{\Gamma(d/2)\Gamma(d/2 - \gamma_v)}{\Gamma(\gamma_v)}$$

$$\times \int \frac{\mathrm{d}^d k}{\pi^{d/2}} \frac{\hat{k} + \hat{p}_1}{[(k + p_1)^2]^{1+\gamma_\chi/2}} \frac{\hat{k} + \hat{p}_2}{[(k + p_2)^2]^{1+\gamma_\chi/2}} \frac{1}{|k|^{d-2+2\gamma_\psi-\gamma_\chi/2}}, \qquad (10.50)$$

where the coefficient in the form of the ratio of the Γ-functions is prescribed by the normalization (10.44) and (10.45), and the indices are related by Eq. (10.47) but can be arbitrary otherwise.*

---

* The only restriction $\gamma_\psi \geq 0$ is imposed by the Källén–Lehmann representation of the propagator, while there is no such restriction on γ_χ since it is a composite field.

Equation (10.50), which results from conformal invariance, unambiguously fixes the function f in Eq. (10.46). In contrast to infinite-dimensional conformal symmetry in $d = 2$, the conformal group in $d > 2$ is less restrictive. It fixes only the tree-point vertex while, say, the four-point vertex remains an unknown function of two variables.

Problem 10.6 Calculate the integral on the RHS of Eq. (10.50) in $d = 3$ to order $1/N$.

Solution The integral on the RHS of Eq. (10.50) looks in $d = 3$ very much like that in Eq. (10.29) and can easily be calculated to the leading order in $1/N$ when only the region of integration over large momenta with $|k| \gtrsim |p|_{\max} \equiv \max\{|p_1|, |p_2|\}$ is essential to this accuracy.

Let us first note that the coefficient in front of the integral is $\propto \gamma_{\rm v} \sim 1/N$, so that one is interested in the term $\sim 1/\gamma_{\rm v}$ in the integral for the vertex to be of order 1. This term comes from the region of integration with $|k| \gtrsim |p|_{\max}$. Recalling that $|p_1 - p_2| \lesssim |p|_{\max}$ in Euclidean space, one obtains

$$\int \frac{\mathrm{d}^3 k}{2\pi} \frac{\hat{k} + \hat{p}_1}{[(k + p_1)^2]^{1 + \gamma_{\rm x}/2}} \frac{\hat{k} + \hat{p}_2}{[(k + p_2)^2]^{1 + \gamma_{\rm x}/2}} \frac{1}{|k|^{1 + 2\gamma_\psi - \gamma_{\rm x}/2}}$$

$$= \int_{p_{\max}^2}^{\infty} \frac{\mathrm{d}k^2}{[k^2]^{1 + \gamma_{\rm v}}} = \frac{1}{\gamma_{\rm v} \, (p_{\max}^2)^{\gamma_{\rm v}}}, \qquad (10.51)$$

where Eq. (10.47) has been used and

$$\Gamma(p_1, p_2) = \left(\frac{\mu^2}{p_{\max}^2} \right)^{\gamma_{\rm v}}. \qquad (10.52)$$

While the integral in Eq. (10.51) is divergent in the ultraviolet for $\gamma_{\rm v} < 0$, this divergence disappears after the renormalization.

Equation (10.30) is reproduced by Eq. (10.51) when expanding in $1/N$. This dependence of the three-vertex solely on the largest momentum is typical for logarithmic theories in the ultraviolet region where one can set, say, $p_1 = 0$ without changing the integral with logarithmic accuracy. This is valid if the integral is quickly convergent in infrared regions which it is in our case.

Remark on broken scale invariance

Scale (and conformal) invariance at a fixed point $g = g_*$ holds only for large momenta $|p| \gg m$. For smaller values of momenta, scale invariance is broken by masses. In fact, any dimensional parameter breaks scale invariance. If the bare coupling g is chosen in the vicinity of g_* according to Eq. (10.39), then scale invariance holds even in the massless case only for $|p| \gg \mu$, while it is broken if $|p| \lesssim \mu$.

10.3 Functional methods for φ^4 theory

The large-N solution of the $O(N)$ vector models, which is given by the sum of the bubble graphs, can be obtained alternatively by evaluating the path integral at large N using the saddle-point method. We shall restrict ourselves to the scalar $O(N)$-symmetric φ^4 theory, while the analysis of the four-Fermi theory is quite analogous.

The action of the $O(N)$-symmetric φ^4 theory is given by

$$S[\varphi^a] = \int d^d x \left[\frac{1}{2} (\partial_\mu \varphi^a)^2 + \frac{1}{2} m^2 \varphi^a \varphi^a + \frac{\lambda}{8} (\varphi^a \varphi^a)^2 \right], \quad (10.53)$$

where $\varphi^a = (\varphi^1, \ldots, \varphi^N)$. The coupling λ in the action (10.53) must be positive for the theory to be well-defined. The vertices of Feynman diagrams are associated with $-\lambda$.

Problem 10.7 Calculate the one-loop Gell-Mann–Low function of the $O(N)$-symmetric φ^4 theory in $d = 4$.

Solution The corresponding diagrams are similar to those of Fig. 10.1, though now the arrows are not essential since the field is real. The diagrams are logarithmically divergent in four dimensions. Each diagram contributes with a positive sign, while the diagram in Fig. 10.1b now has an extra combinatoric factor of $1/2$. The diagrams in Fig. 10.2 result in a mass renormalization and there is no wave-function renormalization of the φ-field in one loop so that one obtains

$$\mathcal{B}(\lambda) = \frac{(N+8)\lambda^2}{16\pi^2}. \quad (10.54)$$

The positive sign in this formula is the same as for QED and is associated with the "triviality" of the φ^4 theory in four dimensions. It is also worth noting that the coefficient $(N+8)$ is large even for $N = 1$.

Introducing the auxiliary field $\chi(x)$ as in Sect. 10.1, the action (10.53) can be rewritten as

$$S[\varphi^a, \chi] = \int d^d x \left[\frac{1}{2} \varphi^a \left(-\partial_\mu^2 + m^2 + \chi \right) \varphi^a - \frac{\chi^2}{2\lambda} \right]. \quad (10.55)$$

The two forms are equivalent owing to the equation of motion

$$\chi = \frac{\lambda}{2} : \varphi^a \varphi^a : . \quad (10.56)$$

In other words, χ is again a composite field.

The correlators of φ and χ are determined by the generating functional

$$Z[J^a, K] = \int_\uparrow \mathcal{D}\chi(x) \int \mathcal{D}\varphi^a(x)$$

$$\times \exp \left\{ -S[\varphi^a, \chi] + \int d^d x \, J^a(x) \varphi^a(x) + \int d^d x \, K(x)\chi(x) \right\},$$

$$(10.57)$$

which is a functional of the sources J^a and K for the fields φ^a and χ and extends Eq. (2.49).

To make the path integral over $\chi(x)$ in Eq. (10.57) convergent, at each point x we integrate over a contour that is parallel to the imaginary axis. This is specific to the Euclidean formulation. The propagator of the χ-field in the Gaussian approximation reads

$$D_0(p) = \langle \chi(-p)\chi(p) \rangle_{\text{Gauss}} = -\lambda, \qquad (10.58)$$

which reproduces the four-boson vertex of perturbation theory.

Since the integral over φ^a is Gaussian, it can be expressed via the Green function

$$G(x, y; \chi) = \left\langle y \left| \frac{1}{-\partial_\mu^2 + m^2 + \chi} \right| x \right\rangle \qquad (10.59)$$

as

$$Z[J^a, K] = \int_\uparrow \mathcal{D}\chi(x) \exp \left\{ \int d^d x \, \frac{\chi^2}{2\lambda} \right.$$
$$+ \frac{1}{2} \int d^d x \, d^d y \, J^a(x) \, G(x, y; \chi) \, J^a(y)$$
$$\left. + \int d^d x \, K(x) \, \chi(x) - \frac{N}{2} \operatorname{Tr} \ln G^{-1}[\chi] \right\}. $$
$$(10.60)$$

Here we have used the obvious notation

$$G^{-1}[\chi] = -\partial_\mu^2 + m^2 + \chi. \qquad (10.61)$$

It will also be convenient to use the short-hand notation

$$g \circ f = \langle g | f \rangle \equiv \int d^d x \, f(x) g(x). \qquad (10.62)$$

Then, Eq. (10.60) can be rewritten as

$$Z[J^a, K] = \int_\uparrow \mathcal{D}\chi(x) \exp \left\{ \frac{\chi \circ \chi}{2\lambda} + \frac{1}{2} J^a \circ G[\chi] \circ J^a \right.$$
$$\left. + K \circ \chi - \frac{N}{2} \operatorname{Tr} \ln G^{-1}[\chi] \right\}. \qquad (10.63)$$

The exponent in Eq. (10.63) is $\mathcal{O}(N)$ at large N so the path integral can be evaluated as $N \to \infty$ by the saddle-point method. The saddle-point field configuration $\chi(x) = \chi_{\text{sp}}(x)$ is determined (implicitly) by the saddle-point equation

$$\chi_{\text{sp}}(x) - \frac{\lambda N}{2} G(x, x; \chi_{\text{sp}})$$

$$+ \frac{\lambda}{2} J^a \circ G(\cdot, x; \chi_{\text{sp}}) \, G(x, \cdot; \chi_{\text{sp}}) \circ J^a + \lambda K(x) \;=\; 0. \qquad (10.64)$$

If $K \sim 1/\lambda$, each term here is $\mathcal{O}(1)$ since

$$\lambda \;\sim\; \frac{1}{N} \qquad\qquad\qquad\qquad (10.65)$$

in analogy with Eq. (10.14).

When the sources J^a and K vanish so that the last two terms on the LHS of Eq. (10.64) equal zero, this equation reduces to

$$\chi_{\text{sp}} - \frac{\lambda N}{2} G(x, x; \chi_{\text{sp}}) \;=\; 0. \qquad (10.66)$$

Its solution is x-independent owing to translational invariance and can be parametrized as

$$\chi_{\text{sp}} \;=\; m_{\text{R}}^2 - m^2, \qquad\qquad (10.67)$$

where m and m_{R} are the bare and renormalized mass, respectively. Equation (10.66) then reduces to the standard formula [Wil73]

$$m^2 \;=\; m_{\text{R}}^2 - \frac{\lambda N}{2} \int^{\Lambda} \frac{d^d k}{(2\pi)^d} \frac{1}{\left(k^2 + m_{\text{R}}^2\right)} \qquad (10.68)$$

for the mass renormalization at large N.

To take into account fluctuations around the saddle point, we expand

$$\chi(x) \;=\; \chi_{\text{sp}} + \delta\chi(x), \qquad\qquad (10.69)$$

where

$$\delta\chi(x) \;\sim\; \sqrt{\lambda} \;\sim\; N^{-1/2}. \qquad\qquad (10.70)$$

The Gaussian integration over $\delta\chi(x)$ determines the pre-exponential factor in (10.63).

To construct the $1/N$ expansion of the generating functional (10.63), it is convenient to use the generating functional for connected Green functions, which was already introduced in Eq. (2.52). It is usually denoted by $W[J^a, K]$ and is related to the partition function (10.57) by

$$Z[J^a, K] \;=\; e^{W[J^a, K]}. \qquad\qquad (10.71)$$

Then we find

$$
\begin{aligned}
W[J^a, K] &= \frac{1}{2\lambda} \chi_{\text{sp}} \circ \chi_{\text{sp}} - \frac{N}{2} \operatorname{Tr} \ln G^{-1}[\chi_{\text{sp}}] \\
&\quad + \frac{1}{2} J^a \circ G[\chi_{\text{sp}}] \circ J^a + K \circ \chi_{\text{sp}} \\
&\quad - \frac{1}{2} \operatorname{Tr} \ln \left(\lambda D^{-1}[\chi_{\text{sp}}] \right) + \mathcal{O}(N^{-1}),
\end{aligned} \tag{10.72}
$$

where

$$
\begin{aligned}
D^{-1}(x, y; \chi) &= -\frac{1}{\lambda} \delta^{(d)}(x - y) - \frac{N}{2} G(x, y; \chi) \, G(y, x; \chi) \\
&\quad + J^a \circ G(\cdot, x; \chi) \, G(x, y; \chi) \, G(y, \cdot; \chi) \circ J^a.
\end{aligned} \tag{10.73}
$$

This operator emerges when integrating over the Gaussian fluctuations around the saddle point. The corresponding (last displayed) term on the RHS of Eq. (10.72) is associated with the pre-exponential factor and, therefore, is ~ 1.

The next terms of the $1/N$ expansion can be calculated in a systematic way by substituting (10.69) in Eq. (10.63) and performing the perturbative expansion in $\delta\chi$.

If the sources J^a and K vanish so that the saddle-point value χ_{sp} is given by the constant (10.67), then the RHS of Eq. (10.73) simplifies to

$$
D^{-1}(x, y; \chi_{\text{sp}}) = -\frac{1}{\lambda} \delta^{(d)}(x - y) - \frac{N}{2} G(x, y; \chi_{\text{sp}}) \, G(y, x; \chi_{\text{sp}}). \tag{10.74}
$$

Remembering the definition (10.59) of G and passing to the momentum-space representation, we obtain

$$
D^{-1}(p) = -\frac{1}{\lambda} - \frac{N}{2} \int \frac{d^d k}{(2\pi)^d} \frac{1}{(k^2 + m_{\text{R}}^2) \left[(k + p)^2 + m_{\text{R}}^2\right]}. \tag{10.75}
$$

The sign of the first term on the RHS is consistent with Eq. (10.58).

Equation (10.75) is analogous to Eq. (10.16) in the fermionic case and can be obtained alternatively by summing bubble graphs of the type shown in Fig. 10.3 for

$$
D(p) = \langle \chi(-p)\chi(p) \rangle. \tag{10.76}
$$

The extra symmetry factor of $1/2$ in Eq. (10.75) is the usual combinatoric one for bosons. Therefore, the large-N saddle-point calculation of the propagator (10.76) results precisely in the zeroth order of the $1/N$-expansion.

We see from Eq. (10.72) the difference between perturbation theory and the $1/N$-expansion. The perturbation theory in λ can be constructed as an expansion (10.69) around the saddle point $\chi_{\rm sp}$ given again by Eq. (10.64), with the omitted second term on the LHS, which is now justified by the fact that λ is small (even for $N \sim 1$). The second term on the RHS of Eq. (10.72), which is associated with a one-loop diagram, appears in perturbation theory as a result of Gaussian fluctuations around this saddle point.

Remark on the effective action

The effective action is a functional of the mean values of fields

$$\varphi_{\rm cl}^a(x) \;=\; \frac{\delta W}{\delta J^a(x)}\,, \qquad \chi_{\rm cl}(x) \;=\; \frac{\delta W}{\delta K(x)} \tag{10.77}$$

in the presence of the external sources. The effective action is defined as the Legendre transformation of $W[J^a, K]$ by

$$\Gamma[\varphi_{\rm cl}^a, \chi_{\rm cl}] \;\equiv\; -W + J^a \circ \varphi_{\rm cl}^a + K \circ \chi_{\rm cl}\,, \tag{10.78}$$

where the sources J^a and K, which are regarded as functionals of $\varphi_{\rm cl}^a$ and $\chi_{\rm cl}$, are to be determined by an inversion of Eq. (10.77). To the leading order in $1/N$ we obtain

$$\left.\begin{array}{rcl} J^a(x) &=& G^{-1}[\chi_{\rm cl}]\,\varphi_{\rm cl}^a(x) + \mathcal{O}(N^{-1})\,, \\[2mm] \chi_{\rm cl}(x) &=& \chi_{\rm sp}(x) + \mathcal{O}(N^{-1})\,. \end{array}\right\} \tag{10.79}$$

When Eq. (10.79) (with the $1/N$ correction included) is substituted into Eq. (10.78) and account is taken of the $1/N$ terms, most of them cancel and we arrive at the relatively simple formula

$$\begin{aligned} \Gamma[\varphi_{\rm cl}^a, \chi_{\rm cl}] \;=\;& -\frac{1}{2\lambda}\chi_{\rm cl} \circ \chi_{\rm cl} + \frac{N}{2}\,{\rm Tr}\ln G^{-1}[\chi_{\rm cl}] \\[2mm] &+ \frac{1}{2}\varphi_{\rm cl}^a \circ G^{-1}[\chi_{\rm cl}] \circ \varphi_{\rm cl}^a + \frac{1}{2}\,{\rm Tr}\ln\left(\lambda D^{-1}[\chi_{\rm cl}]\right) + \mathcal{O}(N^{-1})\,, \end{aligned} \tag{10.80}$$

where

$$\begin{aligned} D^{-1}(x, y; \chi_{\rm cl}) \;=\;& -\frac{1}{\lambda}\delta^{(d)}(x - y) - \frac{N}{2}\,G(x, y; \chi_{\rm cl})\,G(y, x; \chi_{\rm cl}) \\[2mm] &+ \varphi_{\rm cl}^a(x)\,G(x, y; \chi_{\rm cl})\,\varphi_{\rm cl}^a(y) \end{aligned} \tag{10.81}$$

coinciding with (10.73) to the leading order in $1/N$.

The second and fourth terms on the RHS of Eq. (10.80), which involve Tr, are associated with one-loop diagrams of the fields φ^a and χ, respectively, in the classical background fields $\varphi_{\rm cl}^a$ and $\chi_{\rm cl}$. Higher orders in $1/N$

are given by diagrams which are one-particle irreducible with respect to both φ and χ.

It follows immediately from the definitions (10.77) and (10.78) that

$$\frac{\delta \Gamma}{\delta \varphi_{\rm cl}^a(x)} = J^a(x), \qquad \frac{\delta \Gamma}{\delta \chi_{\rm cl}(x)} = K(x). \qquad (10.82)$$

Therefore, $\varphi_{\rm cl}^a(x)$ and $\chi_{\rm cl}(x)$ are determined in the absence of external sources by the equations

$$\frac{\delta \Gamma[\varphi_{\rm cl}^a, \chi_{\rm cl}]}{\delta \varphi_{\rm cl}^b(x)} = 0, \qquad \frac{\delta \Gamma[\varphi_{\rm cl}^a, \chi_{\rm cl}]}{\delta \chi_{\rm cl}(x)} = 0. \qquad (10.83)$$

Substituting (10.80) into Eqs. (10.83), we get to the leading order in $1/N$, respectively, the equations

$$\left[-\partial_\mu^2 + m^2 + \chi_{\rm cl}(x)\right] \varphi_{\rm cl}^a(x) = 0 \qquad (10.84)$$

and

$$\chi_{\rm cl}(x) = \frac{\lambda}{2} \varphi_{\rm cl}^a(x) \varphi_{\rm cl}^a(x) + \frac{\lambda N}{2} G(x, x; \chi_{\rm cl}). \qquad (10.85)$$

The first equation is just a classical equation of motion in an external field $\chi_{\rm cl}(x)$, while the second one is just the average of the (quantum) equation (10.56). Equation (10.85) is often called the *gap equation*.

A solution to Eqs. (10.84) and (10.85) depends on what initial (or boundary) conditions are imposed.

Problem 10.8 Find translationally invariant solutions to Eqs. (10.84) and (10.85) and calculate the corresponding effective potential.

Solution The effective potential $V(\varphi_{\rm cl}^a, \chi_{\rm cl})$ is defined via the integrand in the effective action $\Gamma[\varphi_{\rm cl}^a, \chi_{\rm cl}]$ for translationally invariant

$$\varphi_{\rm cl}^a(x) = \bar\varphi^a, \qquad \chi_{\rm cl} = \bar\chi, \qquad (10.86)$$

i.e. it is given by Γ divided by the volume of Euclidean space. From Eq. (10.80), at large N we find

$$V = -\frac{1}{2\lambda} \bar\chi^2 + \frac{N}{2} \int^\Lambda \frac{d^d k}{(2\pi)^d} \ln\left(k^2 + m^2 + \bar\chi\right) + \frac{1}{2}\left(m^2 + \bar\chi\right) \bar\varphi^2, \qquad (10.87)$$

which obviously recovers Eqs. (10.84) and (10.85) after varying with respect to constant $\bar\varphi^a$ and $\bar\chi$.

It is convenient to perform renormalization by introducing, in $d = 4$, the renormalized coupling $\lambda_{\rm R}$ given by

$$\frac{1}{\lambda_{\rm R}} = \frac{1}{\lambda} + \frac{1}{2} \int^\Lambda \frac{d^4 k}{(2\pi)^4} \frac{1}{k^2 \left(k^2 + m_{\rm R}^2\right)} = \frac{1}{\lambda} + \frac{N}{32\pi^2} \ln \frac{\Lambda^2}{m_{\rm R}^2} \qquad (10.88)$$

and $\bar{\chi}_R = \bar{\chi} + m^2$. Assuming that $\bar{\chi}_R \ll \Lambda^2$ (also $m_R \ll \Lambda$ as usual) and representing Eq. (10.68) in the form

$$\frac{m_R^2}{\lambda_R} = \frac{m^2}{\lambda} - \frac{N}{32\pi^2}\Lambda^2, \tag{10.89}$$

we rewrite Eqs. (10.84) and (10.85) as [Sch74, CJP74]

$$\bar{\chi}_R \bar{\varphi}^a = 0 \tag{10.90}$$

and

$$\bar{\chi}_R \left(1 - \frac{\lambda_R N}{32\pi^2} \ln \frac{\bar{\chi}_R}{m_R^2}\right) = m_R^2 + \frac{\lambda_R}{2}\bar{\varphi}^2. \tag{10.91}$$

Equation (10.87) then gives the renormalized effective potential

$$V_R = -\frac{1}{2\lambda_R}\bar{\chi}_R^2 + \frac{m_R^2 \bar{\chi}_R}{\lambda_R} + \frac{N}{64\pi^2}\bar{\chi}_R^2 \left(-\frac{1}{2} + \ln \frac{\bar{\chi}_R}{m_R^2}\right) + \frac{1}{2}\bar{\chi}_R\bar{\varphi}^2, \tag{10.92}$$

which obviously reproduces Eqs. (10.90) and (10.91).

Equations (10.90) and (10.91) possess the solutions

$$\bar{\varphi}^a = 0, \qquad \bar{\chi}_R = m_R^2 \qquad \text{for } m_R^2 > 0, \tag{10.93}$$

$$\bar{\varphi}^2 = -\frac{2m_R^2}{\lambda_R}, \qquad \bar{\chi}_R = 0 \qquad \text{for } m_R^2 < 0. \tag{10.94}$$

The first of them is associated with an unbroken $O(N)$ symmetry, while the second one corresponds to a spontaneous breaking of $O(N)$ down to $O(N-1)$. Both formulas look like the proper tree-level ones, while the only effect of loop corrections at large N is the renormalization of the coupling constant and mass.

A subtle point is the question of the stability of these solutions. For small deviations of $\bar{\varphi}^2$ from the mean value given by Eqs. (10.93) and (10.94), the effective potential V_R is a monotonically increasing function of $\bar{\varphi}^2$, as can be shown for $\lambda_R N < 32\pi^2$ by eliminating the auxiliary field $\bar{\chi}_R$ from Eq. (10.92) by solving the gap equation (10.91) iteratively in $\bar{\varphi}^2$, and the solutions are locally stable. Both solutions are, however, unstable globally with respect to large fluctuations of the fields. This can be seen by eliminating $\bar{\varphi}^2$ from V_R by solving the gap equation (10.91) for $\bar{\varphi}^2$ which yields

$$V_R = \frac{1}{2}\bar{\chi}_R^2 \left(\frac{1}{\lambda_R} - \frac{N}{32\pi^2} \ln \frac{\bar{\chi}_R}{m_R^2}\right) - \frac{N}{128\pi^2}\bar{\chi}_R^2. \tag{10.95}$$

This function is monotonically decreasing for very large

$$\bar{\chi}_R > m_R^2 \, e^{32\pi^2/\lambda_R N}, \tag{10.96}$$

where the theory becomes unstable. This is related to the usual problem of "triviality" of the φ^4 theory, which makes sense only for small couplings $\lambda_R N$ as an effective theory and cannot be fundamental at very small distances of the order of

$$r \sim m_R^{-1} e^{-16\pi^2/\lambda_R N}. \tag{10.97}$$

Problem 10.9 Find a solution to Eqs. (10.84) and (10.85) which decreases exponentially as

$$\varphi_{\text{cl}}^a(x) = \xi^a m_{\text{R}} \, e^{m_{\text{R}}\tau} \qquad \text{for} \quad \tau \to -\infty, \tag{10.98}$$

where $\tau \equiv x_4$ and ξ^a is an $O(N)$ vector.

Solution The difference with respect to the previous Problem is that φ_{cl} is no longer translationally invariant along the time-variable owing to the initial condition (10.98). Let us denote

$$\varphi_{\text{cl}}^a(x) \equiv \Phi^a(\tau), \qquad \chi_{\text{cl}}(x) \equiv v(\tau). \tag{10.99}$$

The the saddle-point equations (10.84) and (10.85) can be rewritten as

$$\left[-D^2 + m^2 + v(\tau)\right] \Phi^a(\tau) = 0 \tag{10.100}$$

and

$$v(\tau) = \frac{\lambda}{2}\Phi^a(\tau)\Phi^a(\tau) + \frac{\lambda N}{2} \int \frac{d^3k}{(2\pi)^3} G_\omega(\tau, \tau; v), \tag{10.101}$$

where

$$D \equiv \frac{d}{d\tau}, \qquad \omega = \sqrt{k^2 + m^2} \tag{10.102}$$

and we have introduced the Fourier image of the Green function (10.59)

$$G_\omega(\tau, \tau; v) \equiv \int d^3\vec{x} \, e^{i\vec{k}\vec{x}} \, G\big((\tau, \vec{x}), (\tau, \vec{0}); v\big)$$

$$= \left\langle \tau \left| \frac{1}{-D^2 + \omega^2 + v} \right| \tau \right\rangle \tag{10.103}$$

with respect to the spatial coordinate.

The solution to Eqs. (10.100) and (10.101) can be easily found to be

$$\Phi^a(\tau) = \frac{\xi^a m_{\text{R}} \, e^{m_{\text{R}}\tau}}{1 - \frac{\bar{\lambda}_{\text{R}}\xi^2}{16} \, e^{2m_{\text{R}}\tau}}, \qquad v(\tau) = \frac{\bar{\lambda}_{\text{R}}}{2}\Phi^a(\tau)\Phi^a(\tau), \tag{10.104}$$

where the renormalized coupling

$$\bar{\lambda}_{\text{R}} = \frac{\lambda_{\text{R}}}{1 + \frac{\lambda_{\text{R}} N}{16\pi^2}} \tag{10.105}$$

differs from Eq. (10.88) only by an additional final renormalization and the renormalized mass m_{R} is defined in Eq. (10.68). This solution is nontrivial for $\xi^2 \sim N$ and obviously satisfies the initial condition (10.98).

The solution is so simple because the diagonal resolvent (10.103) takes the very simple form

$$G_\omega(\tau, \tau; v) = \frac{1}{2\omega} - \frac{v(\tau)}{4\omega(\omega^2 - m_{\text{R}}^2)} \tag{10.106}$$

for the potential $v(\tau)$ given by Eq. (10.104). This can be verified by substituting into the Gel'fand–Dikii equation (1.127) with $\mathcal{G} = 1$. This is a feature of an integrable potential, which was already discussed in Problem 4.4 on p. 73.

The function $\Phi^a(\tau)$ given by Eq. (10.104) describes large-N amplitudes of multiparticle production at a threshold [Mak94].

10.4 Nonlinear sigma model

The nonlinear $O(N)$ sigma model* in two Euclidean dimensions is defined by the partition function

$$Z = \int \mathcal{D}\vec{n}\, \delta\left(\vec{n}^2 - \frac{1}{g^2}\right) \exp\left[-\frac{1}{2}\int d^2x\, (\partial_\mu \vec{n})^2\right], \qquad (10.107)$$

where $\vec{n} = (n_1, \ldots, n_N)$ is an $O(N)$ vector. While the action is pure Gaussian, the model is not free owing to the constraint

$$\vec{n}^2(x) = \frac{1}{g^2}, \qquad (10.108)$$

which is imposed on the $\vec{n}$ field via the (functional) delta-function.

The sigma model in $d = 2$ is sometimes considered as a toy model for QCD since it possesses:

(1) asymptotic freedom [Pol75];
(2) instantons for $N = 3$ [BP75].

The action in Eq. (10.107) is $\sim N$ as $N \to \infty$ but the entropy, i.e. the contribution from the measure of integration, is also $\sim N$ so that a straightforward saddle point is not applicable.

To overcome this difficulty, we proceed as in the previous section, introducing an auxiliary field $u(x)$, which is ~ 1 as $N \to \infty$, and rewrite the partition function (10.107) as

$$Z \propto \int_\uparrow \mathcal{D}u(x) \int \mathcal{D}\vec{n}(x)\, \exp\left\{-\frac{1}{2}\int d^2x\left[(\partial_\mu \vec{n})^2 - u\left(\vec{n}^2 - \frac{1}{g^2}\right)\right]\right\}, \qquad (10.109)$$

where the contour of integration over $u(x)$ is parallel to the imaginary axis.

Performing the Gaussian integration over $\vec{n}$, we find

$$Z \propto \int_\uparrow \mathcal{D}u(x)\, \exp\left\{-\frac{N}{2}\operatorname{Tr}\ln\left[-\partial_\mu^2 + u(x)\right] + \frac{1}{2g^2}\int d^2x\, u(x)\right\}. \qquad (10.110)$$

---

* The name comes from elementary particle physics where a nonlinear sigma model in four dimensions is used as an effective Lagrangian for describing low-energy scattering of the Goldstone π-mesons.

The first term in the exponent is as before nothing but the sum of one-loop diagrams in two dimensions,

$$\frac{N}{2}\mathrm{Tr}\ln\left[-\partial_\mu^2 + u(x)\right] = \sum_n \frac{1}{n} \quad \text{} \quad , \qquad (10.111)$$

where the auxiliary field u is again denoted by a wavy line. Equation (10.110) looks very much like Eq. (10.63) if we set $J^a = K = 0$. The difference is that the exponent in (10.110) involves the term which is linear in u, while the analogous term in (10.63) is quadratic in χ.

Now the path integral over $u(x)$ in Eq. (10.110) is a typical saddle-point one: the action $\sim N$, while the entropy ~ 1 since only one integration over u is left. The saddle-point equation for the nonlinear sigma model

$$\frac{1}{g^2} - NG(x, x; u_{\mathrm{sp}}) = 0 \qquad (10.112)$$

is quite analogous to Eq. (10.66) for the φ^4 theory, while G is defined by

$$G(x, y; u) = \left\langle y \left| \frac{1}{-\partial_\mu^2 + u} \right| x \right\rangle, \qquad (10.113)$$

which is an analog of Eq. (10.59).

Introducing sources for the $\vec{n}$ and u fields, we can derive the analogs of Eqs. (10.84) and (10.85) for φ^4 theory which are given for the sigma model by

$$\left[-\partial_\mu^2 + u_{\mathrm{cl}}(x)\right]\vec{n}_{\mathrm{cl}}(x) = 0, \qquad (10.114)$$

and

$$\frac{1}{g^2} = \vec{n}_{\mathrm{cl}}^2(x) + NG(x, x; u_{\mathrm{cl}}). \qquad (10.115)$$

For a translationally invariant solution when $\vec{n}_{\mathrm{cl}}(x) = 0$ and $u_{\mathrm{cl}}(x) = u_{\mathrm{sp}}$, we recover Eq. (10.112).

The coupling g^2 in Eq. (10.112) is $\sim 1/N$, as prescribed by the constraint (10.108), which involves a sum over N terms on the LHS. This guarantees that a solution to Eq. (10.112) exists. Next orders of the $1/N$-expansion for the two-dimensional sigma model can be constructed analogously to the previous section.

The $1/N$-expansion of the two-dimensional nonlinear sigma model has many advantages over perturbation theory, which is usually constructed by solving the constraint (10.108) explicitly, say, by choosing

$$n_N = \frac{1}{g}\sqrt{1 - g^2 \sum_{a=1}^{N-1} n_a^2} \tag{10.116}$$

and expanding the square root in g^2. Only $N - 1$ dynamical degrees of freedom are left so that the $O(N)$-symmetry is broken in perturbation theory down to $O(N-1)$. The particles in perturbation theory are massless (like Goldstone bosons) and it suffers from infrared divergences.

In contrast, the solution to Eq. (10.112) has the form

$$u_{\mathrm{sp}} = m_{\mathrm{R}}^2 \equiv \Lambda^2 e^{-4\pi/Ng^2}, \tag{10.117}$$

where Λ is an ultraviolet cutoff. Therefore, all N particles acquire the same mass m_{R} in the $1/N$-expansion so that the $O(N)$ symmetry is restored. This appearance of mass is a result of the dimensional transmutation which says in this case that the parameter m_{R} rather than the renormalized coupling constant g_{R}^2 is observable. The emergence of the mass cures the infrared problem.

Problem 10.10 Show that (10.117) is a solution to Eq. (10.112).

Solution Let us look for a translationally invariant solution $u_{\mathrm{sp}}(x) = m_{\mathrm{R}}^2$. Then Eq. (10.112) in the momentum space gives

$$\frac{1}{g^2} = N \int^{\Lambda} \frac{d^2k}{(2\pi)^2} \frac{1}{k^2 + m_{\mathrm{R}}^2} = \frac{N}{4\pi} \int_0^{\Lambda^2} \frac{dk^2}{k^2 + m_{\mathrm{R}}^2}$$

$$= \frac{N}{4\pi} \ln \frac{\Lambda^2}{m_{\mathrm{R}}^2}. \tag{10.118}$$

The exponentiation results in Eq. (10.117).

Equation (10.118) relates the bare coupling g^2 and the cutoff Λ and allows us to calculate the Gell-Mann–Low function, yielding

$$\mathcal{B}(g^2) \equiv \Lambda \frac{dg^2}{d\Lambda} = -\frac{Ng^4}{2\pi}. \tag{10.119}$$

The analogous one-loop perturbation-theory formula for any N is given by [Pol75]

$$\mathcal{B}(g^2) = -\frac{(N-2)g^4}{2\pi}. \tag{10.120}$$

Thus, the sigma model is asymptotically free in two dimensions for $N > 2$, which is the origin of the dimensional transmutation. There is no asymptotic freedom for $N = 2$ since $O(2)$ is Abelian.

10.5 Large-N factorization in vector models

The fact that a path integral has a saddle point at large N implies a very important feature of large-N theories – the factorization. It is a general property of the large-N limit and holds not only for the $O(N)$ vector models. However, it is useful to illustrate it by these solvable examples.

The factorization at large N holds for averages of *singlet* operators, for example

$$\langle u(x_1) \cdots u(x_k) \rangle$$
$$\equiv Z^{-1} \int_{\uparrow} \mathcal{D}u \, \exp\left[-\frac{N}{2}\mathrm{Tr}\ln\left(-\partial_\mu^2 + u\right) + \frac{1}{2g^2} \int \mathrm{d}^2 x \, u \right]$$
$$\times \, u(x_1) \cdots u(x_k) \tag{10.121}$$

in the two-dimensional sigma model.

Since the path integral has a saddle point at some configuration $u(x) = u_{\mathrm{sp}}(x)$ (which is, in fact, x-independent owing to translational invariance), we obtain to the leading order in $1/N$:

$$\langle u(x_1) \cdots u(x_k) \rangle = u_{\mathrm{sp}}(x_1) \cdots u_{\mathrm{sp}}(x_k) + \mathcal{O}(N^{-1}), \tag{10.122}$$

which can be written in the factorized form

$$\langle u(x_1) \cdots u(x_k) \rangle = \langle u(x_1) \rangle \cdots \langle u(x_k) \rangle + \mathcal{O}(N^{-1}). \tag{10.123}$$

Therefore, u becomes "classical" as $N \to \infty$ in the sense of the $1/N$-expansion. This is an analog of the WKB-expansion in $\hbar = 1/N$. "Quantum" corrections are suppressed as $1/N$.

We shall return to discussing large-N factorization in the next chapter when considering the large-N limit of QCD.

11

Multicolor QCD

The method of $1/N$-expansion can be applied to QCD. This was done by 't Hooft [Hoo74a] using the inverse number of colors for the gauge group $SU(N)$ as an expansion parameter.

For an $SU(N)$ gauge theory without virtual quark loops, the expansion goes in $1/N^2$ and rearranges diagrams of perturbation theory according to their topology. The leading order in $1/N^2$ is given by planar diagrams, which have the topology of a sphere, while the expansion in $1/N^2$ plays the role of a topological expansion. This is similar to an expansion in the string coupling constant in string models of the strong interaction, which also has a topological character.

Virtual quark loops can be easily incorporated in the $1/N$-expansion. One distinguishes between the 't Hooft limit when the number of quark flavors N_f is fixed as $N \to \infty$ and the Veneziano limit [Ven76] when the ratio N_f/N is fixed as $N \to \infty$. Virtual quark loops are suppressed in the 't Hooft limit as $1/N$ and lead in the Veneziano limit to the same topological expansion as dual-resonance models of strong interaction.

The simplification of QCD in the large-N limit arises from the fact that the number of planar graphs grows with the number of vertices only exponentially rather than factorially as do the total number of graphs. Correlators of gauge-invariant operators factorize in the large-N limit, which looks like the leading-order term of a "semiclassical" WKB-expansion in $1/N$.

We begin this chapter with a description of the double-line representation of diagrams of QCD perturbation theory and rearrange it as the topological expansion in $1/N$. Then we discuss some properties of the $1/N$-expansion for a generic matrix-valued field.

11.1 Index or ribbon graphs

In order to describe the $1/N$-expansion of QCD, the extension of which to N colors has already been considered in Sect. 5.1, it is convenient to use the matrix-field representation (5.5).

In this chapter we shall use a slightly different definition

$$[A_\mu(x)]^{ij} = \sum_a A_\mu^a(x)\,[t^a]^{ij}, \tag{11.1}$$

which is similar to that used by 't Hooft [Hoo74a] and differs from (5.5) by a factor of g:

$$\mathcal{A}_\mu^{ij}(x) = g A_\mu^{ij}(x). \tag{11.2}$$

The matrix (11.1) is Hermitian.

The propagator of the matrix field $A^{ij}(x)$, in this notation, takes the form

$$\left\langle A_\mu^{ij}(x)\,A_\nu^{kl}(y) \right\rangle_{\text{Gauss}} = \left(\delta^{il}\delta^{kj} - \frac{1}{N}\delta^{ij}\delta^{kl} \right) D_{\mu\nu}(x-y), \tag{11.3}$$

where we have assumed, as usual, a gauge-fixing to define the gluon propagator in perturbation theory. For instance, one has

$$D_{\mu\nu}(x-y) = \frac{1}{4\pi^2}\frac{\delta_{\mu\nu}}{(x-y)^2} \tag{11.4}$$

in the Feynman gauge.

Equation (11.3) can be derived immediately from the standard formula

$$\left\langle A_\mu^a(x)\,A_\nu^b(y) \right\rangle_{\text{Gauss}} = \delta^{ab} D_{\mu\nu}(x-y) \tag{11.5}$$

multiplying by the generators of the $SU(N)$ gauge group according to the definition (5.5) and using the completeness condition

$$\sum_{a=1}^{N^2-1} (t^a)^{ij}\,(t^a)^{kl} = \left(\delta^{il}\delta^{kj} - \frac{1}{N}\delta^{ij}\delta^{kl} \right) \qquad \boxed{\text{for } SU(N)}. \tag{11.6}$$

We shall explain in Sect. 13.1 how Eq. (11.3) can be derived directly from a path integral over matrices.

We concentrate in this chapter only on the structure of diagrams in the index space, i.e. the space of the indices associated with the $SU(N)$ group. We shall not consider, in most cases, space-time structures of diagrams which are prescribed by Feynman's rules.

Omitting at large N the second term in parentheses on the RHS of Eq. (11.3), we depict the propagator by the double line

$$\left\langle A_\mu^{ij}(x) A_\nu^{kl}(y) \right\rangle_{\text{Gauss}} \propto \delta^{il}\delta^{kj} = \quad \overset{i}{\underset{j}{\rule{2cm}{0pt}}} \overset{l}{\underset{k}{\rule{2cm}{0pt}}} . \qquad (11.7)$$

Each line, often termed the *index line*, represents the Kronecker delta-symbol and has an orientation which is indicated by arrows. This notation is obviously consistent with the space-time structure of the propagator that describes a propagation from x to y.

The arrows are a result of the fact that the matrix A_μ^{ij} is Hermitian and its off-diagonal components are complex conjugate. The independent fields are, say, the complex fields A_μ^{ij} for $i > j$ and the diagonal real fields A_μ^{ii}. The arrow represents the direction of the propagation of the indices of the complex field A_μ^{ij} for $i > j$, while the complex-conjugate field, $A_\mu^{ji} = (A_\mu^{ij})^*$, propagates in the opposite direction. For the real fields A_μ^{ii}, the arrows are not essential.

The double-line notation (11.7) looks similar to that of Sect. 6.5. The reason for that is deep: double lines appear generically in all models describing *matrix* fields in contrast to *vector* (in internal symmetry space) fields, the propagators of which are depicted by single lines as in the previous chapter.

The three-gluon vertex, which is generated by the action (5.13), is depicted in the double-line notation as

$$\propto g\left(\delta^{i_1 j_3}\delta^{i_2 j_1}\delta^{i_3 j_2} - \delta^{i_1 j_2}\delta^{i_2 j_3}\delta^{i_3 j_1}\right),$$

$$(11.8)$$

where the subscripts 1, 2 or 3 refer to each of the three gluons. The relative minus sign arises from the commutator in the cubic-in-A term in the action (5.13). The color part of the three-vertex is antisymmetric under an interchange of gluons. The space-time structure, which is given in the momentum space as

$$\gamma_{\mu_1\mu_2\mu_3}(p_1, p_2, p_3)$$
$$= \delta_{\mu_1\mu_2}(p_1 - p_2)_{\mu_3} + \delta_{\mu_2\mu_3}(p_2 - p_3)_{\mu_1} + \delta_{\mu_1\mu_3}(p_3 - p_1)_{\mu_2},$$
$$(11.9)$$

is antisymmetric as well. We consider all three gluons as incoming so that their momenta obey $p_1 + p_2 + p_3 = 0$. The full vertex is symmetric as prescribed by Bose statistics.

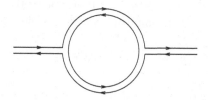

Fig. 11.1. Double-line representation of a one-loop diagram for the gluon propagator. The sum over the N indices is associated with the closed index line. The contribution of this diagram is $\sim g^2 N \sim 1$.

The color structure in Eq. (11.8) can alternatively be obtained by multiplying the standard vertex

$$\Gamma^{a_1 a_2 a_3}_{\mu_1 \mu_2 \mu_3} (p_1, p_2, p_3) = f^{a_1 a_2 a_3} \gamma_{\mu_1 \mu_2 \mu_3} (p_1, p_2, p_3) \qquad (11.10)$$

by $(t^{a_1})^{i_1 j_1} (t^{a_2})^{i_2 j_2} (t^{a_3})^{i_3 j_3}$, with f^{abc} being the structure constants of the $SU(N)$ group, and using the formula

$$f^{a_1 a_2 a_3} (t^{a_1})^{i_1 j_1} (t^{a_2})^{i_2 j_2} (t^{a_3})^{i_3 j_3} = i \left(\delta^{i_1 j_3} \delta^{i_2 j_1} \delta^{i_3 j_2} - \delta^{i_1 j_2} \delta^{i_2 j_3} \delta^{i_3 j_1} \right),$$

$$(11.11)$$

which is a consequence of the completeness condition (11.6).

The four-gluon vertex involves six terms – each of them is depicted by a cross – which differ by interchanging of the color indices. We depict the color structure of the four-gluon vertex for simplicity in the case when $i_1 = j_2 = i$, $i_2 = j_3 = j$, $i_3 = j_4 = k$, $i_4 = j_1 = l$, but i, j, k, l take on different values. Then only the following term is left:

$$\propto g^2, \qquad (11.12)$$

and there are no delta-symbols on the RHS since the color structure is fixed. In other words, we pick up only one color structure by equating indices pairwise.

The diagrams of perturbation theory can now be completely rewritten in the double-line notation [Hoo74a]. The simplest one which describes the one-loop correction to the gluon propagator is depicted in Fig. 11.1. This diagram involves two three-gluon vertices and a sum over the N indices which is associated with the closed index line analogous to Eq. (6.70). Therefore, the contribution of this diagram is $\sim g^2 N$.

In order for the large-N limit to be nontrivial, the bare coupling constant g^2 should satisfy

$$g^2 \sim \frac{1}{N}. \tag{11.13}$$

This dependence on N is similar to Eqs. (10.14) and (10.65) for the vector models and is prescribed by the asymptotic-freedom formula

$$g^2 = \frac{12\pi^2}{11N \ln(\Lambda/\Lambda_{\text{QCD}})} \tag{11.14}$$

of the pure $SU(N)$ gauge theory.

Thus, the contribution of the diagram in Fig. 11.1 is of order

$$\text{Fig. 11.1} \sim g^2 N \sim 1 \tag{11.15}$$

in the large-N limit.

The double lines of the diagram in Fig. 11.1 can be viewed as bounding a piece of a plane. Therefore, these lines represent a two-dimensional object rather than a one-dimensional one as the single lines do in vector models. In mathematics these double-line graphs are often called *ribbon graphs* or *fatgraphs*. In the following we shall see their connection with Riemann surfaces.

Remark on the $U(N)$ gauge group

As was mentioned previously, the second term in the parentheses on the RHS of Eq. (11.6) can be omitted at large N. Such a completeness condition emerges for the $U(N)$ group, the generators of which, T^A $(A = 1, \ldots, N^2)$, are

$$T^A = \left(t^a, \mathbb{I}/\sqrt{N}\right), \qquad \text{tr}\, T^A T^B = \delta^{AB}. \tag{11.16}$$

They obey the completeness condition

$$\sum_{A=1}^{N^2} (T^A)^{ij} (T^A)^{kl} = \delta^{il}\delta^{kj} \qquad \boxed{\text{for } U(N)}. \tag{11.17}$$

The point is that elements of both the $SU(N)$ group and the $U(N)$ group can be represented in the form

$$U = e^{iB}, \tag{11.18}$$

where B is a general Hermitian matrix for $U(N)$ and a traceless Hermitian matrix for $SU(N)$.

Therefore, the double-line representation of the perturbation-theory diagrams which is described in this chapter holds, strictly speaking, only for the $U(N)$ gauge group. However, the large-N limit of both the $U(N)$ group and the $SU(N)$ group is the same.

Fig. 11.2.   Double-line representation of a four-loop diagram for the gluon propagator. The sum over the N indices is associated with each of the four closed
index lines, the number of which is equal to the number of loops. The contribution of this diagram is $\sim g^8 N^4 \sim 1$.

11.2 Planar and nonplanar graphs

The double-line representation of perturbation-theory diagrams in the
index space is very convenient to estimate their orders in $1/N$. Each
three- or four-gluon vertex contributes a factor of g or g^2, respectively.
Each closed index line contributes a factor of N. The order of g in $1/N$
is given by Eq. (11.13).

Let us consider a typical diagram for the gluon propagator depicted in
Fig. 11.2. It has eight three-gluon vertices and four closed index lines,
which coincides with the number of loops. Therefore, the order of this
diagram in $1/N$ is

$$\text{Fig. 11.2} \sim \left(g^2 N\right)^4 \sim 1. \tag{11.19}$$

The diagrams of the type in Fig. 11.2, which can be drawn on a sheet
of paper without crossing any lines, are called *planar* diagrams. For such
diagrams, the addition of a loop inevitably results in the addition of two
three-gluon (or one four-gluon) vertices. A planar diagram with n_2 loops
has n_2 closed index lines. It is of order

$$n_2\text{-loop planar diagram} \sim \left(g^2 N\right)^{n_2} \sim 1, \tag{11.20}$$

so that all planar diagrams survive in the large-N limit.

Let us now consider a *nonplanar* diagram of the type depicted in
Fig. 11.3. This diagram is a three-loop one and has six three-gluon vertices. The crossing of the two lines in the middle does not correspond to
a four-gluon vertex and is merely a result of the fact that the diagram
cannot be drawn on a sheet of paper without crossing the lines. The diagram has only one closed index line. The order of this diagram in $1/N$
is

$$\text{Fig. 11.3} \sim g^6 N \sim \frac{1}{N^2}. \tag{11.21}$$

It is therefore suppressed at large N by $1/N^2$.

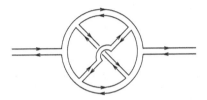

Fig. 11.3. Double-line representation of a three-loop nonplanar diagram for the gluon propagator. The diagram has six three-gluon vertices but only one closed index line (although it has three loops!). The order of this diagram is $\sim g^6 N \sim 1/N^2$.

The nonplanar diagram in Fig. 11.3 can be drawn without line-crossing on a surface with one handle which in mathematics is usually called a torus or a surface of genus one. A plane is then equivalent to a sphere and has genus zero. Adding a handle to a surface produces a hole according to mathematical terminology. A general Riemann surface with h holes has genus h.

The above evaluations of the order of the diagrams in Figs. 11.1–11.3 can be described by the single formula

$$\text{genus-}h \text{ diagram} \;\sim\; \left(\frac{1}{N^2}\right)^{\text{genus}}. \qquad (11.22)$$

Thus, the expansion in $1/N$ rearranges perturbation-theory diagrams according to their topology [Hoo74a]. For this reason, it is referred to as the *topological expansion* or the *genus expansion*. The general proof of Eq. (11.22) for an arbitrary diagram is given in Sect. 11.4.

Only planar diagrams, which are associated with genus zero, survive in the large-N limit. This class of diagrams is an analog of the bubble graphs in the vector models. However, the problem of summing the planar graphs is much more complicated than that of summing the bubble graphs. Nevertheless, it is simpler than the problem of summing all the graphs, since the number of planar graphs with n_0 vertices grows geometrically at large n_0 [Tut62, KNN77]

$$\#_{\mathrm{p}}(n_0) \;\equiv\; \text{no of planar graphs} \;\sim\; \text{const}^{n_0}, \qquad (11.23)$$

while the total number of graphs grows factorially with n_0. There is no dependence in Eq. (11.23) on the number of external lines of a planar graph which is assumed to be much less than n_0.

It is instructive to see the difference between the planar diagrams and, for instance, the ladder diagrams which describe $e^+ e^-$ elastic scattering in QED. Let the ladder have n rungs. Then there are $n!$ ladder diagrams, but

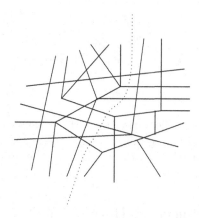

Fig. 11.4. Cutting a planar graph into two graphs. The cutting is along the dotted line. The numbers of vertices of each part and of the whole graph obey Eq. (11.24).

only one of them is planar. This simple example shows why the number of planar graphs is much smaller than the total number of graphs, most of which are nonplanar.

In the rest of this book, we shall discuss what is known concerning solving the problem of summing the planar graphs.

Problem 11.1 Show that Eq. (11.23) for the number of planar graphs is consistent with its independence of the number of external lines.

Solution Let us split a planar graph into two parts by cutting along some line as depicted in Fig. 11.4. The numbers of vertices of each part, n_0' and n_0'', are obviously related to that of the original graph, n_0, by

$$n_0' + n_0'' = n_0. \tag{11.24}$$

We assume that both n_0' and n_0'' are large.

The number of cut lines is $\sim \sqrt{n_0}$ for a planar graph in contrast to that for a generic nonplanar one, when it would be $\sim n_0$. Disregarding the cut lines, we obtain

$$\#_{\mathrm{p}}(n_0) = \#_{\mathrm{p}}(n_0') \cdot \#_{\mathrm{p}}(n_0''), \tag{11.25}$$

which is obviously satisfied by the formula (11.23) accounting for Eq. (11.24).

Problem 11.2 Cutting all loops of a planar graph, obtain the upper bound

$$\#_{\mathrm{p}} \leq (1024)^{n_2} \tag{11.26}$$

for the number of planar graphs with n_2 loops.

Solution Since $\#_{\mathrm{p}}$ does not depend on the number of external lines (see Problem 11.1), let us consider a one-particle irreducible planar graph with one external line and cut all the loops as depicted in Fig. 11.5a. By a continuous

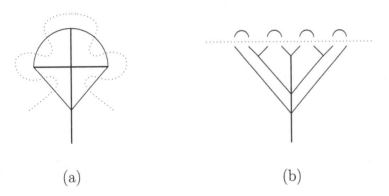

(a) (b)

Fig. 11.5. Cutting a planar graph into trees and arches. The line of cutting is depicted in (a) by a dotted line. The combination of tree and arches in (b) is obtained from (a) by a continuous distortion.

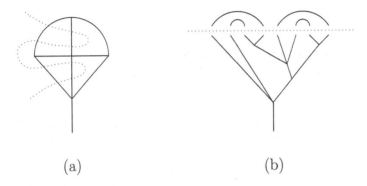

(a) (b)

Fig. 11.6. Alternative cutting of the same planar graph as in Fig. 11.5 into trees and arches.

distortion, it can be depicted as in Fig. 11.5b, where below the dotted line we have a tree with n_0 vertices and above the dotted line we have n_2 arches. The latter number coincides with the number of loops of the planar graph. The number of tips of the tree is $2n_2$.

Since each planar graph can be cut in several ways, $\#_p$ is bounded from above by

$$\#_p \leq \#_A(n_2)\,\#_T(n_0, 2n_2)\,, \tag{11.27}$$

where $\#_A(n_2)$ denotes the number of arches and $\#_T(n_0, 2n_2)$ denotes the number of trees with n_0 vertices and $2n_2$ tips. An alternative way of cutting the same planar graph, which leads to a different combination of arches and trees, is depicted in Fig. 11.6.

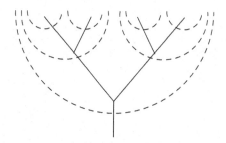

Fig. 11.7. A tree graph (the solid lines) and its dual (the dashed arches).

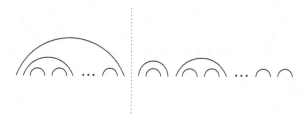

Fig. 11.8.   Recurrence relation for the number of arches. The dotted line separates a configuration of n arches into two pieces: n' to the left and $n - n'$ to the right.

The number of arches is well-known in mathematics and is given by the Catalan number of order n:

$$\#_{\mathrm{A}}(n) \;=\; \frac{2n!}{n!(n+1)!} \;\overset{n\to\infty}{\longrightarrow}\; 4^n \,. \tag{11.28}$$

The number of trees is not independent since a graph, dual to a tree graph, consists of arches as is illustrated by Fig. 11.7. The number of arches of this dual graph equals the sum of the number n_0 of vertices and the number $2n_2$ of tips, i.e. equals $n_0 + 2n_2$. Given the number n_2 of loops, the number n_0 of vertices is maximal when all vertices are trivalent, so that

$$n_0 \;\leq\; 2n_2 - 1 \tag{11.29}$$

($n_0 = 2n_2$ for trivalent and $n_0 = n_2$ for fourvalent vertices when n_2 is large). Therefore, the number of arches of the dual graph is bounded by $4n_2$, so that

$$\#_{\mathrm{T}}(n_0, 2n_2) \;\leq\; \#_{\mathrm{A}}(4n_2)\,. \tag{11.30}$$

Substituting in (11.27), we obtain [KNN77] the inequality (11.26).
Finally, Eq. (11.23) can be obtained by noting that $n_0 \sim n_2$ for large n_2.

Problem 11.3 Derive Eq. (11.28) for the number of arches.

Solution Let us consider a general configuration of n arches as depicted in Fig. 11.8. Let us pick up the leftmost arch, splitting the configuration into two

Fig. 11.9. Recurrence relation for the number of trees. The trees inside the left and right dotted circles have n' and $n - n'$ tips, respectively.

pieces: n' arches to the left and $n - n'$ arches to the right of the dotted line. The number of arches obviously satisfies the recurrence relation

$$\#_{\mathrm{A}}(n) \; = \; \sum_{n'=1}^{n} \#_{\mathrm{A}}(n' - 1)\, \#_{\mathrm{A}}(n - n'), \tag{11.31}$$

where the number of arches to the left of the dotted line is described by $\#_{\mathrm{A}}(n' - 1)$ because one arch encircles $n' - 1$ others. Equation (11.31) expresses $\#_{\mathrm{A}}(n)$ recurrently via $\#_{\mathrm{A}}(0) = 1$.

Introducing the generating function

$$f_{\mathrm{A}}(g) \; = \; \sum_{n=0}^{\infty} g^{n+1} \#_{\mathrm{A}}(n), \tag{11.32}$$

we rewrite Eq. (11.31) as the quadratic equation

$$f_{\mathrm{A}}(g) - g \; = \; f_{\mathrm{A}}^2(g). \tag{11.33}$$

Its solution

$$f_{\mathrm{A}}(g) \; = \; \frac{1 - \sqrt{1 - 4g}}{2} = \sum_{n=0}^{\infty} g^{n+1} \frac{(2n)!}{n!(n + 1)!} \tag{11.34}$$

gives Eq. (11.28) for the number of arches.

Problem 11.4 Improve the inequality (11.26), calculating the number of trivalent tree graphs with n tips.

Solution Let us first note that the number of vertices of a trivalent tree graph with n tips equals $n - 1$. Hence, we are interested in

$$\#_{\mathrm{T}}(n) \; \equiv \; \#_{\mathrm{T}}(n - 1, n) \tag{11.35}$$

in the notation of Problem 11.2. Picking up the first vertex in a tree as depicted in Fig. 11.9, we obtain the following recursion relation for the number of trivalent tree graphs:

$$\#_{\mathrm{T}}(n) \;=\; \sum_{n'=1}^{n-1} \#_{\mathrm{T}}(n')\,\#_{\mathrm{T}}(n-n')\,, \tag{11.36}$$

which expresses $\#_{\mathrm{T}}(n)$ via $\#_{\mathrm{T}}(1) = 1$.

Introducing the generating function

$$f_{\mathrm{T}}(g) \;=\; \sum_{n=1}^{\infty} g^{n-1}\#_{\mathrm{T}}(n)\,, \tag{11.37}$$

where g^{n-1} corresponds to $n-1$ vertices of each tree, we rewrite Eq. (11.36) as the quadratic equation

$$f_{\mathrm{T}}(g) - 1 \;=\; g f_{\mathrm{T}}^2(g)\,. \tag{11.38}$$

Its solution

$$f_{\mathrm{T}}(g) \;=\; \frac{1 - \sqrt{1-4g}}{2g} \;=\; \sum_{n=1}^{\infty} g^{n-1}\frac{(2n-2)!}{n!\,(n-1)!} \tag{11.39}$$

gives

$$\#_{\mathrm{T}}(n) \;=\; \frac{(2n-2)!}{n!(n-1)!}\,. \tag{11.40}$$

Returning to Problem 11.2, it is shown that

$$\#_{\mathrm{T}}(n_0, 2n_2) \;\leq\; \#_{\mathrm{T}}(2n_2 - 1, 2n_2) \;=\; \#_{\mathrm{A}}(2n_2 - 1)\,. \tag{11.41}$$

The inequality here is a result of (11.29) and the equality is because of the explicit formulas (11.28) and (11.40). Thus we have improved the estimate (11.30) having calculated the number of tree graphs. The inequality (11.26) is now improved as

$$\#_{\mathrm{p}} \;\leq\; (64)^{n_2}\,. \tag{11.42}$$

The actual number of planar graphs was first evaluated by Tutte [Tut62]. In Sect. 13.2 we shall obtain the estimate

$$\#_{\mathrm{p}} \;\approx\; \left(12\sqrt{3}\right)^{n_2} \tag{11.43}$$

for the number of trivalent planar graphs at asymptotically large n_2.

11.3 Planar and nonplanar graphs (the boundaries)

Equation (11.22) holds, strictly speaking, only for the gluon propagator, while the contribution of all planar diagrams to a connected n-point Green function is $\sim g^{n-2}$, which is its natural order in $1/N$. The three-gluon Green function is $\sim g$, the four-gluon one is $\sim g^2$ and so on. In order to make contributions of all planar diagrams to be of the same order ~ 1 in the large-N limit, independently of the number of external lines, it is convenient to contract the Kronecker delta-symbols associated with external lines.

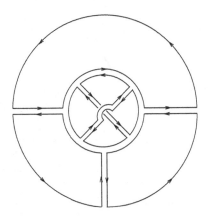

Fig. 11.10.   Generic index diagram with $n_0 = 10$ vertices, $n_1 = 10$ gluon prop-agators, $n_2 = 4$ closed index lines, and $B = 1$ boundary. The color indices of the external lines are contracted by the Kronecker delta-symbols (represented by the single lines) in a cyclic order. The extra factor of $1/N$ arises from the normalization (11.44). Its order in $1/N$ is $\sim 1/N^2$ in accord with Eq. (11.22).

Let us do this in a cyclic order as depicted in Fig. 11.10 for a generic con-nected diagram with three external gluon lines. The extra delta-symbols, which are added to contract the color indices, are depicted by the single lines. They can be viewed as a *boundary* of the given diagram. The actual size of the boundary is not essential – it can be shrunk to a point. Then a bounded piece of a plane will be topologically equivalent to a sphere with a puncture. I shall prefer to draw planar diagrams in a plane with an extended boundary (boundaries) rather than in a sphere with a puncture (punctures).

It is clear from the graphical representation that the diagram in Fig. 11.10 is associated with the trace over the color indices of the three-point Green function

$$G^{(3)}_{\mu_1\mu_2\mu_3}(x_1, x_2, x_3) \;\equiv\; \frac{g^3}{N} \left\langle \operatorname{tr}\left[A_{\mu_1}(x_1)\, A_{\mu_2}(x_2)\, A_{\mu_3}(x_3) \right] \right\rangle. \qquad (11.44)$$

Here we have introduced the factor of g^3/N to make $G^{(3)}$ of $\mathcal{O}(1)$ in the large-N limit. Therefore, the contribution of the diagram in Fig. 11.10 having one boundary should be divided by N, while the factor of g^3 is naturally associated with three extra vertices which appear after the contraction of color indices.

The extension of Eq. (11.44) to multipoint Green functions is obvious:

$$G^{(n)}_{\mu_1\cdots\mu_n}(x_1, \ldots, x_n) \;\equiv\; \frac{g^n}{N} \left\langle \operatorname{tr}\left[A_{\mu_1}(x_1)\cdots A_{\mu_n}(x_n) \right] \right\rangle. \qquad (11.45)$$

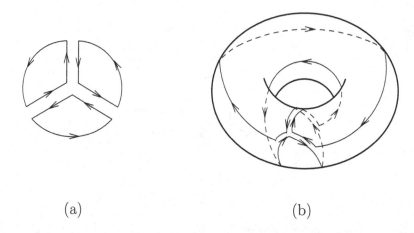

(a)   (b)

Fig. 11.11.   Planar (a) and nonplanar (b) contributions of the two color structures in Eq. (11.8) for three-gluon vertex to $G^{(3)}$ in the lowest order of perturbation theory.

The factor of $1/N$, which normalizes the trace, provides the natural normalization $G^{(0)} = 1$ of the averages.

Though the two terms in the index-space representation (11.8) of the
three-gluon vertex look very similar, their fate in the topological expansion is quite different. When the color indices are contracted anticlockwise, the first term leads to the planar contributions to $G^{(3)}$, the simplest
of which is depicted in Fig. 11.11a. The anticlockwise contraction of the
color indices in the second term leads to a nonplanar graph in Fig. 11.11b
which can be drawn without a crossing of lines only on a torus. Therefore,
the two color structures of the three-gluon vertex contribute to different
orders of the topological expansion. The same is true for the four-gluon
vertex.

Remark on oriented Riemann surfaces

Each line of an index graph of the type depicted in Fig. 11.10 is oriented.
This orientation continues along a closed index line, while the pairs of
index lines of each double line have opposite orientations. The overall
orientation of the lines is prescribed by the orientation of the external
boundary which we choose to be, say, anticlockwise. Since the lines are
oriented, the faces of the Riemann surface associated with a given graph
are also oriented – all in the same way – anticlockwise. Vice versa, such an
orientation of the Riemann surfaces unambiguously fixes the orientation
of all the index lines. This is the reason why we shall often omit the

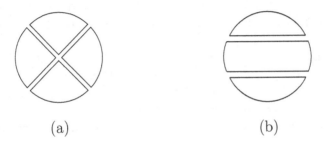

(a)   (b)

Fig. 11.12. Example of (a) connected and (b) disconnected planar graphs.

arrows associated with the orientation of the index lines: their directions are obvious.

Remark on cyclic-ordered Green functions

The cyclic-ordered Green functions (11.45) arise naturally in the expansion of the trace of the non-Abelian phase factor for a closed contour, which was considered in Problem 5.2 on p. 89. One obtains

$$\left\langle \frac{1}{N} \operatorname{tr} \boldsymbol{P} \, e^{ig \oint_\Gamma dx^\mu A_\mu(x)} \right\rangle$$

$$= \sum_{n=0}^{\infty} i^n \oint_\Gamma dx_1^{\mu_1} \int_{x_1}^{x_1} dx_2^{\mu_2} \cdots \int_{x_1}^{x_{n-1}} dx_n^{\mu_n} \, G^{(n)}_{\mu_1 \cdots \mu_n}(x_1, \ldots, x_n) \, .$$

$$(11.46)$$

The reason for this is that the ordering along a closed path implies cyclic-ordering in the index space.

Remark on generating functionals for planar graphs

By connected or disconnected planar graphs we mean, respectively, the graphs which were connected or disconnected before the contraction of the color indices as illustrated by Fig. 11.12. The graph in Fig. 11.12a is connected-planar, while the graph in Fig. 11.12b is disconnected-planar.

The usual relation (2.52) between the generating functionals $W[J]$ and $Z[J]$ for connected graphs and all graphs, which is discussed in the Remark on p. 44, does not hold for the planar graphs. The reason for this is that exponentiation of such a connected planar diagram for the cyclic-ordered Green functions (11.45) can give disconnected nonplanar diagrams.

The generating functionals for all and connected planar graphs can be constructed [Cvi81] by means of introducing noncommutative sources $j_\mu(x)$. "Noncommutative" means that there is no way to transform $j_{\mu_1}(x_1)\,j_{\mu_2}(x_2)$ into $j_{\mu_2}(x_2)\,j_{\mu_1}(x_1)$. This noncommutativity of the sources reflects the cyclic-ordered structure of the Green functions (11.45) which possess only cyclic symmetry.

Using the shorthand notation (10.62) where the symbol $\circ$ includes the sum over the d-vector (or whatever is available) indices except for the color ones:

$$j \circ \mathcal{A} \equiv \sum_\mu \int d^d x\, j_\mu(x)\, \mathcal{A}_\mu(x)\,, \tag{11.47}$$

we write down the definitions of the generating functionals for all planar and connected planar graphs, respectively, as

$$Z[j] \equiv \sum_{n=0}^{\infty} i^n \left\langle \frac{1}{N} \operatorname{tr} (j \circ \mathcal{A})^n \right\rangle \tag{11.48}$$

and

$$W[j] \equiv \sum_{n=0}^{\infty} i^n \left\langle \frac{1}{N} \operatorname{tr} (j \circ \mathcal{A})^n \right\rangle_{\text{conn}}. \tag{11.49}$$

The planar contribution to the Green functions (11.45) and their connected counterparts can be obtained, respectively, from the generating functionals $Z[j]$ and $W[j]$ by applying the noncommutative derivative which is defined by

$$\frac{\delta}{\delta j_\mu(x)} j_\nu(y) f(j) = \delta_{\mu\nu}\, \delta^{(d)}(x-y)\, f(j)\,, \tag{11.50}$$

where f is an arbitrary function of j_μ. In other words, the derivative picks up only the leftmost variable.

The relation which replaces Eq. (2.52) for planar graphs is

$$Z[j] = W[j Z[j]]\,, \tag{11.51}$$

while the cyclic symmetry gives

$$W[j Z[j]] = W[Z[j]\,j]\,. \tag{11.52}$$

A graphical derivation of Eqs. (11.51) and (11.52) is given in Fig. 11.13. In other words, given $W[j]$, one should construct an inverse function as the solution to the equation

$$j_\mu(x) = J_\mu(x)\, W[j]\,, \tag{11.53}$$

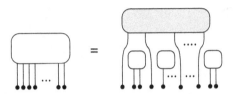

Fig. 11.13. Graphical derivation of Eq. (11.51): $Z[j]$ is denoted by an empty box, $W[j]$ is denoted by a shaded box, j is denoted by a filled circle. By picking the leftmost external line of a planar graph, we end up with a connected planar graph, whose remaining external lines are somewhere to the right interspersed by disconnected planar graphs. It is evident that $jZ[j]$ plays the role of a new source for the connected planar graph. If we instead pick up the rightmost external line, we obtain the inverse order $Z[j]\,j$, which results in Eq. (11.52).

after which Eq. (11.51) gives

$$Z[j] \;=\; W[J]\,. \tag{11.54}$$

More concerning this approach to the generating functionals for planar graphs can be found in [CLS82].

Problem 11.5 Solve Eq. (11.51) iteratively for the Gaussian case.

Solution In the Gaussian case, only $G^{(2)}$ is nonvanishing which yields

$$W[j] \;=\; 1 - g^2 j \circ D \circ j\,, \tag{11.55}$$

where the propagator D is given by Eq. (11.4). Using Eq. (11.51), we find explicitly

$$Z[j] \;=\; 1 - g^2 \int \mathrm{d}^d x\, \mathrm{d}^d y\, D_{\mu\nu}(x-y)\, j_\mu(x)\, Z[j]\, j_\nu(y)\, Z[j]\,. \tag{11.56}$$

While this equation for $Z[j]$ is quadratic, its solution can be written only as a continued fraction owing to the noncommutative nature of the variables. In order to find it, we rewrite Eq. (11.56) as

$$Z[j] \;=\; \frac{1}{1 + g^2 \displaystyle\int \mathrm{d}^d x\, \mathrm{d}^d y\, D_{\mu\nu}(x-y)\, j_\mu(x)\, Z[j]\, j_\nu(y)}\,, \tag{11.57}$$

the iterative solution of which is given by [Cvi81]

$$Z[j] \;=\; \cfrac{1}{1 + g^2 j\,\cfrac{\circ D \circ}{1 + g^2 j\,\cfrac{\circ D \circ}{1 + g^2 j\,\cfrac{\circ D \circ}{\vdots}\,j}\,j}\,j}\,. \tag{11.58}$$

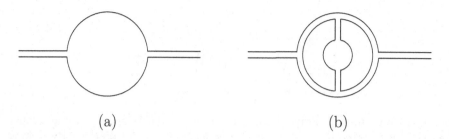

(a)   (b)

Fig. 11.14.   Diagrams for the gluon propagator with a quark loop which is represented by the single lines. Diagram (a) involves one quark loop and has no closed index lines so that its order is $\sim g^2 \sim 1/N$. Diagram (b) involves three loops, one of which is a quark loop. Its order is $\sim g^6 N^2 \sim 1/N$.

11.4 Topological expansion and quark loops

It is easy to incorporate quarks in the topological expansion. A quark field belongs to the fundamental representation of the gauge group $SU(N)$ and its propagator is represented by a single line

$$\langle \psi_i \bar{\psi}_j \rangle \ \propto \ \delta_{ij} \ = \ i \longrightarrow j \, . \tag{11.59}$$

The arrow indicates, as usual, the direction of propagation of a (complex) field ψ. We shall omit these arrows for simplicity.

The diagram for the gluon propagator which involves one quark loop is depicted in Fig. 11.14a. It has two three-gluon vertices and no closed index lines so that its order in $1/N$ is

$$\text{Fig. 11.14a} \ \sim \ g^2 \ \sim \ \frac{1}{N} \, . \tag{11.60}$$

Analogously, the order of a more complicated tree-loop diagram in Fig. 11.14b, which involves one quark loop and two closed index lines, is

$$\text{Fig. 11.14b} \ \sim \ g^6 N^2 \ \sim \ \frac{1}{N} \, . \tag{11.61}$$

It is evident from this consideration that quark loops are not accompanied by closed index lines. One should add a closed index line for each quark loop in order for a given diagram with L quark loops to have the same double-line representation as for pure gluon diagrams. Therefore, given Eq. (11.22), diagrams with L quark loops are suppressed at large N by

$$L \text{ quark loops} \ \sim \ \left(\frac{1}{N} \right)^{L + 2 \cdot \text{genus}} \, . \tag{11.62}$$

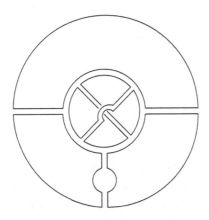

Fig. 11.15. Generic diagram in the index space which has $L = 1$ quark loop and $B = 1$ loop associated with the external boundary. Its order in $1/N$ is described by Eq. (11.70).

The single-line representation of the quark loops is similar to that of the external boundary in Fig. 11.10. Moreover, such a diagram emerges when one calculates perturbative gluon corrections to the vacuum expectation value of the quark operator

$$O = \frac{1}{N}\bar{\psi}\psi, \qquad (11.63)$$

where the factor of $1/N$ is introduced to make it $\mathcal{O}(1)$ in the large-N limit. Therefore, the external boundary can be viewed as a single line associated with valence quarks. The difference between virtual quark loops and external boundaries is that each of the latter has a factor of $1/N$ owing to the definitions (11.45) and (11.63).

In order to prove Eqs. (11.22) and its quark counterpart (11.62), let us consider a generic diagram in the index space which has $n_0^{(3)}$ three-point vertices (either three-gluon or quark–gluon ones), $n_0^{(4)}$ four-gluon vertices, n_1 propagators (either gluon or quark ones), n_2 closed index lines, L virtual quark loops and B external boundaries. A typical such diagram is depicted in Fig. 11.15. Its order in $1/N$ is

$$\frac{1}{N^B} g^{n_0^{(3)}+2n_0^{(4)}} N^{n_2} \sim N^{n_2 - n_0^{(3)}/2 - n_0^{(4)} - B} \qquad (11.64)$$

as has already been explained. The extra factor of $1/N^B$ arises from the extra normalization factor of $1/N$ in operators associated with external boundaries.

The number of propagators and vertices are related by

$$2n_1 = 3n_0^{(3)} + 4n_0^{(4)}, \qquad (11.65)$$

since three- and four-point vertices emit three or four propagators, respectively, and each propagator connects two vertices. Using the relation (11.65), we rewrite the RHS of (11.64) as

$$N^{n_2-n_0^{(3)}/2-n_0^{(4)}-B} = N^{n_2-n_1+n_0-B}, \tag{11.66}$$

where the total number of vertices

$$n_0 = n_0^{(3)} + n_0^{(4)} \tag{11.67}$$

is introduced.

The exponent on the RHS of Eq. (11.66) can be expressed via the Euler characteristic χ of a given graph of genus h. Let us first mention that an appropriate Riemann surface, which is associated with a given graph, is open and has $B+L$ boundaries (represented by single lines). This surface can be closed by attaching a cap to each boundary. The single lines then become double-lines together with the lines of the boundary of each cap. We have already considered this procedure when deducing Eq. (11.62) from Eq. (11.22).

The number of faces for a closed Riemann surface constructed in such a manner is $n_2 + L + B$, while the number of edges and vertices are n_1 and n_0, respectively. Euler's theorem states that

$$\chi \equiv 2 - 2h = n_2 + L + B - n_1 + n_0. \tag{11.68}$$

Therefore the RHS of Eq. (11.66) can be rewritten as

$$N^{n_2-n_1+n_0-B} = N^{2-2h-L-2B}. \tag{11.69}$$

We have thus proven that the order in $1/N$ of a generic graph does not depend on its order in the coupling constant and is completely expressed via the genus h and the number of virtual quark loops L and external boundaries B by

$$\text{generic graph} \sim \left(\frac{1}{N}\right)^{2h+L+2(B-1)}. \tag{11.70}$$

For $B = 1$, we recover Eqs. (11.22) and (11.62).

Remark on the order of gauge action

We see from Eq. (11.45) that the natural variables for the large-N limit are the calligraphic matrices $\mathcal{A}_\mu$ which include the extra factor of g with respect to A_μ (see Eq. (11.2)). For these matrices

$$\frac{1}{N} \langle \operatorname{tr} [\mathcal{A}_{\mu_1}(x_1) \cdots \mathcal{A}_{\mu_n}(x_n)] \rangle = G^{(n)}_{\mu_1\cdots\mu_n}(x_1,\ldots,x_n) \tag{11.71}$$

so that they are $\mathcal{O}(1)$ in the large-N limit since the trace is $\mathcal{O}(N)$.

In these variables, the gluon part of the QCD action (5.13) takes the simple form

$$S \;=\; \frac{1}{4g^2} \int \mathrm{d}^d x \, \mathrm{tr}\, \mathcal{F}_{\mu\nu}^2(x)\,. \tag{11.72}$$

Since g^2 in this formula is $\sim 1/N$ and the trace is $\sim N$, the action is $\mathcal{O}(N^2)$ at large N.

This result can be anticipated from the free theory because the kinetic part of the action involves the sum over N^2-1 free gluons. Therefore, the non-Abelian field strength (3.62) is ~ 1 for $g^2 \sim 1/N$.

The fact that the action is $\mathcal{O}(N^2)$ in the large-N limit is a generic property of the models describing matrix fields. It will be crucial for developing saddle-point approaches at large N which are considered below.

Problem 11.6 Rederive the formula (11.70) using the calligraphic notation.

Solution The propagator of the $\mathcal{A}$-field is $\sim g^2$, while both three- and four-gluon vertices are now $\sim g^{-2}$ as a consequence of Eq. (11.72). The contribution of a generic graph is now of the order

$$\left(g^2\right)^{n_1-n_0} N^{n_2-B} \;\sim\; N^{n_2-n_1+n_0-B} \tag{11.73}$$

for $g^2 \sim 1/N$. This coincides with the RHS of Eq. (11.66) which results in Eq. (11.70).

11.5 't Hooft versus Veneziano limits

In QCD there are several species or flavors of quarks (u-, d-, s- and so on). We denote the number of flavors by N_f and associate a Greek letter α or β with a flavor index of the quark field.

The quark propagator then has the Kronecker delta-symbol with respect to the flavor indices in addition to Eq. (11.59):

$$\left\langle \psi_i^\alpha \bar\psi_j^\beta \right\rangle \;\propto\; \delta^{\alpha\beta}\delta_{ij}\,. \tag{11.74}$$

Their contraction results in

$$\sum_{\alpha=1}^{N_\mathrm{f}} \delta_{\alpha\alpha} \;=\; N_\mathrm{f}\,. \tag{11.75}$$

Therefore, an extra factor of N_f corresponds to each closed quark loop for the N_f flavors.

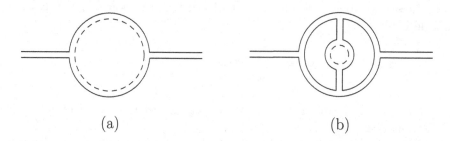

(a)   (b)

Fig. 11.16.   Diagrams with quark loops in the Veneziano limit. Color and flavor indices of a quark loop are represented by the solid and dashed single lines, respectively.   Diagram (a) is $\sim g^2 N_\mathrm{f} \sim N_\mathrm{f}/N$. Diagram (b) is $\sim g^6 N^2 N_\mathrm{f} \sim N_\mathrm{f}/N$.

The limit when N_f is fixed as $N \to \infty$, as was considered in the original paper by 't Hooft [Hoo74a], is called the *'t Hooft limit*. Only valence quarks are left in the 't Hooft limit. Hence, it is associated with the quenched approximation which was discussed in the Remark on p. 158. In order for a meson to decay into other mesons built out of quarks, say for a ρ-meson to decay into a pair of π-mesons, a quark–antiquark pair must be produced out of the vacuum. Consequently, the ratios of meson widths to their masses are

$$\frac{\Gamma_\text{total}}{M} \sim \frac{N_\mathrm{f}}{N} \tag{11.76}$$

in the 't Hooft limit. The ratio on the LHS of Eq. (11.76) is 10–15% experimentally for the ρ-meson. The hope of solving QCD in the 't Hooft limit is the hope to describe QCD with this accuracy.

An alternative large-N limit of QCD when $N_\mathrm{f} \sim N$ as $N \to \infty$ was proposed by Veneziano [Ven76]. Some diagrams for the gluon propagator, which involve one quark loop, are depicted in Fig. 11.16. The dashed single line represents propagation of the flavor index. Each closed loop of the dashed line is associated with the factor of N_f according to Eq. (11.75). This is analogous to the vector models which exactly describe the $O(N_\mathrm{f})$ flavor symmetry in this notation.

The diagrams in Fig. 11.16 contribute, respectively,

$$\text{Fig. 11.16a} \sim g^2 N_\mathrm{f} \sim \frac{N_\mathrm{f}}{N} \tag{11.77}$$

and

$$\text{Fig. 11.16b} \sim g^6 N^2 N_\mathrm{f} \sim \frac{N_\mathrm{f}}{N} \tag{11.78}$$

in the limit (11.13).

Likewise, a more general diagram with L quark loops will contribute

$$L \text{ quark loops } \sim \left(\frac{N_f}{N}\right)^L \left(\frac{1}{N^2}\right)^{\text{genus}}. \tag{11.79}$$

This formula obviously follows from Eq. (11.62) since each quark loop results in N_f.

We see from Eq. (11.79) that quark loops are not suppressed at large N in the Veneziano limit

$$N_f \sim N \to \infty \tag{11.80}$$

if the diagram is planar. Furthermore, the representation of a flavored quark by one solid and one dashed line is obviously similar to the double-line representation of a gluon. All that is said above concerning the topological expansion of pure gluodynamics holds for QCD with quarks in the Veneziano limit.

It is the Veneziano limit (11.80) that is related to the hadronic topological expansion in the dual-resonance models. In the Veneziano limit hadrons can have finite widths according to Eq. (11.76). I refer the reader to the original paper by Veneziano [Ven76] for further details.

There is an alternative way to show why virtual quarks are suppressed in the 't Hooft limit and survive in the Veneziano limit. Let us integrate over the quark fields which yields

$$\int \mathcal{D}\bar{\psi}\,\mathcal{D}\psi\, e^{-\int d^4x \left(\bar{\psi}\widehat{\nabla}\psi + m\bar{\psi}\psi\right)} = e^{\mathrm{Tr}\ln(\widehat{\nabla}+m)} \tag{11.81}$$

as is discussed in Sect. 2.2. The trace in the exponent involves summation both over color and flavor indices, so that

$$\mathrm{Tr}\ln\left(\widehat{\nabla}+m\right) \sim N N_f. \tag{11.82}$$

The order in N of the pure gluon action is $\mathcal{O}(N^2)$ as was discussed in the Remark on p. 232. Hence, the quark contribution to the action is $\sim N_f/N$ in comparison with the gluon one. The quark determinant can be disregarded in the 't Hooft limit, but is essential in the Veneziano limit.

The consideration of the previous paragraph also explains why each quark loop contributes a factor of $\sim N_f/N$. The exponent on the RHS of Eq. (11.81) is associated with one-loop diagrams. A diagram with L quark loops corresponds to the Lth term of the expansion of the exponential. This explains the factor of $(N_f/N)^L$ in Eq. (11.79). A diagram with two quark loops, which appears in the second order of this expansion, is depicted in Fig. 11.17.

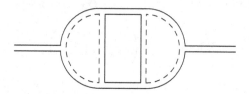

Fig. 11.17.   Diagram with two quark loops in the Veneziano limit. The diagram
is $\sim g^6 N N_{\rm f}^2 \sim (N_{\rm f}/N)^2$.

Remark on asymptotic freedom in the Veneziano limit

Though the number of flavors becomes large in the Veneziano limit, this
does not mean that asymptotic freedom is lost. The leading-order coefficient of the $\mathcal{B}$-function of QCD with N colors and $N_{\rm f}$ flavors is given
by

$$b \;=\; \frac{1}{4\pi^2}\left(-\frac{11}{3}N + \frac{2}{3}N_{\rm f}\right) \tag{11.83}$$

which reproduces Eq. (9.71) for $N = 3$. It is still negative if $N_{\rm f}/N < 11/2$
in the Veneziano limit.

Remark on phenomenology of multicolor QCD

While $N = 3$ in the real world, there are phenomenological indications
that $1/N$ may be considered as a small parameter. We have already
mentioned some of them in the text – the simplest one is that the ratio of
the ρ-meson width to its mass, which is $\sim 1/N$, is small. Considering $1/N$
as a small parameter immediately leads to qualitative phenomenological
consequences which are preserved by the planar diagrams associated with
multicolor QCD, but are violated by the nonplanar diagrams.

The most important consequence is the relation of the $1/N$-expansion to
the topological expansion in the dual-resonance model of hadrons. Vast
numbers of properties of hadrons are explained by the dual-resonance
model. A very clear physical picture behind this model is that hadrons
are excitations of a string with quarks at the ends.

I shall briefly list some consequences of multicolor QCD.

(1) The "naive" quark model of hadrons emerges at $N = \infty$. Hadrons
    are built out of a pair of (valence or constituent) quark and anti-
    quark $q\bar{q}$, while exotic states like $qq\bar{q}\bar{q}$ do not appear.

(2) The partial width of decay of the ϕ-meson, which is built out of $s\bar{s}$
    (the strange quark and antiquark), into K^+K^- is $\sim 1/N$, while that

into $\pi^+\pi^-\pi^0$ is $\sim 1/N^2$. This explains Zweig's rule. The masses of the ρ- and ω-mesons are degenerate at $N = \infty$.

(3) The coupling constant of the meson–meson interaction is small at large N.

(4) The widths of glueballs are $\sim 1/N^2$, i.e. they should be even narrower than mesons built out of quarks. The glueballs do not interact or mix with mesons at $N = \infty$.

All of these hadron properties (except the last one) agree approximately with experiment, and were well-known even before 1974 when multicolor QCD was introduced. Glueballs have not yet been detected experimentally (possibly because of their property listed in item (4)).

11.6 Large-N factorization

The vacuum expectation values of several colorless or white operators, which are singlets with respect to the gauge group, factorize in the large-N limit of QCD (or other matrix models). This property is similar to that already discussed in Sect. 10.5 for the vector models.

The simplest gauge-invariant operator in a pure $SU(N)$ gauge theory is the square of the non-Abelian field strength:

$$O(x) \;=\; \frac{1}{N^2}\,\mathrm{tr}\,F_{\mu\nu}^2(x)\,. \tag{11.84}$$

The normalizing factor provides the natural normalization

$$\left\langle \frac{1}{N^2}\,\mathrm{tr}\,F_{\mu\nu}^2(x) \right\rangle \;=\; \left\langle \frac{1}{N^2}\,F_{\mu\nu}^a(x)\,F_{\mu\nu}^a(x) \right\rangle \;\sim\; 1\,. \tag{11.85}$$

In order to verify the factorization in the large-N limit, let us consider the index-space diagrams for the average of two colorless operators $O(x_1)$ and $O(x_2)$, which are depicted in Fig. 11.18.

The graph in Fig. 11.18a represents the zeroth order of perturbation theory. It involves four closed index lines (the factor of N^4) and the normalization factor of $1/N^4$ according to the definition (11.84). Its contribution is

$$\text{Fig. 11.18a} \;\sim\; \frac{1}{N^2}N^2 \cdot \frac{1}{N^2}N^2 \;\sim\; 1\,, \tag{11.86}$$

i.e. $\mathcal{O}(1)$ in accord with the general estimate (11.85).

The graph in Fig. 11.18b involves a gluon line which is emitted and absorbed by the same operator $O(x_1)$. It has five closed index lines (the

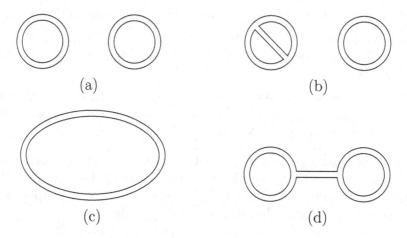

Fig. 11.18. Demonstration of the large-N factorization to the lowest orders of perturbation theory. The closed double line represents the average of the operator (11.84) to the zeroth order in g. Diagrams (a) and (b), which are associated with the factorized part of the average on the LHS of Eq. (11.90), are $\mathcal{O}(1)$. Diagrams (c) and (d), which would violate the factorization, are suppressed by $1/N^2$.

factor of N^5), the normalization factor of $1/N^4$, and g^2 owing to two three-gluon vertices. Its contribution is

$$\text{Fig. 11.18b} \sim g^2 N \sim 1, \tag{11.87}$$

i.e. also $\mathcal{O}(1)$ in the limit (11.13).

The graph in Fig. 11.18c is of the same type as the graph in Fig. 11.18a, but the double lines now connect two different operators. It has two closed index lines (the factor of N^2) and the normalization factor of $1/N^4$, so that its contribution

$$\text{Fig. 11.18c} \sim \frac{1}{N^2} \tag{11.88}$$

is suppressed by $1/N^2$.

The graph in Fig. 11.18d, which is of the same order in the coupling constant as the graph in Fig. 11.18b, involves only three closed index lines (the factor of N^3) and is of order $1/N^2$:

$$\text{Fig. 11.18d} \sim g^2 \frac{1}{N} \sim \frac{1}{N^2}. \tag{11.89}$$

Therefore, it is suppressed by $1/N^2$ in the large-N limit. For this graph, the gluon line is emitted and absorbed by different operators $O(x_1)$ and $O(x_2)$.

This lowest-order example illustrates the general property that only (planar) diagrams with gluon lines emitted and absorbed by the same operators survive as $N \to \infty$. Hence, correlations between the colorless operators $O(x_1)$ and $O(x_2)$ are of order $1/N^2$, so that the *factorization* property holds as $N \to \infty$:

$$
\left\langle \frac{1}{N^2} \operatorname{tr} F^2(x_1) \frac{1}{N^2} \operatorname{tr} F^2(x_2) \right\rangle
$$

$$
= \left\langle \frac{1}{N^2} \operatorname{tr} F^2(x_1) \right\rangle \left\langle \frac{1}{N^2} \operatorname{tr} F^2(x_2) \right\rangle + \mathcal{O}(N^{-2}) . \tag{11.90}
$$

For a general set of gauge-invariant operators $O_1, \ldots, O_n$, the factorization property can be represented by

$$
\langle O_1 \cdots O_n \rangle = \langle O_1 \rangle \cdots \langle O_n \rangle + \mathcal{O}(N^{-2}) . \tag{11.91}
$$

This is analogous to Eq. (10.123) for the vector models.

The factorization in large-N QCD was first discovered by A.A. Migdal in the late 1970s. The important observation that the factorization implies a semiclassical nature of the large-N limit of QCD was made by Witten [Wit79]. We shall discuss this in the next two sections.

The factorization property also holds for gauge-invariant operators constructed from quarks as in Eq. (11.63). For the case of several flavors N_f, we normalize these quark operators by

$$
O_\Gamma = \frac{1}{N_\mathrm{f} N} \bar{\psi} \Gamma \psi . \tag{11.92}
$$

Here Γ denotes one of the combination of the γ-matrices:

$$
\Gamma = \mathbb{I}, \ \gamma_5, \ \gamma_\mu, \ i\gamma_\mu\gamma_5, \ \Sigma_{\mu\nu} = \frac{1}{2i} [\gamma_\mu, \gamma_\nu] , \ldots . \tag{11.93}
$$

The lowest-order diagrams of perturbation theory for the average of two quark operators (11.92) are depicted in Fig. 11.19. The estimation of their order in $1/N$ is analogous to that for the pure gluon graphs in Fig. 11.18.

The graph in Fig. 11.19a represents the zeroth order of perturbation theory for the average of two quark operators. It involves two closed color and two closed flavor index lines (the factor of $N_\mathrm{f}^2 N^2$) and the normalization factor of $1/(N_\mathrm{f} N)^2$ according to the definition (11.92). Its contribution is

$$
\text{Fig. 11.19a} \sim \frac{1}{N_\mathrm{f}^2 N^2} N_\mathrm{f}^2 N^2 \sim 1 . \tag{11.94}
$$

This justifies the normalization factor in Eq. (11.92).

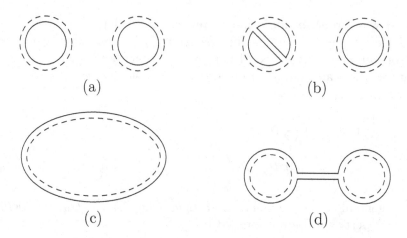

Fig. 11.19.   Same as in Fig. 11.18 but for quark operators (11.92). The solid and dashed lines are associated with color and flavor indices, respectively. Diagrams (a) and (b), which contribute to the factorized part of the average on the LHS of Eq. (11.98), are $\mathcal{O}(1)$. Diagrams (c) and (d), which would violate the factorization, are suppressed by $1/(N_f N)$ and $1/N^2$, respectively.

The graph in Fig. 11.19b involves a gluon line which is emitted and absorbed by the same quark operator. It has three closed color and two closed flavor index lines (the factor of $N_f^2 N^3$), the normalization factor of $1/(N_f N)^2$, and g^2 arising from two quark–gluon vertices. Its contribution is

$$\text{Fig. 11.19b} \sim \frac{1}{N_f^2 N^2} N_f^2 g^2 N^3 \sim g^2 N \sim 1 \qquad (11.95)$$

in full analogy with the pure gluon diagram in Fig. 11.18b.

The graph in Fig. 11.19c is similar to the graph in Fig. 11.18c – the lines connect two different quark operators. It has one closed color and one closed flavor index lines (the factor of $N_f N$) and the normalization factor of $1/(N_f N)^2$, so that its contribution

$$\text{Fig. 11.19c} \sim \frac{1}{N_f N} \qquad (11.96)$$

is suppressed by $1/(N_f N)$.

Finally, the graph in Fig. 11.19d involves a gluon line which is emitted by one quark operator and absorbed by the other. It has one closed color and two closed flavor index lines (factor of $N_f^2 N$), the normalization factor of $1/(N_f N)^2$, and g^2 owing to two quark–gluon vertices. Its contribution

$$\text{Fig. 11.19d} \sim \frac{1}{N_f^2 N^2} N_f^2 g^2 N \sim \frac{g^2}{N} \sim \frac{1}{N^2} \qquad (11.97)$$

is suppressed by $1/N^2$ in the limit (11.13).

We see that the factorization of the gauge-invariant quark operators holds both in the 't Hooft and Veneziano limits:

$$\langle O_{\Gamma_1} \cdots O_{\Gamma_n} \rangle = \langle O_{\Gamma_1} \rangle \cdots \langle O_{\Gamma_n} \rangle + \mathcal{O}(1/(N_{\mathrm{f}}N)). \qquad (11.98)$$

The nonfactorized part, which is associated with connected diagrams, is $\sim 1/N$ in the 't Hooft limit. This leads, in particular, to the coupling constant of meson–meson interaction of order $1/N$, clarifying the property of multicolor QCD listed in item (3) on p. 237. The Veneziano limit is analogous to pure gluodynamics as has already been mentioned.

It is worth noting that the factorization can be seen alternatively (at all orders of perturbation theory) from Eq. (11.70) for the contribution of a generic connected graph of genus h with B external boundaries which are precisely associated with the quark operators O_{Γ}, as is explained in Sect. 11.4. The diagrams with gluon lines emitted and absorbed by the same operator as in Fig. 11.19b are products of diagrams having only one boundary. Hence, their contribution is of order one. Otherwise, the diagrams with gluon lines emitted and absorbed by two different operators as in Fig. 11.19d have two boundaries. According to Eq. (11.70), their contribution is suppressed by $1/N^2$. Alternatively, the diagrams as in Fig. 11.19c (including its planar dressing by gluons) have one boundary.[*] Their contribution is $\mathcal{O}(1)$ times $1/(N_{\mathrm{f}}N)$ coming from the normalization of the operator (11.92). This proves the factorization property (11.98) at all orders of perturbation theory.

Remark on factorization beyond perturbation theory

The large-N factorization can also be verified at all orders of the strong-coupling expansion in the $SU(N)$ lattice gauge theory. A nonperturbative proof of the factorization will be given in the next chapter using quantum equations of motion (the loop equations).

Problem 11.7 Prove the factorization of the Wilson loop operators within the strong-coupling expansion of the $SU(N)$ lattice gauge theory as $N \to \infty$.

Solution Let us first estimate the order in N of the Wilson loop average (6.42). The explicit result to the leading order in β is given by Eqs. (6.73) and (6.72), where

$$\beta \sim N^2 \qquad (11.99)$$

in the limit (11.13) as prescribed by Eq. (6.32). Therefore, $W(C) \sim 1$ in the large-N limit.

---

[*] In the dual-resonance model, they are associated with the meson–meson interaction arising from an exchange of constituent quarks.

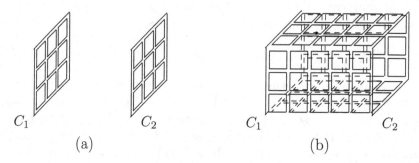

Fig. 11.20. Factorization of the Wilson loop operators in the strong-coupling expansion as $N \to \infty$. The surfaces are constructed from plaquettes which come from the expansion of the exponential of the lattice action. Each link in the surface is passed at least twice: otherwise the result vanishes. Diagram (a) involves two separate surfaces enclosed by Wilson loops. It contributes to the factorized part of the average on the LHS of Eq. (11.101). The surface in diagram (b) connects two different Wilson loops and would violate factorization. It has two boundaries and its contribution is suppressed by $1/N^2$ according to the general formula (11.100).

To be precise, we first perform the strong-coupling expansion in β and then set $N \to \infty$ in each term of the strong coupling expansion. As we shall see in a moment, the actual parameter is β/N^2, so that the limits $\beta \to 0$ and $N \to \infty$ are interchangeable.

It is easy to estimate the order in N of any graph of the strong coupling expansion for $W(C)$, which looks like that in Fig. 6.8 on p. 116. Let the plaquettes fill an arbitrary surface enclosed by the loop C, with n_2, n_1, and n_0 being the number of plaquettes, links, and sites which belong to the surface. Each plaquette contributes β/N since it comes from the expansion of the exponential of the lattice action, each link contributes $1/N$ owing to Eq. (6.60), and each site contributes N since it is associated with summing over the color indices owing to Eq. (6.70). Accounting for the normalization factor of $1/N^B$, where $B = 1$ is the number of boundaries, the contribution is of order

$$
\left(\frac{\beta}{N}\right)^{n_2} N^{-n_1+n_0-B} \sim \left(\frac{\beta}{N^2}\right)^{n_2} N^{n_2-n_1+n_0-B} \sim \left(\frac{\beta}{N^2}\right)^{n_2} \left(\frac{1}{N}\right)^{2h+2(B-1)},
$$

$$(11.100)$$

where we have used Euler's theorem (11.68). In the limit (11.99), the contribution does not depend on the order of the strong-coupling expansion and is completely determined by the number B of boundaries and the genus h of the surface. This is analogous to the perturbation theory. For the minimal surface, we reproduce previous results.

We are now in a position to analyze the order in N of different terms in the strong-coupling expansion of the average of two Wilson loop operators. The factorized part results from the surfaces of the type depicted in Fig. 11.20a, which are spanned by each individual loop. Its contribution is $\mathcal{O}(1)$ as $N \to \infty$. A nonfactorized part emerges from surfaces of the type depicted in Fig. 11.20b, which connect two different Wilson loops. They look like a cylinder and have two boundaries. Their contribution is suppressed by $1/N^2$ according to the general formula (11.100).

Thus, we have proven the factorization property

$$\left\langle \frac{1}{N} \mathrm{tr}\, U(C_1) \frac{1}{N} \mathrm{tr}\, U(C_2) \right\rangle = \left\langle \frac{1}{N} \mathrm{tr}\, U(C_1) \right\rangle \left\langle \frac{1}{N} \mathrm{tr}\, U(C_2) \right\rangle + \mathcal{O}(N^{-2})$$

(11.101)

at all orders of the strong-coupling expansion.

Problem 11.8 Find the relation between the Wilson loop averages in the fundamental and adjoint representations for an $SU(N)$ pure gauge theory at large N.

Solution The characters in the fundamental and adjoint representations are related by Eq. (6.28). Using the factorization formula (11.101) with coinciding contours C_1 and C_2, we obtain

$$W_{\mathrm{adj}}(C) = [W_{\mathrm{fun}}(C)]^2 + \mathcal{O}(N^{-2}).$$

(11.102)

As was discussed in Part 2, the Wilson loop average in the fundamental representation obeys the area law (6.75). The same is true at $N = \infty$ for the Wilson loop average in the adjoint representation owing to Eq. (11.102). In particular, the string tensions in the fundamental and adjoint representations at $N = \infty$ are related by

$$K_{\mathrm{adj}} = 2K_{\mathrm{fun}}.$$

(11.103)

On the other hand, the adjoint test quark can be screened at finite N by a gluon produced out of the vacuum. This is similar to the breaking of the flux tube in the fundamental representation by a quark–antiquark pair, which is discussed in Sect. 9.5. Therefore, the perimeter law (6.79) must dominate for large contours. The point is that the perimeter law appears owing to connected diagrams which are suppressed as $1/N^2$:

$$W_{\mathrm{adj}}(C) \xrightarrow{\text{large } C} \mathrm{e}^{-2KA_{\min}(C)} + \frac{1}{N^2} \mathrm{e}^{-\mu \cdot L(C)}.$$

(11.104)

These properties of the adjoint representation were first pointed out in [KM81].

11.7 The master field

The large-N factorization in QCD assumes that gauge-invariant objects behave as c-numbers, rather than as operators. Likewise for vector models, this suggests that the path integral is dominated by a saddle point.

We have already seen in Sect. 10.5 that the factorization in the vector models does not mean that the fundamental field itself, for instance $\vec{n}$ in the sigma-model, becomes "classical". It is the case, instead, for a singlet composite field.

We are now going to apply a similar idea to the Yang–Mills theory, the partition function of which is

$$Z = \int \mathcal{D}A_\mu^a \, e^{-\int d^4x \, \frac{1}{4} F_{\mu\nu}^a F_{\mu\nu}^a} .$$ (11.105)

The action, $\sim N^2$, is large as $N \to \infty$, but the entropy is also $\sim N^2$ as a result of the $N^2 - 1$ integrations over A_μ^a:

$$\mathcal{D}A_\mu^a \sim e^{N^2} .$$ (11.106)

Consequently, the saddle-point equation of the large-N Yang–Mills theory is *not* the classical one which is given by[*]

$$\frac{\delta S}{\delta A_\nu^a} = -(\nabla_\mu F_{\mu\nu})^a = 0 .$$ (11.107)

The idea is to rewrite the path integral over A_μ for the Yang–Mills theory as that over a colorless composite field $\Phi[A]$, likewise this was done in Sect. 10.4 for the sigma-model. The expected new path-integral representation of the partition function (11.105) would be something like

$$Z \propto \int \mathcal{D}\Phi \, \frac{1}{\left| \frac{\partial \Phi[A]}{\partial A_\mu^a} \right|} \, e^{-N^2 S[\Phi]} .$$ (11.108)

The Jacobian

$$\left| \frac{\partial \Phi[A]}{\partial A_\mu^a} \right| \equiv e^{-N^2 J[\Phi]}$$ (11.109)

in Eq. (11.108) is related to the old entropy factor, so that $J[\Phi] \sim 1$ in the large-N limit.

The original partition function (11.105) can be then rewritten as

$$Z \propto \int \mathcal{D}\Phi \, e^{N^2 J[\Phi] - N^2 S[\Phi]} ,$$ (11.110)

where $S[\Phi]$ represents the Yang–Mills action in the new variables. The new entropy factor of $\mathcal{D}\Phi$ is $\mathcal{O}(1)$ because the variable $\Phi[A]$ is a color singlet. The large parameter N enters Eq. (11.110) only in the exponent. Therefore, the saddle-point equation can be immediately written as

$$\frac{\delta S}{\delta \Phi} = \frac{\delta J}{\delta \Phi} .$$ (11.111)

---

[*] It was already discussed in Problem 5.1 on p. 87.

Remembering that Φ is a functional of A_μ, $\Phi \equiv \Phi[A]$, we rewrite the saddle-point equation (11.111) as

$$\frac{\delta S}{\delta A_\nu^a} = -(\nabla_\mu F_{\mu\nu})^a = \frac{\delta J}{\delta A_\nu^a}. \tag{11.112}$$

It differs from the classical Yang–Mills equation (11.107) by the term on the RHS coming from the Jacobian (11.109).

Given $J[\Phi]$, which depends on the precise from of the variable $\Phi[A]$, Eq. (11.112) has a solution

$$A_\mu(x) = A_\mu^{\rm cl}(x). \tag{11.113}$$

Let us first assume that there exists only one solution to Eq. (11.112). Then the path integral is saturated by a single configuration (11.113), so that the vacuum expectation values of gauge-invariant operators are given by their values at this configuration:

$$\langle O \rangle = O\left(A_\mu^{\rm cl}(x)\right). \tag{11.114}$$

The factorization property (11.91) will obviously be satisfied.

The existence of such a classical field configuration in multicolor QCD was conjectured by Witten [Wit79]. It was discussed in the lectures by Coleman [Col79] who called it the *master field*. Equation (11.112) which determines the master field is often referred to as the master-field equation.

A subtle point with the master field is that a solution to Eq. (11.112) is determined only up to a gauge transformation. To preserve gauge invariance, it is more reasonable to speak about the whole gauge orbit as a solution of Eq. (11.112). However, this will not change Eq. (11.114) since the operator O is gauge invariant.

The conjecture concerning the existence of the master field has surprisingly rich consequences. Since vacuum expectation values are Poincaré invariant, the RHS of Eq. (11.114) is also. This implies that $A_\mu^{\rm cl}(x)$ must itself be Poincaré invariant up to a gauge transformation: a change of $A_\mu^{\rm cl}(x)$ under translations or rotations can be compensated by a gauge transformation. Moreover, there must exist a gauge in which $A_\mu^{\rm cl}(x)$ is space-time-independent: $A_\mu^{\rm cl}(x) = A_\mu^{\rm cl}(0)$. In this gauge, rotations must be equivalent to a global gauge transformation, so that $A_\mu^{\rm cl}(0)$ transforms as a Lorentz vector.

In fact, the idea concerning such a master field in multicolor QCD may not be correct as was pointed out by Haan [Haa81]. The conjecture concerning the existence of only one solution to the master-field equation (11.112) seems to be too strong. If several solutions exist, one needs an additional averaging over these solutions. This is a very delicate matter, since this additional averaging must still preserve the factorization property. One might be better to think about this situation as if $A^{\text{cl}}_\mu(0)$ were an operator in some Hilbert space rather than a c-valued function. This is simply because $A^{\text{cl}}_\mu(0)$ is, in the matrix notation (11.1), an $N \times N$ matrix which becomes, as $N \to \infty$, an infinite matrix, or an operator in Hilbert space. Such an operator-valued master field is sometimes called the master field in the *weak* sense, while the above conjecture concerning a single classical configuration of the gauge field, which saturates the path integral, is called the master field in the *strong* sense.

The concept of the master field is rather vague until a precise form of the composite field $\Phi[A]$, and consequently the Jacobian $\Phi[A]$ that enters Eq. (11.112), is defined. However, what is important is that the master field (in the weak sense) is space-time-independent. This looks like a simplification of the problem of solving large-N QCD. A Hilbert space, in which the operator $A^{\text{cl}}_\mu(0)$ acts, should be specified by $\Phi[A]$. In the next section we shall consider a realization of these ideas for the case of $\Phi[A]$ given by the trace of the non-Abelian phase factor for closed contours.

Remark on noncommutative probability theory

An adequate mathematical language for describing the master field in multicolor QCD (and, generically, in matrix models at large N) was found by I. Singer in 1994. It is based on the concept of free random variables of noncommutative probability theory, introduced by Voiculescu [VDN95]. How to describe the master field in this language and some other applications of noncommutative free random variables to the problems of planar quantum field theory are discussed in [Dou95, GG95].

11.8 $1/N$ as semiclassical expansion

A natural candidate for the composite operator $\Phi[A]$ from the previous section is given by the trace of the non-Abelian phase factor for closed contours – the Wilson loop. It is labeled by the loop C in the same sense as the field $A_\mu(x)$ is labeled by the point x, so we shall use the notation

$$\Phi(C) \;\equiv\; \Phi[A] \;=\; \frac{1}{N} \operatorname{tr} \boldsymbol{P} \, e^{ig \oint_C dx^\mu A_\mu(x)}. \qquad (11.115)$$

Nobody up to now has managed to reformulate QCD at finite N in terms of $\Phi(C)$ in the language of the path integral. This is due to the fact that self-intersecting loops are not independent (they are related by the so-called Mandelstam relations [Man79]),[*] and the Jacobian is huge.

---

[*] See, for example, Appendix C of the review [Mig83].

The reformulation was performed [MM79] in the language of Schwinger–Dyson or loop equations which will be described in the next chapter.

Schwinger–Dyson equations are a convenient way of performing the semiclassical expansion, which is an alternative to the path integral. Let us illustrate an idea of how to do this by an example of the φ^3 theory, the Schwinger–Dyson equations of which are given by Eq. (2.47).

The RHS of Eq. (2.47) is proportional to Planck's constant $\hbar$ as is explained in Sect. 2.5. In the semiclassical limit $\hbar \to 0$, we obtain

$$\left(-\partial_1^2 + m^2\right) \langle \varphi(x_1) \cdots \varphi(x_n) \rangle + \frac{\lambda}{2} \langle \varphi^2(x_1) \cdots \varphi(x_n) \rangle = 0, \qquad (11.116)$$

the solution of which is of the factorized form

$$\langle \varphi(x_1) \cdots \varphi(x_n) \rangle = \langle \varphi(x_1) \rangle \cdots \langle \varphi(x_n) \rangle + \mathcal{O}(\hbar) \qquad (11.117)$$

provided that

$$\langle \varphi(x) \rangle \equiv \varphi_{\rm cl}(x) \qquad (11.118)$$

obeys

$$\left(-\partial^2 + m^2\right) \varphi_{\rm cl}(x) + \frac{\lambda}{2} \varphi_{\rm cl}^2(x) = 0. \qquad (11.119)$$

Equation (11.119) is nothing but the classical equation of motion for the φ^3 theory, which specifies extrema of the action (2.22) entering the path integral (2.2). Thus, we have reproduced, using the Schwinger–Dyson equations, the well-known fact that the path integral is dominated by a classical trajectory as $\hbar \to 0$. It is also clear how to perform the semiclassical expansion in $\hbar$ in the language of the Schwinger–Dyson equations: one should solve Eq. (2.47) by iteration.

The reformulation of multicolor QCD in terms of the loop functionals $\Phi(C)$ is, in a sense, a realization of the idea of the master field in the weak sense, when the master field acts as an operator in the space of loops. The loop equation of the next chapter will be a sort of master-field equation in the loop space.

Remark on the large-N limit as statistical averaging

There is yet another, purely statistical, explanation why the large-N limit is a "semiclassical" limit for the collective variables $\Phi(C)$. The matrix $U^{ij}[C_{xx}]$, that describes the parallel transport along a closed contour C_{xx}, can be reduced by the unitary transformation to

$$U[C_{xx}] = \Omega[C_{xx}] \, {\rm diag} \left(e^{ig\alpha_1(C)}, \ldots, e^{ig\alpha_N(C)} \right) \Omega^{\dagger}[C_{xx}]. \qquad (11.120)$$

Then $\Phi(C)$ is given by

$$\Phi(C) = \frac{1}{N} \sum_{j=1}^{N} e^{ig\alpha_j(C)}. \tag{11.121}$$

The phases $\alpha_j(C)$ are gauge invariant modulo permutations and normal-ized so that $\alpha_j(C) \sim 1$ as $N \to \infty$. For simplicity we omit below all the indices (including space ones) except for color.

The commutator of Φs can be estimated using the representation (11.121). Since

$$\left[\alpha_i(C),\, \alpha_j(C')\right] \propto \delta_{ij}, \tag{11.122}$$

one obtains

$$\left[\Phi(C),\, \Phi(C')\right] \sim g^2 \frac{1}{N} \sim \frac{1}{N^2} \tag{11.123}$$

in the limit (11.13), i.e. the commutator can be neglected as $N \to \infty$, and the field $\Phi(C)$ becomes classical.

Note that the commutator (11.123) is of order $1/N^2$. One factor of $1/N$ is because of g in the definition (11.121) of $\Phi(C)$, while the other has a deeper reason. Let us image the summation over j in Eq. (11.121) as some statistical averaging. It is well-known in statistics that such averages fluctuate weakly as $N \to \infty$, so that the dispersion is of order $1/N$. It is this factor that emerges in the commutator (11.123).

The factorization is valid only for the gauge-invariant quantities which involve the averaging over the color indices, such as that in Eq. (11.121). There is no reason to expect factorization for gauge invariants which do not involve this averaging and therefore fluctuate strongly even at $N = \infty$. An explicit example of such strongly fluctuating gauge-invariant quanti-ties was first constructed in [Haa81].

This Remark may be summarized to give that the factorization arises from the additional statistical averaging in the large-N limit. There is no reason to assume the existence of a master field in the strong sense in order to explain the factorization.

12

QCD in loop space

QCD can be entirely reformulated in terms of the colorless composite field $\Phi(C)$ – the trace of the Wilson loop for closed contours. This fact involves two main steps:

(1) all of the observables are expressed via $\Phi(C)$;

(2) the dynamics is entirely reformulated in terms of $\Phi(C)$.

This approach is especially useful in the large-N limit where everything is expressed via the vacuum expectation value of $\Phi(C)$ – the Wilson loop average. Observables are given by summing the Wilson loop average over paths with the same weight as in free theory. The Wilson loop average itself obeys a close functional equation – the loop equation.

We begin this chapter by presenting the formulas which relate observables to Wilson loops. Then we translate the quantum equation of motion of Yang–Mills theory into loop space. We derive the closed equation for the Wilson loop average as $N \to \infty$ and discuss its various properties, including a nonperturbative regularization. Finally, we briefly comment on what is known concerning solutions of the loop equation.

12.1 Observables in terms of Wilson loops

All observables in QCD can be expressed via the Wilson loops $\Phi(C)$ defined by Eq. (11.115). This property was first advocated by Wilson [Wil74] on a lattice. Calculation of QCD observables can be divided into two steps:

(1) calculation of the Wilson loop averages for arbitrary contours;

(2) summation of the Wilson loop averages over the contours with some weight depending on a given observable.

249

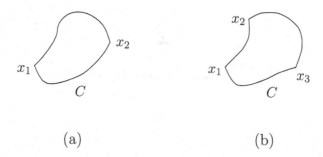

Fig. 12.1.   Contours in the sum over paths representing observables: (a) in Eq. (12.3) and (b) in Eq. (12.4). The contour (a) passes x_1 and x_2. The contour (b) passes x_1, x_2, and x_3.

At finite N, observables are expressed via the n-loop averages

$$W_n(C_1, \ldots, C_n) \;\; = \;\; \langle\, \Phi(C_1) \cdots \Phi(C_n) \,\rangle, \qquad (12.1)$$

which are analogous to the n-point Green functions (2.45). The appropriate formulas for the continuum theory can be found in [MM81].

Great simplifications occur in these formulas at $N = \infty$, when all observables are expressed only via the one-loop average

$$W(C) \;\; = \;\; \langle\, \Phi(C) \,\rangle \;\; \equiv \;\; \left\langle\, \frac{1}{N}\, \mathrm{tr}\, \boldsymbol{P}\, \mathrm{e}^{ig\oint_C \mathrm{d}x^\mu A_\mu} \,\right\rangle. \qquad (12.2)$$

This is associated with the quenched approximation discussed in the Remark on p. 158.

For example, the average of the product of two colorless quark vector currents (11.92) is given at large N by

$$\langle \bar\psi\gamma_\mu\psi(x_1)\, \bar\psi\gamma_\nu\psi(x_2) \rangle \;\; = \;\; \sum_{C \ni x_1, x_2} J_{\mu\nu}(C)\, \langle\, \Phi(C) \,\rangle, \qquad (12.3)$$

where the sum runs over contours C passing through the points x_1 and x_2 as is depicted in Fig. 12.1a. An analogous formula for the (connected) correlators of three quark scalar currents can be written as

$$\langle \bar\psi\psi(x_1)\, \bar\psi\psi(x_2)\, \bar\psi\psi(x_3) \rangle_{\mathrm{conn}} \;\; = \;\; \sum_{C \ni x_1, x_2, x_3} J(C)\, \langle\, \Phi(C) \,\rangle, \qquad (12.4)$$

where the sum runs over contours C passing through the three points x_1, x_2, and x_3 as depicted in Fig. 12.1b. A general (connected) correlator of n quark currents is given by a similar formula with C passing through n points $x_1, \ldots, x_n$ (some of them may coincide).

The weights $J_{\mu\nu}(C)$ in Eq. (12.3) and $J(C)$ in Eq. (12.4) are completely determined by free theory. If quarks were scalars rather than spinors, then we would have

$$J(C) = e^{-\frac{1}{2}m^2\tau - \frac{1}{2}\int_0^\tau dt\, \dot{z}_\mu^2(t)} = e^{-mL(C)} \quad \boxed{\text{scalar quarks}}, \quad (12.5)$$

where $L(C)$ is the length of the (closed) contour C, as was shown in Sect. 1.6. Using the notation (1.156), we can rewrite Eq. (12.4) for scalar quarks as

$$\left\langle \psi^\dagger\psi(x_1)\,\psi^\dagger\psi(x_2)\,\psi^\dagger\psi(x_3) \right\rangle_{\text{conn}} = \sum_{C\ni x_1,x_2,x_3}^{\prime} \left\langle \Phi(C) \right\rangle. \quad (12.6)$$

Therefore, we obtain the sum over paths of the Wilson loop, likewise in Sect. 1.7 and Problem 5.4 on p. 91.

For spinor quarks, an additional disentangling of the γ-matrices is needed. This can be done in terms of a path integral over the momentum variable, with $k_\mu(t)$ $(0 \leq t \leq \tau)$ being an appropriate trajectory. The result is given by [BNZ79]

$$J(C) = \int \mathcal{D}k_\mu(t)\, \text{sp}\, \boldsymbol{P}\, e^{-\int_0^\tau dt\, \{ik_\mu(t)[\dot{x}_\mu(t) - \gamma_\mu(t)] + m\}} \quad (12.7)$$

and

$$J_{\mu\nu}(C) = \int \mathcal{D}k_\mu(t)\, \text{sp}\, \boldsymbol{P}\, \left[\gamma_\mu(t_1)\, \gamma_\nu(t_2)\, e^{-\int_0^\tau dt\, \{ik_\mu(t)[\dot{x}_\mu(t) - \gamma_\mu(t)] + m\}} \right], \quad (12.8)$$

where the values t_1 and t_2 of the parameter t are associated with the points x_1 and x_2 in Eq. (12.3), and the symbol of P-ordering puts the matrices γ_μ and γ_ν at a proper order.

Problem 12.1 Derive Eqs. (12.7) and (12.8).

Solution Since the spinor field ψ enters the QCD action quadratically, it can be integrated out in the correlators (12.3) and (12.4), so that they can be represented, in the first quantized language, via the resolvent of the Dirac operator in the external field $\mathcal{A}_\mu$, with subsequent averaging over $\mathcal{A}_\mu$. Proceeding as in Chapter 1, we express the resolvent by

$$\left\langle y \left| \frac{1}{\widehat{\nabla} + m} \right| x \right\rangle = \int_0^\infty d\tau\, \left\langle y \left| e^{-\tau(\widehat{\nabla} + m)} \right| x \right\rangle \quad (12.9)$$

and represent the matrix element of the exponential of the Dirac operator as

$$\left\langle y \left| e^{-\tau(\widehat{\nabla} + m)} \right| x \right\rangle = e^{-\tau m}\, \boldsymbol{P}\, e^{-\int_0^\tau dt\, \widehat{\nabla}(t)} \delta^{(d)}(x - y). \quad (12.10)$$

In order to disentangle the RHS, we insert unity, represented by

$$1 = \int_{z_\mu(0)=x_\mu} \mathcal{D}z_\mu(t) \int \mathcal{D}p_\mu(t)\, e^{-i\int_0^\tau dt\, p_\mu(t)\dot{z}_\mu(t)}, \tag{12.11}$$

where the path integration over $p_\mu(t)$ is unrestricted, i.e. the integrals over $p_\mu(0)$ and $p_\mu(\tau)$ are included. Then we obtain

$$\left\langle y \left| e^{-\tau(\hat{\nabla}+m)} \right| x \right\rangle = e^{-\tau m} \int_{z_\mu(0)=x_\mu} \mathcal{D}z_\mu(t) \int \mathcal{D}p_\mu(t)$$
$$\times \boldsymbol{P}\, e^{-\int_0^\tau dt\, \{ip_\mu(t)\dot{z}_\mu(t)-[ip_\mu(t)+i\mathcal{A}_\mu(t)]\gamma_\mu(t)+\partial_\mu(t)\dot{z}_\mu(t)\}}\, \delta^{(d)}(x-y), \tag{12.12}$$

the equivalence of which to the original expression is obvious since everything commutes under the sign of the P-ordering (so that we can substitute $p_\mu(t) = -i\partial_\mu(t)$ in the integrand).

By making the change of the integration variable, $p_\mu(t) = k_\mu(t) - \mathcal{A}_\mu(t)$, and proceeding as in Problem 1.13 on p. 29, we represent the RHS of Eq. (12.12) by

$$\left\langle y \left| e^{-\tau(\hat{\nabla}+m)} \right| x \right\rangle = e^{-\tau m} \int_{z_\mu(0)=x_\mu} \mathcal{D}z_\mu(t) \int \mathcal{D}k_\mu(t)$$
$$\times \boldsymbol{P}\, e^{-\int_0^\tau dt\, \{ik_\mu(t)[\dot{z}_\mu(t)-\gamma_\mu(t)]-i\dot{z}_\mu(t)\mathcal{A}_\mu(t)+\partial_\mu(t)\dot{z}_\mu(t)\}}\, \delta^{(d)}(x-y)$$
$$= e^{-\tau m} \int_{\substack{z_\mu(0)=x_\mu \\ z_\mu(\tau)=y_\mu}} \mathcal{D}z_\mu(t) \int \mathcal{D}k_\mu(t)\, \boldsymbol{P}\, e^{i\int_x^y dz_\mu\, \mathcal{A}_\mu(z)}\, \boldsymbol{P}\, e^{-i\int_0^\tau dt\, k_\mu(t)[\dot{z}_\mu(t)-\gamma_\mu(t)]}, \tag{12.13}$$

where the first P-exponential on the RHS depends only on color matrices (it is nothing but the non-Abelian phase factor), and the second one depends only on spinor matrices. In [BNZ79], Eq. (12.13) is derived by discretizing paths.

Equation (12.13) leads to Eqs. (12.7) and (12.8).

Remark on renormalization of Wilson loops

Perturbation theory for $W(C)$ can be obtained by expanding the path-ordered exponential in the definition (12.2) in g (see Eq. (11.46)) and averaging over the gluon field A_μ. Because of ultraviolet divergences, we need a (gauge-invariant) regularization. After such a regularization has been introduced, the Wilson loop average for a smooth contour C of the type in Fig. 12.2a reads as

$$W(C) = \exp\left[-g^2 \frac{(N^2-1)}{4\pi N} \frac{L(C)}{a}\right] W_{\text{ren}}(C), \tag{12.14}$$

where a is the cutoff, $L(C)$ is the length of C, and $W_{\text{ren}}(C)$ is finite when expressed via the renormalized charge g_{R}. The exponential factor is a

(a)   (b)

Fig. 12.2. Examples of (a) a smooth contour and (b) a contour with a cusp. The tangent vector to the contour jumps through an angle γ at the cusp.

result of the renormalization of the mass of a heavy test quark, which was already discussed in the Remark on p. 113. This factor does not emerge in the dimensional regularization where $d = 4 - \varepsilon$. The multiplicative renormalization of the smooth Wilson loop was shown in [GN80, Pol80, DV80].

If the contour C has a cusp (or cusps) but no self-intersections as is illustrated by Fig. 12.2b, then $W(C)$ is still multiplicatively renormalizable [BNS81]:

$$W(C) \;=\; Z(\gamma)\, W_{\text{ren}}(C)\,, \tag{12.15}$$

while the (divergent) factor of $Z(\gamma)$ depends on the cusp angle (or angles) γ (or γs) and $W_{\text{ren}}(C)$ is finite when expressed via the renormalized charge g_{R}.

Problem 12.2 Calculate the divergent parts of the Wilson loop average (12.2) for contours without self-intersections to order g^2. Consider the cases of a smooth contour C and a contour with a cusp.

Solution Expanding the Wilson loop average (12.2) in g^2 (see Eq. (11.46) and Problem 5.2 on p. 89), we obtain

$$W(C) \;=\; 1 + W^{(2)}(C) + \mathcal{O}\big(g^4\big) \tag{12.16}$$

with

$$W^{(2)}(C) \;=\; -g^2 \frac{(N^2 - 1)}{2N} \oint_C dx_\mu \oint_C dy_\nu\, D_{\mu\nu}(x - y)\,, \tag{12.17}$$

where $D_{\mu\nu}(x - y)$ is the gluon propagator (11.4).

Since the contour integral in Eq. (12.17) diverges for $x = y$, we introduce the regularization by

$$D_{\mu\nu}(x - y) \;\overset{\text{reg.}}{\Longrightarrow}\; \frac{1}{4\pi^2} \frac{\delta_{\mu\nu}}{[(x - y)^2 + a^2]} \tag{12.18}$$

with a being the ultraviolet cutoff. Parametrizing the contour C using the function $z_\mu(\sigma)$, we rewrite the contour integral in Eq. (12.17) as

$$\oint_C dx_\mu \oint_C dy_\mu \frac{1}{(x-y)^2 + a^2} = \int ds \int dt \frac{\dot{z}_\mu(s)\dot{z}_\mu(s+t)}{[z(s+t) - z(s)]^2 + a^2}. \qquad (12.19)$$

Choosing the proper-length parametrization (1.101) when $\dot{z}_\mu(s)\ddot{z}_\mu(s) = 0$, expanding in powers of t, and assuming that the contour C is smooth as is depicted in Fig. 12.2a, we obtain for the integral (12.19)

$$\int ds\, \dot{x}^2(s) \int dt \frac{1}{\dot{x}^2(s)t^2 + a^2} = \frac{\pi}{a} \int ds\, \sqrt{\dot{x}^2(s)} = \frac{\pi}{a} L(C). \qquad (12.20)$$

Typical values of t in the last integral are $\sim a$, which justifies the expansion in t: the next terms lead to a finite contribution as $a \to 0$.

Thus, we find

$$W^{(2)}(C) = -g^2 \frac{(N^2 - 1)}{4\pi N} \frac{L(C)}{a} + \text{finite term as } a \to 0 \qquad (12.21)$$

for a smooth contour. This is precisely the renormalization of the mass of a heavy test quark owing to the interaction.

If the contour C is not smooth and has a cusp at some value s_0 of the parameter, as depicted in Fig. 12.2b, then an extra divergent contribution in the integral (12.19) emerges when $s \approx s_0$, $t \approx t_0$. Introducing $\Delta s = s - s_0$ and $\Delta t = t - t_0$, we represent this extra divergent term by

$$\dot{x}_\mu(s_0 + 0)\dot{x}_\mu(s_0 - 0) \int d\Delta s \int d\Delta t \frac{1}{[\dot{x}_\mu(s_0 + 0)\Delta s - \dot{x}_\mu(s_0 - 0)\Delta t]^2 + a^2}$$

$$= (\gamma \cot \gamma - 1) \ln \frac{L(C)}{a}, \qquad (12.22)$$

where γ is the angle of the cusp ($\cos \gamma \equiv \dot{x}_\mu(s_0+0)\dot{x}_\mu(s_0-0)$) and the upper limit of the integrations is chosen to be $L(C)$ with logarithmic accuracy. Collecting all of this together, we obtain finally for the divergent part of $W^{(2)}(C)$:

$$W^{(2)}(C) = -g^2 \frac{(N^2 - 1)}{4\pi N} \left[\frac{L(C)}{a} + \frac{1}{\pi}(\gamma \cot \gamma - 1) \ln \frac{L(C)}{a} \right]$$

$$+ \text{ finite term as } a \to 0. \qquad (12.23)$$

The second term in square brackets is associated with the bremsstrahlung radiation of a particle changing its velocity when passing the cusp. The answers in the Abelian and non-Abelian cases coincide to this order in g^2.

Problem 12.3 Obtain Coulomb's law of interaction in Maxwell's theory by calculating the average of a rectangular Wilson loop.

Solution Performing the Gaussian averaging over A_μ in Maxwell's theory, we obtain from Eqs. (6.50) and (6.51)

$$
\begin{aligned}
-\ln W(C) &= \frac{1}{2} \int \mathrm{d}^4 x \int \mathrm{d}^4 y \, J^\mu(x) \, D_{\mu\nu}(x-y) \, J^\nu(y) \\
&= \frac{e^2}{2} \oint_C \mathrm{d}x^\mu \oint_C \mathrm{d}y^\nu \, D_{\mu\nu}(x-y) .
\end{aligned}
\tag{12.24}
$$

The interaction potential is now determined by Eq. (6.43) for a rectangular contour depicted in Fig. 6.6 on p. 111 as $T \gg R$. The contribution to the interaction potential arises when the photon line is emitted by the upper part of the rectangular contour and absorbed by the lower part. Otherwise, we obtain singular terms associated with the renormalization of the Wilson loop as discussed in the previous Problem.

Choosing the parametrization with $x_\mu = (R, \ldots, s)$ for the upper part and $x_\mu = (0, \ldots, t)$ for the lower part of the rectangular contour with $0 \le s, t \le T$, we have

$$
V(R)\,T = \frac{e^2}{4\pi^2} \int_0^T \mathrm{d}s \int_0^T \mathrm{d}t \, \frac{1}{(s-t)^2 + R^2} .
\tag{12.25}
$$

Introducing $u = (s+t)/2$ and $v = s - t$, we obtain

$$
V(R)\,T = \frac{e^2}{4\pi^2} \int_0^T \mathrm{d}u \int_{-T}^T \mathrm{d}v \, \frac{1}{v^2 + R^2} = \frac{e^2}{4\pi R} T
\tag{12.26}
$$

which reproduces Coulomb's law.

12.2 Schwinger–Dyson equations for Wilson loop

The dynamics of (quantum) Yang–Mills theory is described by the quantum equation of motion

$$
-\nabla_\mu^{ab} F_{\mu\nu}^b(x) \overset{\text{w.s.}}{=} \hbar \frac{\delta}{\delta A_\nu^a(x)}
\tag{12.27}
$$

which is analogous to Eq. (2.27) for the scalar field, and is again understood in the weak sense, i.e. for the averages

$$
-\left\langle \nabla_\mu^{ab} F_{\mu\nu}^b(x) \, Q\,[A] \right\rangle = \hbar \left\langle \frac{\delta}{\delta A_\nu^a(x)} Q\,[A] \right\rangle .
\tag{12.28}
$$

The standard set of Schwinger–Dyson equations of Yang–Mills theory emerges when the functional $Q[A]$ is chosen in the form of the product of A_{μ_i} as in Eq. (11.45).

Strictly speaking, the last statement is incorrect, since in Eqs. (12.27) and (12.28) we have not added contributions coming from the variation of gauge-fixing and ghost terms in the Yang–Mills action. However, these two contributions are mutually canceled for gauge-invariant functionals $Q[A]$. We shall deal only with such gauge-invariant functionals (the Wilson loops). This is why we have not considered the contribution of the gauge-fixing and ghost terms.

It is also convenient to use the matrix notation (5.5), when Eq. (12.27) for the Wilson loop takes the form

$$-\left\langle \frac{1}{N} \operatorname{tr} \boldsymbol{P} \nabla_\mu \mathcal{F}_{\mu\nu}(x) \, e^{i \oint_C d\xi^\mu A_\mu} \right\rangle = \left\langle \frac{g^2}{N} \operatorname{tr} \frac{\delta}{\delta A_\nu(x)} \boldsymbol{P} \, e^{i \oint_C d\xi^\mu A_\mu} \right\rangle,$$

(12.29)

where we have restored the units with $\hbar = 1$.

The variational derivative on the RHS can be calculated by virtue of the formula

$$\frac{\delta A_\mu^{ij}(y)}{\delta A_\nu^{kl}(x)} = \delta_{\mu\nu} \, \delta^{(d)}(x-y) \left(\delta^{il} \delta^{kj} - \frac{1}{N} \delta^{ij} \delta^{kl} \right)$$

(12.30)

which is a consequence of

$$\frac{\delta A_\mu^a(y)}{\delta A_\nu^b(x)} = \delta_{\mu\nu} \, \delta^{(d)}(x-y) \, \delta^{ab}.$$

(12.31)

The second term in the parentheses in Eq. (12.30) – same as in Eq. (11.6) – is because $\mathcal{A}_\mu$ is a matrix from the adjoint representation of $SU(N)$.

By using Eq. (12.30), we obtain for the variational derivative on RHS of Eq. (12.29):

$$\operatorname{tr} \frac{\delta}{\delta A_\nu(x)} \boldsymbol{P} \, e^{i \oint_C d\xi^\mu A_\mu} = i \oint_C dy_\nu \, \delta^{(d)}(x-y)$$

$$\times \left[\frac{1}{N} \operatorname{tr} \boldsymbol{P} \, e^{i \int_{C_{yx}} d\xi^\mu A_\mu} \frac{1}{N} \operatorname{tr} \boldsymbol{P} \, e^{i \int_{C_{xy}} d\xi^\mu A_\mu} - \frac{1}{N^3} \operatorname{tr} \boldsymbol{P} \, e^{i \int_C d\xi^\mu A_\mu} \right].$$

(12.32)

The contours C_{yx} and C_{xy}, which are depicted in Fig. 12.3, are the parts of the loop C: from x to y and from y to x, respectively. They are always closed owing to the presence of the delta-function. It implies that x and y should be the same points of *space* but not necessarily of the *contour* (i.e. they may be associated with different values of the parameter σ).

Finally, we rewrite Eq. (12.29) as

$$i \left\langle \frac{1}{N} \operatorname{tr} \boldsymbol{P} \nabla_\mu \mathcal{F}_{\mu\nu}(x) \, e^{i \oint_C d\xi^\mu A_\mu} \right\rangle$$

$$= \lambda \oint_C dy_\nu \, \delta^{(d)}(x-y) \left[\langle \Phi(C_{yx}) \Phi(C_{xy}) \rangle - \frac{1}{N^2} \langle \Phi(C) \rangle \right],$$

(12.33)

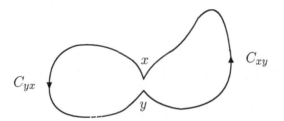

Fig. 12.3.   Contours C_{yx} and C_{xy} which enter the RHSs of Eqs. (12.29) and (12.33).

where we have introduced the *'t Hooft coupling*

$$\lambda = g^2 N. \tag{12.34}$$

Note that the RHS of Eq. (12.33) is completely represented via the (closed) Wilson loops.

Problem 12.4 Prove the cancellation of the contributions of the gauge-fixing and ghost terms in the Lorentz gauge.

Solution   The Yang–Mills action, associated with the Lorentz gauge, is given by

$$S_{\text{gf}} = \frac{1}{g^2} \int \mathrm{d}^d x \left[\frac{1}{4} \operatorname{tr} \mathcal{F}_{\mu\nu}^2 + \frac{1}{2\alpha} \operatorname{tr} (\partial_\mu \mathcal{A}_\mu)^2 \right]. \tag{12.35}$$

Since $\Phi(C)$ is gauge invariant, the (infinite) group-volume factors, in the numerator and denominator in the definition of the average, cancel when fixing the gauge (see the Remark on p. 109), and we obtain

$$W(C) \equiv \frac{\int \mathcal{D}A_\mu \, \mathrm{e}^{-S} \, \Phi(C)}{\int \mathcal{D}A_\mu \, \mathrm{e}^{-S}} = \frac{\int \mathcal{D}A_\mu \, \det(\partial_\mu \nabla_\mu) \, \mathrm{e}^{-S_{\text{gf}}} \, \Phi(C)}{\int \mathcal{D}A_\mu \, \det(\partial_\mu \nabla_\mu) \, \mathrm{e}^{-S_{\text{gf}}}}, \tag{12.36}$$

where $\det(\partial_\mu \nabla_\mu)$ is associated with ghosts.

The Schwinger–Dyson equation for the Yang–Mills theory in the Lorentz gauge is

$$-\nabla_\mu^{ab} F_{\mu\nu}^b(x) \overset{\text{w.s.}}{=} \hbar \frac{\delta}{\delta A_\nu^a(x)} + \frac{1}{\alpha} \partial_\nu \partial_\mu A_\mu^a(x) + \frac{\partial}{\partial x_\nu} f^{abc} G^{bc}(x' = x, x; A), \tag{12.37}$$

where $G^{bc}(x', x; A)$ is the Green function of the ghost in an external field A_μ. Applying this equation to the Wilson loop and using the gauge Ward identity (the Slavnov–Taylor identity), we transform the contribution from the second term on the RHS to

$$\frac{i}{\alpha}\left\langle \frac{1}{N} \operatorname{tr} \partial_\nu \partial_\mu A_\mu U(C_{xx}) \right\rangle_{\text{gf}}$$

$$= g^2 \oint_C d\xi_\mu \left\langle \frac{1}{N} \operatorname{tr} \left[U(C_{\xi x}) t^a U(C_{x\xi}) t^b \right] \frac{\partial}{\partial x_\nu} \nabla_\mu^{bc}(\xi) \, G^{ca}(\xi, x; A) \right\rangle_{\text{gf}}$$

$$= g^2 \oint_C d\xi_\mu \frac{\partial}{\partial \xi_\mu} \left\langle \frac{1}{N} \operatorname{tr} \left[U(C_{\xi x}) t^a U(C_{x\xi}) t^b \right] \frac{\partial}{\partial x_\nu} G^{ba}(\xi, x; A) \right\rangle_{\text{gf}}$$

$$= g^2 \left\langle \frac{1}{N} \operatorname{tr} \left\{ U(C_{xx}) \left[t^a, t^b \right] \right\} \frac{\partial}{\partial x_\nu} G^{ba}(x' = x, x; A) \right\rangle_{\text{gf}} \tag{12.38}$$

which exactly cancels the contribution from the ghost term in Eq. (12.37).

We have thus proven that the contribution of gauge-fixing and ghost terms in Eq. (12.37) are mutually canceled, when applied to the Wilson loop (and, in fact, to any gauge-invariant functional).

12.3 Path and area derivatives

As we already mentioned, the RHS of Eq. (12.33) is completely represented via the (closed) Wilson loops. It is crucial for the loop-space formulation of QCD that the LHS of Eq. (12.33) can also be represented in loop space as some operator applied to the Wilson loop. To do this we need to develop a differential calculus in loop space.

Loop space consists of arbitrary continuous closed loops, C. They can be described in a parametric form by the functions $x_\mu(\sigma) \in L_2$,[*] where $\sigma_0 \leq \sigma \leq \sigma_1$ and $\mu = 1, \ldots, d$, which take on values in a d-dimensional Euclidean space. The functions $x_\mu(\sigma)$ can be discontinuous, generally speaking, for an arbitrary choice of the parameter σ. The continuity of the loop C implies a continuous dependence on parameters of the type of proper length

$$s(\sigma) = \int_{\sigma_0}^{\sigma} d\sigma' \sqrt{\dot{x}_\mu^2(\sigma')}, \tag{12.39}$$

where $\dot{x}_\mu(\sigma) = dx_\mu(\sigma)/d\sigma$.

The functions $x_\mu(\sigma) \in L_2$ which are associated with the elements of loop space obey the following restrictions.

(1) The points $\sigma = \sigma_0$ and $\sigma = \sigma_1$ are identified: $x_\mu(\sigma_0) = x_\mu(\sigma_1)$ – the loops are closed.

---

[*] Let us remind the reader that L_2 denotes the Hilbert space of functions $x_\mu(\sigma)$, the square of which is integrable over the Lebesgue measure: $\int_{\sigma_0}^{\sigma_1} d\sigma \, x_\mu^2(\sigma) < \infty$. We have already mentioned this in the Remark on p. 19.

(2) The functions $x_\mu(\sigma)$ and $\Lambda_{\mu\nu} x_\nu(\sigma) + \alpha_\mu$, with $\Lambda_{\mu\nu}$ and α_μ independent of σ, represent the same element of the loop space – rotational and translational invariance.

(3) The functions $x_\mu(\sigma)$ and $x_\mu(\sigma')$ with $\sigma' = f(\sigma)$, $f'(\sigma) \geq 0$ describe the same loop – reparametrization invariance.

An example of functionals which are defined on the elements of loop space is the Wilson loop average (12.2) or, more generally, the n-loop average (12.1).

The differential calculus in loop space is built out of the path and area derivatives.

The *area derivative* of a functional $\mathcal{F}(C)$ is defined by the difference

$$\frac{\delta \mathcal{F}(C)}{\delta \sigma_{\mu\nu}(x)} \equiv \frac{1}{\delta \sigma_{\mu\nu}} \left[\mathcal{F}\left(\begin{array}{c} \end{array} \right) - \mathcal{F}\left(\begin{array}{c} \end{array} \right) \right],$$

(12.40)

where an infinitesimal loop $\delta C_{\mu\nu}(x)$ is attached to a given loop at the point x in the (μ, ν)-plane and $\delta \sigma_{\mu\nu}$ denotes the area enclosed by $\delta C_{\mu\nu}(x)$. For a rectangular loop $\delta C_{\mu\nu}(x)$, one finds

$$\delta \sigma_{\mu\nu} = dx_\mu \wedge dx_\nu,$$

(12.41)

where the symbol $\wedge$ implies antisymmetrization. The sign of $\delta \sigma_{\mu\nu}$ is determined by the orientation of $\delta C_{\mu\nu}(x)$.

Analogously, the *path derivative* is defined by

$$\partial_\mu^x \mathcal{F}(C_{xx}) \equiv \frac{1}{\delta x_\mu} \left[\mathcal{F}\left(\begin{array}{c} \end{array} \right) - \mathcal{F}\left(\begin{array}{c} \end{array} \right) \right],$$

(12.42)

where the point x is shifted from the loop along an infinitesimal path $\delta \Gamma_\mu$ and δx_μ denotes the length of $\delta \Gamma_\mu$. The sign of δx_μ is determined by the direction of $\delta \Gamma_\mu$.

As is usual in quantum field theory, the typical size of $\delta C_{\mu\nu}$ in the definition of the area derivative as well as the length of $\delta \Gamma_\mu$ in the definition of the path derivative should be smaller than the size of an ultraviolet cutoff.

These two differential operations are well-defined for so-called functionals of the Stokes type which satisfy the backtracking condition – they do

not change when a small path passing back and forth is added to the loop
at some point x:

$$\mathcal{F}\left(\!\!\!\begin{array}{c}\includegraphics\end{array}\!\!\!x\right) = \mathcal{F}\left(\!\!\!\begin{array}{c}\includegraphics\end{array}\!\!\!\right). \tag{12.43}$$

This condition is equivalent to the Bianchi identity of Yang–Mills theory
and is obviously satisfied by the Wilson loop (12.2) owing to the properties
of the non-Abelian phase factor (see Eq. (5.47)). Such functionals are
known in mathematics as Chen integrals.*

A simple example of the Stokes functional is the area of the minimal
surface, $A_{\min}(C)$. It obviously satisfies Eq. (12.43). Otherwise, the length
$L(C)$ of the loop C is not a Stokes functional, since the lengths of contours
on the LHS and RHS of Eq. (12.43) are different.

For the Stokes functionals, the variation on the RHS of Eq. (12.40) is
proportional to the area enclosed by the infinitesimally small loop $\delta C_{\mu\nu}(x)$
and does not depend on its shape. Analogously, the variation on the RHS
of Eq. (12.42) is proportional to the length of the infinitesimal path $\delta\Gamma_\mu$
and does not depend on its shape.

If x is a regular point (such as any point of the contour for the functional (12.2)), the RHS of Eq. (12.42) vanishes owing to the backtracking
condition (12.43). In order for the result to be nonvanishing, the point x
should be a *marked* (or irregular) point. A simple example of the functional with a marked point x is

$$\Phi^a[C_{xx}] \equiv \frac{1}{N}\,\mathrm{tr}\left(t^a \boldsymbol{P}\,\mathrm{e}^{\mathrm{i}\int_{C_{xx}}\mathrm{d}\xi^\mu A_\mu(\xi)}\right) \tag{12.44}$$

with the $SU(N)$ generator t^a being inserted in the path-ordered product
at the point x.

The area derivative of the Wilson loop is given by the Mandelstam
formula

$$\frac{\delta}{\delta\sigma_{\mu\nu}(x)}\frac{1}{N}\,\mathrm{tr}\,\boldsymbol{P}\,\mathrm{e}^{\mathrm{i}\oint_C\mathrm{d}\xi^\mu A_\mu} = \frac{\mathrm{i}}{N}\,\mathrm{tr}\,\boldsymbol{P}\,\mathcal{F}_{\mu\nu}(x)\,\mathrm{e}^{\mathrm{i}\oint_C\mathrm{d}\xi^\mu A_\mu}. \tag{12.45}$$

In order to prove this, it is convenient to choose $\delta C_{\mu\nu}(x)$ to be a rectangle
in the (μ,ν)-plane, as was done in Problem 5.8 on p. 94, and use straightforwardly the definition (12.40). The sense of Eq. (12.45) is very simple:

---

* See, for example, [Tav93] which contains definitions of path and area derivatives in
this language.

Table 12.1. Vocabulary for translation of Yang–Mills theory from ordinary space into loop space.

Ordinary space		Loop space	
$\Phi[A]$	Phase factor	$\Phi(C)$	Loop functional
$F_{\mu\nu}(x)$	Field strength	$\dfrac{\delta}{\delta\sigma_{\mu\nu}(x)}$	Area derivative
∇^x_μ	Covariant derivative	∂^x_μ	Path derivative
$\nabla \wedge F = 0$	Bianchi identity		Stokes functionals
$-\nabla_\mu F_{\mu\nu}$ $= \delta/\delta A_\nu$	Schwinger–Dyson equations		Loop equations

$\mathcal{F}_{\mu\nu}$ is a curvature associated with the connection $\mathcal{A}_\mu$, as we discussed in the Remark on p. 95.

The functional on the RHS of Eq. (12.45) has a marked point x, and is of the same type as in Eq. (12.44). When the path derivative acts on such a functional according to the definition (12.42), the result is given by

$$\partial^x_\mu \frac{1}{N} \operatorname{tr} \boldsymbol{P} \, B(x) \, \mathrm{e}^{\mathrm{i}\oint_C \mathrm{d}\xi^\mu \mathcal{A}_\mu} \;=\; \frac{1}{N} \operatorname{tr} \boldsymbol{P} \, \nabla_\mu B(x) \, \mathrm{e}^{\mathrm{i}\oint_C \mathrm{d}\xi^\mu \mathcal{A}_\mu}, \qquad (12.46)$$

where

$$\nabla_\mu B \;=\; \partial_\mu B - \mathrm{i}\,[\mathcal{A}_\mu, B] \qquad (12.47)$$

is the covariant derivative (5.10) in the adjoint representation (see also Problem 5.7 on p. 93).

Combining Eqs. (12.45) and (12.46), we finally represent the expression on the LHS of Eq. (12.29) (or Eq. (12.33)) as

$$\frac{\mathrm{i}}{N} \operatorname{tr} \boldsymbol{P} \, \nabla_\mu \mathcal{F}_{\mu\nu}(x) \, \mathrm{e}^{\mathrm{i}\oint_C \mathrm{d}\xi^\mu \mathcal{A}_\mu} \;=\; \partial^x_\mu \frac{\delta}{\delta\sigma_{\mu\nu}(x)} \frac{1}{N} \operatorname{tr} \boldsymbol{P} \, \mathrm{e}^{\mathrm{i}\oint_C \mathrm{d}\xi^\mu \mathcal{A}_\mu}, \qquad (12.48)$$

i.e. via the action of the path and area derivatives on the Wilson loop. It is therefore rewritten in loop space.

A summary of the results of this section is presented in Table 12.1 as a vocabulary for translation of Yang–Mills theory from the language of ordinary space in the language of loop space.

Remark on Bianchi identity for Stokes functionals

The backtracking condition (12.43) can be represented equivalently as

$$\epsilon_{\mu\nu\lambda\rho}\, \partial_\mu^x \frac{\delta}{\delta\sigma_{\nu\lambda}(x)}\, \Phi(C) \;=\; 0\,, \tag{12.49}$$

by choosing the small path in Eq. (12.43) to be an infinitesimal straight line in the ρ-direction and applying Stokes' theorem geometrically. Using Eqs. (12.45) and (12.46), Eq. (12.49) can in turn be rewritten as

$$\epsilon_{\mu\nu\lambda\rho} \frac{1}{N}\, \mathrm{tr}\, \boldsymbol{P}\, \nabla_\mu \mathcal{F}_{\nu\lambda}(x)\, e^{\,i \oint_C \mathrm{d}\xi^\mu A_\mu} \;=\; 0\,. \tag{12.50}$$

Therefore, Eq. (12.49) represents the Bianchi identity (5.18) in loop space.

Remark on the regularized length

The length $L(C)$ can be approximated by the Stokes functional

$$L_a(C) \;\overset{\text{def}}{=}\; \oint_C \mathrm{d}x_\mu \oint_C \mathrm{d}y_\mu \frac{1}{\sqrt{2\pi a}}\, e^{-(x-y)^2/2a^2} \;\overset{a\to 0}{\longrightarrow}\; L(C)\,. \tag{12.51}$$

This works for the contours, the size of which is much larger than the ultraviolet cutoff a. The area derivative of the functional $L_a(C)$ is finite at finite a but does not commute with taking the limit $a \to 0$. This illustrates the above statement that the size of the variation should be much smaller than the ultraviolet cutoff.

Problem 12.5 Prove Eq. (12.51).

Solution The calculation is similar to that in Problem 12.2 on p. 253. We have

$$\int \mathrm{d}s\, \dot{x}_\mu(s) \int \mathrm{d}t\, \dot{x}_\mu(s+t) \frac{1}{\sqrt{2\pi a}}\, e^{-(x(s+t)-x(s))^2/2a^2}$$

$$\overset{a\to 0}{\longrightarrow} \int \mathrm{d}s\, \dot{x}^2(s) \int_{-\infty}^{+\infty} \mathrm{d}\tau\, \frac{1}{\sqrt{2\pi}}\, e^{-\dot{x}^2(s)\tau^2/2}$$

$$= \int \mathrm{d}s\, \sqrt{\dot{x}^2(s)} \;=\; L(C)\,, \tag{12.52}$$

where $\tau = t/a$. This proves Eq. (12.51).

Remark on the relation with the variational derivative

The standard variational derivative, $\delta/\delta x_\mu(\sigma)$, can be expressed via the path and area derivatives using the formula

$$\frac{\delta}{\delta x_\mu(\sigma)} \;=\; \dot{x}_\nu(\sigma) \frac{\delta}{\delta\sigma_{\mu\nu}(x(\sigma))} + \sum_{i=1}^{m} \partial_\mu^{x_i}\delta(\sigma-\sigma_i)\,, \tag{12.53}$$

where the sum on the RHS is present for the case of a functional having m marked (irregular) points $x_i \equiv x(\sigma_i)$. The simplest example of the functional with m marked points is just a function of m variables $x_1, \ldots, x_m$.

Using Eq. (12.53), the path derivative can be calculated as the limiting procedure

$$\partial_\mu^{x(\sigma)} = \int\limits_{\sigma-0}^{\sigma+0} d\sigma' \frac{\delta}{\delta x_\mu(\sigma')} . \tag{12.54}$$

The result is obviously nonvanishing only when ∂_μ^x is applied to a functional with $x(\sigma)$ being a marked point.

It is nontrivial that the area derivative can also be expressed via the variational derivative [Pol80]:

$$\frac{\delta}{\delta\sigma_{\mu\nu}(x(\sigma))} = \int\limits_{\sigma-0}^{\sigma+0} d\sigma'(\sigma' - \sigma) \frac{\delta}{\delta x_\mu(\sigma')} \frac{\delta}{\delta x_\nu(\sigma)} . \tag{12.55}$$

The point is that the six-component quantity, $\delta/\delta\sigma_{\mu\nu}(x(\sigma))$, is expressed via the four-component one, $\delta/\delta x_\mu(\sigma)$, which is possible because the components of $\delta/\delta\sigma_{\mu\nu}(x(\sigma))$ are dependent owing to the loop-space Bianchi identity (12.49).

12.4 Loop equations

By virtue of Eq. (12.48), Eq. (12.33) can be represented completely in loop space:

$$\partial_\mu^x \frac{\delta}{\delta\sigma_{\mu\nu}(x)} \langle \Phi(C) \rangle$$
$$= \lambda \oint\limits_C dy_\nu\, \delta^{(d)}(x - y) \left\langle \left[\Phi(C_{yx}) \Phi(C_{xy}) - \frac{1}{N^2} \Phi(C) \right] \right\rangle , \tag{12.56}$$

or, using the definitions (12.1) and (12.2) of the loop averages, as

$$\partial_\mu^x \frac{\delta}{\delta\sigma_{\mu\nu}(x)} W(C) = \lambda \oint\limits_C dy_\nu\, \delta^{(d)}(x - y) \left[W_2(C_{yx}, C_{xy}) - \frac{1}{N^2} W(C) \right] . \tag{12.57}$$

This equation is not closed. Having started from $W(C)$, we obtain another quantity, $W_2(C_1, C_2)$, so that Eq. (12.57) connects the one-loop average with a two-loop one. This is similar to the case of the (quantum)

φ^3-theory, whose Schwinger–Dyson equations (2.47) connect the n-point Green functions with different n. We shall derive this complete set of equations for the n-loop averages later in this section.

However, the two-loop average factorizes in the large-N limit:

$$W_2(C_1, C_2) \;=\; W(C_1)\,W(C_2) + \mathcal{O}(N^{-2}), \qquad (12.58)$$

as was discussed in Sect. 11.6. Keeping the constant λ (defined by Eq. (12.34)) fixed in the large-N limit as prescribed by Eq. (11.13), we obtain [MM79]

$$\partial_\mu^x \frac{\delta}{\delta\sigma_{\mu\nu}(x)} W(C) \;=\; \lambda \oint_C \mathrm{d}y_\nu\, \delta^{(d)}(x-y)\, W(C_{yx})\, W(C_{xy}) \qquad (12.59)$$

as $N \to \infty$.

Equation (12.59) is a closed equation for the Wilson loop average in the large-N limit. It is referred to as the *loop equation* or the Makeenko–Migdal equation.

To find $W(C)$, Eq. (12.59) should be solved in the class of Stokes functionals with the initial condition

$$W(0) \;=\; 1 \qquad (12.60)$$

for loops which are shrunk to points. This is a consequence of the obvious property of the Wilson loop

$$\mathrm{e}^{\mathrm{i}\oint_0 \mathrm{d}\xi^\mu \mathcal{A}_\mu} \;=\; 1 \qquad (12.61)$$

and the normalization $\langle\,1\,\rangle = 1$ of the averages.

The factorization (12.58) can itself be derived from the chain of loop equations. Proceeding as before, we obtain

$$\frac{1}{\lambda}\partial_\mu^x \frac{\delta}{\delta\sigma_{\mu\nu}(x)} W_n(C_1, \ldots, C_n)$$

$$= \oint_{C_1} \mathrm{d}y_\nu\, \delta^{(d)}(x-y) \left[W_{n+1}(C_{xy}, C_{yx}, \ldots, C_n) - \frac{1}{N^2} W_n(C_1, \ldots, C_n) \right]$$

$$+ \sum_{j\geq 2} \frac{1}{N^2} \oint_{C_j} \mathrm{d}y_\nu\, \delta^{(d)}(x-y) \left[W_{n-1}\Big(C_1 C_j, \ldots, \underline{C_j}, \ldots, C_n\Big) \right.$$

$$\left. - W_n(C_1, \ldots, C_n) \right]. \qquad (12.62)$$

Here x belongs to C_1; $C_1 C_j$ denotes the joining of C_1 and C_j; $\underline{C_j}$ denotes that C_j is omitted.

Equation (12.62) looks like Eq. (2.47) for φ^3-theory. Moreover, the number of colors N enters Eq. (12.62) simply as a scalar factor, N^{-2}, likewise Planck's constant $\hbar$ enters Eq. (2.47). It is the major advantage of the use of loop space. What was mentioned in Sect. 11.8 concerning the "semiclassical" nature of the $1/N$-expansion of QCD is realized explicitly in Eq. (12.62). Its expansion in $1/N$ is straightforward.

At $N = \infty$, Eq. (12.62) is simplified to

$$\partial_\mu^x \frac{\delta}{\delta\sigma_{\mu\nu}(x)} W_n(C_1, \ldots) = \lambda \oint_{C_1} dy_\nu \, \delta^{(d)}(x - y) \, W_{n+1}(C_{yx}, C_{xy}, \ldots).$$

$$(12.63)$$

This equation possesses [Mig80] a factorized solution

$$W_n(C_1, \ldots, C_n) = \langle\Phi(C_1)\rangle \cdots \langle\Phi(C_n)\rangle + \mathcal{O}(N^{-2})$$

$$\equiv W(C_1) \cdots W(C_n) + \mathcal{O}(N^{-2}) \qquad (12.64)$$

provided $W(C)$ obeys Eq. (12.59) which plays the role of a "classical" equation in the large-N limit. Thus, we have given a nonperturbative proof of the large-N factorization of the Wilson loops.

Problem 12.6 Derive a lattice analog of the loop equation.

Solution The derivation is similar to that in Problem 6.3 on p. 105 for the classical case. We perform the shift (6.22) in the definition (6.42) of the lattice Wilson loop average. Similarly to Eqs. (6.24) and (12.59), we obtain

$$\frac{\beta}{2N^2} \sum_p \left[W(C \, \partial p) - W(C \, \partial p^{-1}) \right] = \sum_{l \in C} \delta_{xy} \tau_\nu(l) \, W(C_{yx}) \, W(C_{xy}).$$

$$(12.65)$$

Here we use the notations of Problem 5.6 on p. 92 so that the contours $C \, \partial p$ and $C \, \partial p^{-1}$ are obtained from C_{xx} by adding the boundary of the plaquette p (∂p^{-1} denotes that the orientation of the boundary is opposite) and the sum over p goes over the $2(d-1)$ plaquettes involving the link at which the shift of $U_\nu(x)$ is performed. These contours are depicted in Fig. 12.4.

The sum on the RHS goes over the links belonging to the contour C. The unit vector $\tau_\nu(l) = 0, \pm 1$ denotes the projection of the (oriented) link $l \in C$ on the axis ν ($\tau_\nu(l) = 1, -1$ or 0 when the directions are parallel, antiparallel, or perpendicular, respectively). The point y is defined as the beginning of the link l if it has positive direction, or as the end of l if it has negative direction. Such an asymmetry arises from the fact that we have performed the right shift (6.22) of $U_\nu(x)$. The Kronecker symbol δ_{xy} guarantees that C_{yx} and C_{xy} are always closed.

Equation (12.65) is a lattice regularization of the continuum loop equation (12.59). The loop equation on the lattice was first discussed in [Foe79, Egu79] and with quarks in [Wei79].

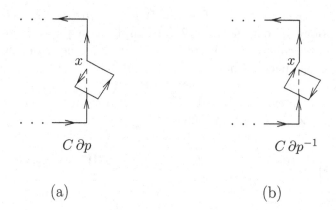

$C\,\partial p$ $C\,\partial p^{-1}$

(a)   (b)

Fig. 12.4. Contours (a) $C\,\partial p$ and (b) $C\,\partial p^{-1}$ on the RHS of the lattice loop equation (12.65).

Problem 12.7 Find a solution to the lattice loop equation (12.65) at small β/N^2.

Solution A strong-coupling solution to Eq. (12.65) can be obtained iteratively in β/N^2. Let us choose the contour C to be the boundary ∂p_0 of a plaquette p_0. Since δ_{xy} on the RHS of Eq. (12.65) is nonvanishing only when y coincides with x, we rewrite Eq. (12.65) as

$$W(\partial p_0) \;=\; \frac{\beta}{2N^2}\sum_p\left[W(\partial p_0\,\partial p) - W(\partial p_0\,\partial p^{-1})\right]. \tag{12.66}$$

One of the terms on the RHS is

$$W(\partial p_0\,\partial p) \;=\; W\!\left(\;\boxed{}\;\right) \;=\; W(0) \;=\; 1\,, \tag{12.67}$$

when p and p_0 have opposite orientations as depicted in Fig. 6.7 on p. 115, owing to the backtracking condition (12.43) and the initial condition (12.60). We thus obtain

$$W(\partial p) \;=\; \frac{\beta}{2N^2} \tag{12.68}$$

to the leading order in β/N^2, which reproduces Eq. (6.72). The other terms on the RHS of Eq. (12.66) are of the next order in β/N^2.

Analogously, Eq. (6.73) is reproduced for a general contour C to the leading order in β/N^2, since

$$\min\{A\,(C\,\partial p)\} \;=\; A_{\min}\,(C) - 1 \tag{12.69}$$

in the lattice units.

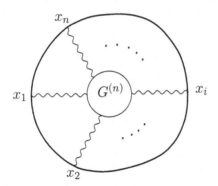

Fig. 12.5. Graphical representation of the terms on the RHS of Eq. (12.70).

12.5 Relation to planar diagrams

The perturbation-theory expansion of the Wilson loop average can be calculated from Eq. (11.46), which we represent in the form

$$
W(C) = 1 + \sum_{n=2}^{\infty} i^n \oint_C dx_1^{\mu_1} \oint_C dx_2^{\mu_2} \cdots \oint_C dx_n^{\mu_n}
$$
$$
\times \theta_c(1, 2, \ldots, n) \, G_{\mu_1 \mu_2 \cdots \mu_n}^{(n)}(x_1, x_2, \ldots, x_n), \qquad (12.70)
$$

where $\theta_c(1, 2, \ldots, n)$ orders the points $x_1, \ldots, x_n$ along the contour in cyclic order and $G_{\mu_1 \cdots \mu_n}^{(n)}$ is given by Eq. (11.71). This θ-function has the meaning of the propagator of a test heavy particle on contour C (see Problem 5.3 on p. 90).

We assume, for definiteness, dimensional regularization throughout this section to make all the integrals well-defined.

Each term on the RHS of Eq. (12.70) can be conveniently represented by the diagram in Fig. 12.5, where the integration over contour C is associated with each point x_i lying on contour C.

These diagrams are analogous to those discussed in Sect. 11.3 with one external boundary – the Wilson loop in the given case. This was already mentioned in the Remark on p. 227. In the large-N limit, only planar diagrams survive. Some of them, which are of the lowest order in λ, are depicted in Fig. 12.6. The diagram in Fig. 12.6a has already been considered in Problem 12.2 (see Eq. (12.17)).

The large-N loop equation (12.59) describes the sum of the planar diagrams. Its iterative solution in λ reproduces the set of planar diagrams for $W(C)$ provided the initial condition (12.60) and some boundary conditions for asymptotically large contours are imposed.

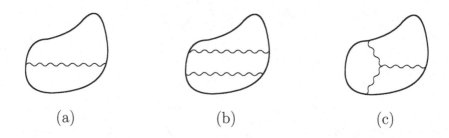

$$(a) \qquad\qquad (b) \qquad\qquad (c)$$

Fig. 12.6.   Planar diagrams for $W(C)$: (a) of order λ with a gluon propagator, and of order λ^2 (b) with two noninteracting gluons and (c) with the three-gluon vertex. Diagrams of order λ^2 with one-loop insertions to the gluon propagator are not shown.

Equation (12.70) can be viewed as an ansatz for $W(C)$ with some unknown functions $G^{(n)}_{\mu_1 \cdots \mu_n}(x_1, \ldots, x_n)$ to be determined by substitution into the loop equation. To preserve symmetry properties of $W(C)$, the functions $G^{(n)}$ must be symmetric under a cyclic permutation of the points $1, \ldots, n$ and depend only on $x_i - x_j$ (translational invariance). The main advantage of this ansatz is that it corresponds automatically to a Stokes functional, owing to the properties of vector integrals, and the initial condition (12.60) is satisfied.

The action of the area and path derivatives on the ansatz (12.70) is easily calculable. For instance, the area derivative is given by

$$
\begin{aligned}
\frac{\delta W(C)}{\delta \sigma_{\mu\nu}(z)} \;=\; & \sum_{n=1}^{\infty} \mathrm{i}^{(n+1)} \oint_C \mathrm{d}x_1^{\mu_1} \ldots \oint_C \mathrm{d}x_n^{\mu_n}\, \theta_{\mathrm{c}}(1, 2, \ldots, n) \\
& \times \Big[\left(\partial_\mu^z \delta_{\nu\alpha} - \partial_\nu^z \delta_{\mu\alpha}\right) G^{(n+1)}_{\alpha\mu_1\cdots\mu_n}(z, x_1, \ldots, x_n) \\
& \quad + \mathrm{i}\, \left(\delta_{\mu\beta}\delta_{\nu\alpha} - \delta_{\mu\alpha}\delta_{\nu\beta}\right) G^{(n+2)}_{\alpha\beta\mu_1\cdots\mu_n}(z, z, x_1, \ldots, x_n) \Big].
\end{aligned}
$$

$$(12.71)$$

The analogy with the Mandelstam formula (12.45) is obvious.

More concerning solving the loop equation by the ansatz (12.70) can be found in [MM81, BGS82, Mig83].

Problem 12.8 Solve Eq. (12.59) to order λ using the ansatz (12.70).

Solution   To order λ, we can restrict ourselves by the $n = 2$ term in the ansatz (12.70). For the θ-function, we have

$$
\theta_{\mathrm{c}}(1, 2) \;\equiv\; \frac{1}{2}\left[\theta(1, 2) + \theta(2, 1)\right] \;=\; \frac{1}{2}.
$$

$$(12.72)$$

The meaning of this formula is obvious: there is no cyclic ordering for two points. We therefore rewrite the ansatz as

$$W(C) \;=\; 1 - \frac{\lambda}{2} \oint_C dx_\mu \oint_C dy_\nu \, D_{\mu\nu}(x-y) + \mathcal{O}(\lambda^2) \tag{12.73}$$

with some unknown function $D_{\mu\nu}(x-y)$. Its tensor structure reads

$$D_{\mu\nu}(x-y) \;=\; \delta_{\mu\nu} D(x-y) + \partial_\mu \partial_\nu f(x-y). \tag{12.74}$$

The second (longitudinal) term in this formula does not contribute to $W(C)$ since the contour integral of this term vanishes in Eq. (12.73). We can thus write

$$W(C) \;=\; 1 - \frac{\lambda}{2} \oint_C dx_\mu \oint_C dy_\mu \, D(x-y) + \mathcal{O}(\lambda^2). \tag{12.75}$$

The area derivative can be calculated easily using Stokes' theorem, which gives

$$\frac{\delta}{\delta\sigma_{\mu\nu}(z)} \oint_C dx_\rho \oint_C dy_\rho \, D(x-y) \;=\; 2\left[\oint_C dy_\nu \, \partial_\mu D(z-y) - \oint_C dy_\mu \, \partial_\nu D(z-y) \right] \tag{12.76}$$

and

$$\partial_\mu^z \frac{\delta}{\delta\sigma_{\mu\nu}(z)} \oint_C dx_\rho \oint_C dy_\rho \, D(x-y) \;=\; 2 \oint_C dy_\nu \, \partial^2 D(z-y) \tag{12.77}$$

since

$$\partial_\mu \oint dy_\mu \, \partial_\nu D(x-y) \;=\; 0. \tag{12.78}$$

Substituting into the loop equation (12.59), we find

$$-\oint_C dy_\nu \, \partial^2 D(x-y) \;=\; \oint_C dy_\nu \, \delta^{(d)}(x-y) \tag{12.79}$$

which is equivalent to

$$-\partial^2 D(x-y) \;=\; \delta^{(d)}(x-y) \tag{12.80}$$

since the contour C is arbitrary. The solution to Eq. (12.80) is unique, provided $D(x-y)$ decreases for large $x-y$, and recovers the propagator (11.4).

12.6 Loop-space Laplacian and regularization

The loop equation (12.59) is *not* yet entirely formulated in loop space. It is a d-vector equation, both sides of which depend explicitly on the point x which does not belong to loop space. The fact that we have a d-vector equation for a scalar quantity means, in particular, that Eq. (12.59) is overspecified.

A practical difficulty in solving Eq. (12.59) is that the area and path derivatives, $\delta/\delta\sigma_{\mu\nu}(x)$ and ∂_μ^x, which enter the LHS are complicated, generally speaking, noncommutative operators. They are intimately related to the Yang–Mills perturbation theory where they correspond to the non-Abelian field strength $F_{\mu\nu}$ and the covariant derivative ∇_μ. However, it is not easy to apply these operators to a generic functional $W(C)$ which is defined on elements of loop space.

A much more convenient form of the loop equation can be obtained by integrating both sides of Eq. (12.59) over dx_ν along the same contour C, which yields

$$\oint_C dx_\nu\, \partial_\mu^x\, \frac{\delta}{\delta\sigma_{\mu\nu}(x)} W(C) \;=\; \lambda \oint_C dx_\mu \oint_C dy_\mu\, \delta^{(d)}(x-y)\, W(C_{yx})\, W(C_{xy}).$$

$$(12.81)$$

Now both the operator on the LHS and the functional on the RHS are scalars without labeled points and are well-defined in loop space. The operator on the LHS of Eq. (12.81) can be interpreted as an infinitesimal variation of elements of loop space.

Equations (12.59) and (12.81) are completely equivalent. A proof of equivalence of the scalar Eq. (12.81) and original d-vector Eq. (12.59) is based on the important property of Eq. (12.59), for which both sides are identically annihilated by the operator ∂_ν^x. It is a consequence of the identity (see Sect. 5.1)

$$\nabla_\mu \nabla_\nu\, \mathcal{F}_{\mu\nu} \;=\; -\frac{1}{2}[\mathcal{F}_{\mu\nu}, \mathcal{F}_{\mu\nu}] \;=\; 0 \qquad (12.82)$$

in ordinary space. Owing to this property, the vanishing of the contour integral of some vector is equivalent to the vanishing of the vector itself, so that Eq. (12.59) can in turn be deduced from Eq. (12.81).

Equation (12.81) is associated with the second-order Schwinger–Dyson equation

$$-\int d^d x\, \nabla_\mu F_{\mu\nu}^a(x)\, \frac{\delta}{\delta A_\nu^a(x)} \;\overset{\text{w.s.}}{=}\; \hbar \int d^d x\, d^d y\, \delta^{(d)}(x-y)\, \frac{\delta}{\delta A_\nu^a(y)}\, \frac{\delta}{\delta A_\nu^a(x)}$$

$$(12.83)$$

in the same sense as Eq. (12.59) is associated with Eq. (12.27). It is called "second order" since the RHS involves two variational derivatives with respect to A_ν.

The operator on the LHS of Eq. (12.81) is a well-defined object in loop space. When applied to regular functionals which do not have marked points, it can be represented, using Eqs. (12.54) and (12.55), in an equivalent form

$$\Delta \equiv \oint_C \mathrm{d}x_\nu \, \partial_\mu^x \frac{\delta}{\delta\sigma_{\mu\nu}(x)} = \int_{\sigma_0}^{\sigma_1} \mathrm{d}\sigma \int_{\sigma-0}^{\sigma+0} \mathrm{d}\sigma' \frac{\delta}{\delta x_\mu(\sigma')} \frac{\delta}{\delta x_\mu(\sigma)}. \qquad (12.84)$$

As was first pointed out by Gervais and Neveu [GN79b], this operator is nothing but a functional extension of the Laplace operator, which is known in mathematics as the Lévy operator.* Equation (12.81) can be represented in turn as an (inhomogeneous) functional Laplace equation

$$\Delta W(C) = \lambda \oint_C \mathrm{d}x_\mu \oint_C \mathrm{d}y_\mu \, \delta^{(d)}(x-y) \, W(C_{yx}) \, W(C_{xy}). \qquad (12.85)$$

We shall refer to this equation as the loop-space Laplace equation.

The form (12.85) of the loop equation is convenient for a nonperturbative ultraviolet regularization.

The idea is to start from the regularized version of Eq. (12.83), replacing the delta-function on the RHS by the kernel of the regularizing operator:

$$\delta^{ab}\delta^{(d)}(x-y) \overset{\text{reg.}}{\Longrightarrow} \left\langle y \left| R^{ab} \right| x \right\rangle = R^{ab}\delta^{(d)}(x-y) \qquad (12.86)$$

with

$$R^{ab} = \left(e^{a^2\nabla^2/2} \right)^{ab}, \qquad (12.87)$$

where ∇_μ is the covariant derivative in the adjoint representation. The regularized version of Eq. (12.83) is

$$-\int \mathrm{d}^d x \, \nabla_\mu F_{\mu\nu}^a(x) \frac{\delta}{\delta A_\nu^a(x)} \overset{\text{w.s.}}{=} \hbar \int \mathrm{d}^d x \, \mathrm{d}^d y \left\langle y \left| R^{ab} \right| x \right\rangle \frac{\delta}{\delta A_\nu^a(y)} \frac{\delta}{\delta A_\nu^b(x)}. \qquad (12.88)$$

To translate Eq. (12.88) in loop space, we use the path-integral representation (see Problem 5.5 on p. 91)

$$\left\langle y \left| R^{ab} \right| x \right\rangle = \int_{\substack{r_\mu(0)=x_\mu \\ r_\mu(a^2)=y_\mu}} \mathcal{D}r_\mu(t) \, e^{-\frac{1}{2}\int_0^{a^2} \mathrm{d}t \, \dot{r}_\mu^2(t)} \, \mathrm{tr} \left[t^a U(r_{yx}) t^b U(r_{xy}) \right] \qquad (12.89)$$

with

$$U(r_{yx}) = P \, e^{\mathrm{i}\int_x^y \mathrm{d}r^\mu A_\mu(r)}, \qquad (12.90)$$

---

* See the book by Lévy [Lev51] and the review [Fel86].

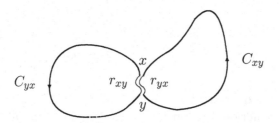

Fig. 12.7. Contours $C_{yx}r_{xy}$ and $C_{xy}r_{yx}$ which enter the RHSs of Eqs. (12.92) and (12.93).

where the integration is over regulator paths $r_\mu(t)$ from x to y, for which the typical length is $\sim a$. The conventional measure is implied in (12.89) so that

$$\int_{\substack{r_\mu(0)=x_\mu \\ r_\mu(a^2)=y_\mu}} \mathcal{D}r_\mu(t)\, \mathrm{e}^{-\frac{1}{2}\int_0^{a^2} dt\, \dot{r}_\mu^2(t)} \, \mathrm{tr}\left[t^a t^b\right] = \delta^{ab}\frac{1}{(2\pi a^2)^{d/2}}\, \mathrm{e}^{-(x-y)^2/2a^2}.$$

$$(12.91)$$

Calculating the variational derivatives on the RHS of Eq. (12.88), using Eq. (12.89) and the completeness condition (11.6), we obtain as $N \to \infty$

$$\int d^d x\, d^d y\, \left\langle y\left|R^{ab}\right|x\right\rangle \frac{\delta}{\delta A_\nu^a(y)}\frac{\delta}{\delta A_\nu^b(x)}\Phi(C)$$

$$= \lambda \oint_C dx_\mu \oint_C dy_\mu \int_{\substack{r_\mu(0)=x_\mu \\ r_\mu(a^2)=y_\mu}} \mathcal{D}r_\mu(t)\, \mathrm{e}^{-\frac{1}{2}\int_0^{a^2} dt\, \dot{r}_\mu^2(t)}\, \Phi(C_{yx}r_{xy})\, \Phi(C_{xy}r_{yx}),$$

$$(12.92)$$

where the contours $C_{yx}r_{xy}$ and $C_{xy}r_{yx}$ are depicted in Fig. 12.7. Averaging over the gauge field and using the large-N factorization, we arrive at the regularized loop-space Laplace equation [HM89]

$$\Delta W(C)$$

$$= \lambda \oint_C dx_\mu \oint_C dy_\mu \int_{\substack{r_\mu(0)=x_\mu \\ r_\mu(a^2)=y_\mu}} \mathcal{D}r_\mu(t)\, \mathrm{e}^{-\frac{1}{2}\int_0^{a^2} dt\, \dot{r}_\mu^2(t)}\, W(C_{yx}r_{xy})\, W(C_{xy}r_{yx})$$

$$(12.93)$$

which manifestly recovers Eq. (12.85) when $a \to 0$.

The constructed regularization is nonperturbative, while perturbatively it reproduces regularized Feynman diagrams. An advantage of this regularization of the loop equation is that the contours $C_{yx}r_{xy}$ and $C_{xy}r_{yx}$ on the RHS of Eq. (12.93) are both closed and do not have marked points if C does not have one. Therefore, Eq. (12.93) is written entirely in loop space.

Remark on functional Laplacian

It is worth noting that the representation of the functional Laplacian on the RHS of Eq. (12.84) is defined for a wider class of functionals than Stokes functionals. The point is that the standard definition of the functional Laplacian from the book by Lévy [Lev51] uses solely the concept of the second variation of a functional $U[x]$, namely the term in the second variation which is proportional to $[\delta x_\mu(\sigma)]^2$:

$$\delta^2 U[x] \;=\; \frac{1}{2} \int_{\sigma_0}^{\sigma_1} \mathrm{d}\sigma \, [\delta x_\mu(\sigma)]^2 \, U''_{xx}[x] + \cdots . \qquad (12.94)$$

The functional Laplacian Δ is then defined by the formula

$$\Delta U[x] \;=\; \int_{\sigma_0}^{\sigma_1} \mathrm{d}\sigma \, U''_{xx}[x] . \qquad (12.95)$$

Here $U[x]$ can be an arbitrary, not necessarily parametric invariant, functional. To emphasize this obstacle, we use the notation $U[x]$ for generic functionals which are defined on L_2 space in comparison with $U(C)$ for the functionals which are defined on elements of loop space. It is easier to deal with the whole operator Δ, rather than separately with the area and path derivatives.

The functional Laplacian is parametric invariant and possesses a number of remarkable properties. While a finite-dimensional Laplacian is a second-order operator, the functional Laplacian is of first order and satisfies the Leibnitz rule

$$\Delta (UV) \;=\; (\Delta U) V + U (\Delta V) . \qquad (12.96)$$

The functional Laplacian can be approximated [Mak88] in loop space by a (second-order) partial differential operator in such a way as to preserve these properties in the continuum limit. This loop-space Laplacian can be inverted to determine a Green function $G(C, C')$ in the form of a sum over surfaces $S_{C,C'}$ connecting two loops:

$$G(C, C') \;=\; \sum_{S_{C,C'}} \cdots , \qquad (12.97)$$

which is analogous to the representation (1.102) of the Green function of the ordinary Laplacian. The standard perturbation theory can then be recovered by iterating Eq. (12.85) (or its regularized version (12.93)) in λ with the Green function (12.97).

12.7 Survey of nonperturbative solutions

While the loop equations were proposed long ago, not much is known concerning their nonperturbative solutions except in two dimensions. We briefly list some of the available results.

It was shown [MM80] that the area law

$$W(C) \equiv \langle \Phi(C) \rangle \quad \propto \quad e^{-K \cdot A_{\min}(C)} \tag{12.98}$$

satisfies the large-N loop equation for asymptotically large C. However, a self-consistency equation for K, which should relate it to the bare charge and the cutoff, was not investigated. In order to do this, one needs more detailed information concerning the behavior of $W(C)$ for intermediate loops.

The *free* bosonic Nambu–Goto string which is defined as a sum over surfaces spanned by C

$$W(C) \quad = \quad \sum_{S:\partial S=C} e^{-K \cdot A(S)}, \tag{12.99}$$

with the action being the area $A(S)$ of the surface S, is *not* a solution for intermediate loops. Consequently, QCD does not reduce to this kind of string, as was expected originally in [GN79a, Nam79, Pol79]. Roughly speaking, the ansatz (12.99) is not consistent with the factorized structure on the RHS of Eq. (12.59).

Nevertheless, it was shown that if a free string satisfies Eq. (12.59), then the same interacting string satisfies the loop equations for finite N. Here "free string" means, as is usual in string theory, that only surfaces of genus zero are present in the sum over surfaces, while surfaces or higher genera are associated with a string interaction. The coupling constant of this interaction is $\mathcal{O}(N^{-2})$.

A formal solution of Eq. (12.59) for all loops was found by Migdal [Mig81] in the form of a fermionic string

$$W(C) \quad = \quad \sum_{S:\partial S=C} \int \mathcal{D}\psi \, e^{-\int d^2\xi \, [\bar{\psi}\sigma_k \partial_k \psi + \bar{\psi}\psi m \sqrt[4]{g}]}, \tag{12.100}$$

where the world sheet of the string is parametrized by the coordinates ξ_1 and ξ_2 for which the two-dimensional metric is conformal, i.e. diagonal. The field $\psi(\xi)$ describes two-dimensional elementary fermions (elves) living in the surface S, and m denotes their mass. Elves were introduced to provide a factorization which now holds owing to some remarkable properties of two-dimensional fermions. For large loops, the internal fermionic structure becomes frozen, so that the empty string behavior (12.98) is recovered. For small loops, the elves are necessary for asymptotic freedom.

However, it is unclear whether or not the string solution (12.100) is practically useful for a study of multicolor QCD, since the methods of dealing with the string theory in four dimensions have not yet been developed.

A very interesting solution of the large-N loop equation on a lattice, found by Eguchi and Kawai [EK82], shows that the $SU(N)$ gauge theory on an infinite lattice and a unit hypercube are equivalent at $N = \infty$. With slight modifications this large-N reduction holds in the continuum theory as well, so that the space-time can be absorbed by the internal symmetry group. More concerning the large-N reduction will be said in Part 4.

12.8 Wilson loops in QCD$_2$

Two-dimensional QCD (QCD$_2$) has been popular since the paper by 't Hooft [Hoo74b] as a simplified model of QCD$_4$.

One can always choose the axial gauge

$$A_1 = 0, \tag{12.101}$$

so that the commutator in the non-Abelian field strength (5.14) vanishes in two dimensions. Therefore, there is no gluon self-interaction in this gauge and the theory looks, at first glance, like the Abelian one.

The Wilson loop average in QCD$_2$ can be calculated straightforwardly via the expansion (12.70) where only disconnected (free) parts of the correlators $G^{(n)}$ for even n should be left, since there is no interaction. Only the planar structure of color indices contributes at $N = \infty$. Diagrammatically, the diagrams of the type depicted in Figs. 12.6a and b are relevant for contours without self-intersections, while that in Fig. 12.6c should be omitted in two dimensions.

The color structure of the relevant planar diagrams can be reduced by virtue of the formula

$$\sum_a (t^a)^{ik} (t^a)^{kj} = N\delta^{ij}, \tag{12.102}$$

which is a consequence of the completeness condition (11.6) at large N. We have

$$W(C) = 1 + \sum_k^\infty (-\lambda)^k \oint_C dx_1^{\mu_1} \oint_C dx_2^{\nu_1} \cdots \oint_C dx_{2k-1}^{\mu_k} \oint_C dx_{2k}^{\nu_k}$$
$$\times \theta_c(1, 2, \ldots, 2k) D_{\mu_1\nu_1}(x_1 - x_2) \cdots D_{\mu_k\nu_k}(x_{2k-1} - x_{2k}), \tag{12.103}$$

where the points $x_1, \ldots, x_{2k}$ are still cyclic ordered along the contour. Similarly to Problem 5.2 on p. 89, we can exponentiate the RHS of

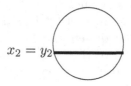

$x_2 = y_2$

Fig. 12.8.   Graphical representation of the contour integral on the LHS of Eq. (12.108) in the axial gauge. The bold line represents the gluon propagator (12.105) with $x_2 = y_2$ owing to the delta-function.

Eq. (12.103) to obtain finally

$$W(C) \;=\; \exp\left[-\frac{\lambda}{2}\oint_C \mathrm{d}x^\mu \oint_C \mathrm{d}y^\nu D_{\mu\nu}(x-y)\right]. \qquad (12.104)$$

This is the same formula as in the Abelian case if λ denotes e^2.

The propagator $D_{\mu\nu}(x,y)$ is, strictly speaking, the one in the axial gauge (12.101) which is given by

$$D_{\mu\nu}(x-y) \;=\; \frac{1}{2}\delta_{\mu 2}\delta_{\nu 2}\,|x_1 - y_1|\,\delta^{(1)}(x_2 - y_2)\,. \qquad (12.105)$$

However, the contour integral on the RHS of Eq. (12.104) is gauge invariant, and we can simply choose

$$D_{\mu\nu}(x-y) \;=\; \delta_{\mu\nu}\,D(x-y)\,. \qquad (12.106)$$

In two dimensions* we have

$$D(x-y) \;=\; \frac{1}{4\pi}\ln\frac{\ell^2}{(x-y)^2}\,, \qquad (12.107)$$

where ℓ is an arbitrary parameter with dimension of length. Nothing depends on it because the contour integral of a constant vanishes.

The propagator (12.106) is usually associated with the Feynman gauge. The explicit form (12.104) indicates that a contribution of diagrams with vertices, which are present in the Feynman gauge, vanishes in two dimensions.

The contour integral in the exponent on the RHS of Eq. (12.104) can be represented graphically as depicted in Fig. 12.8, where $x_2 = y_2$ owing

---

* In d dimensions

$$D(x-y) \;=\; \frac{1}{4\pi^{d/2}}\,\Gamma\left(\frac{d}{2}-1\right)\frac{1}{\left[(x-y)^2\right]^{d/2-1}}\,.$$

to the delta-function in Eq. (12.105) and the bold line represents $|x_1 - y_1|$. This gives

$$\oint_C \mathrm{d}x^\mu \oint_C \mathrm{d}y^\nu D_{\mu\nu}(x - y) = A(C), \qquad (12.108)$$

where $A(C)$ is the area enclosed by the contour C. Finally, we obtain

$$W(C) = \mathrm{e}^{-\frac{\lambda}{2}A(C)} \qquad (12.109)$$

for contours without self-intersections.

Therefore, the area law holds in two dimensions both in the non-Abelian and Abelian cases. This is, roughly speaking, because of the form of the two-dimensional propagator (12.107), which decreases with distance only logarithmically in the Feynman gauge.

Problem 12.9 Prove Eq. (12.109) in the Feynman gauge.

Solution To prove Eq. (12.108) in the Feynman gauge (12.106) and (12.107), we note that the area element in two dimensions can be represented by

$$\mathrm{d}\sigma^{\mu\nu}(x) \equiv \mathrm{d}x^\mu \wedge \mathrm{d}x^\nu = \varepsilon^{\mu\nu}\mathrm{d}^2x, \qquad (12.110)$$

where $\varepsilon^{\mu\nu}$ is the antisymmetric tensor $\varepsilon^{12} = -\varepsilon^{21} = 1$. Therefore, the area can be represented by the double integral

$$A(C) = \frac{1}{2} \int_{S(C)} \mathrm{d}\sigma^{\mu\nu}(x) \int_{S(C)} \mathrm{d}\sigma^{\mu\nu}(y)\, \delta^{(2)}(x - y) \qquad (12.111)$$

which goes along the surface $S(C)$ enclosed by the (nonintersecting) loop C.

Applying Stokes' theorem, we obtain

$$\oint_C \mathrm{d}x^\mu \oint_C \mathrm{d}y^\mu D(x - y) = \int_{S(C)} \mathrm{d}\sigma^{\mu\nu}(x)\, \partial_\nu \oint_C \mathrm{d}y^\mu D(x - y)$$

$$= -\int_{S(C)} \mathrm{d}\sigma^{\mu\nu}(x) \int_{S(C)} \mathrm{d}\sigma^{\mu\rho}(y)\, \partial_\nu\partial_\rho D(x - y)$$

$$= -\frac{1}{2} \int_{S(C)} \mathrm{d}\sigma^{\mu\nu}(x) \int_{S(C)} \mathrm{d}\sigma^{\mu\nu}(y)\, \partial^2 D(x - y)$$

$$= \frac{1}{2} \int_{S(C)} \mathrm{d}\sigma^{\mu\nu}(x) \int_{S(C)} \mathrm{d}\sigma^{\mu\nu}(y)\, \delta^{(2)}(x - y). \qquad (12.112)$$

Using Eq. (12.111) we prove Eq. (12.109) in the Feynman gauge.

It is worth noting that Eq. (12.112) is based only on Stokes' theorem and holds for contours with arbitrary self-intersections. In contrast, Eq. (12.111) itself is valid only for nonintersecting loops.

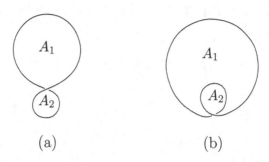

Fig. 12.9. Contours with one self-intersection: A_1 and A_2 denote the areas of the proper windows. The total area enclosed by the contour in (a) is $A_1 + A_2$. The areas enclosed by the exterior and interior loops in (b) are $A_1 + A_2$ and A_2, respectively, while the total area of the surface with the folding is $A_1 + 2A_2$.

The difference between the Abelian and non-Abelian cases shows up for the contours with self-intersections.

We first note that the simple formula (12.108) does *not* hold for contours with arbitrary self-intersections.

The simplest contours with one self-intersection are depicted in Fig. 12.9. There is nothing special about the contour in Fig. 12.9a. Equation (12.108) still holds in this case with $A(C)$ being the total area, $A(C) = A_1 + A_2$.

The Wilson loop average for the contour in Fig. 12.9a coincides both for the Abelian and non-Abelian cases and equals

$$W(C) \;=\; e^{-\frac{\lambda}{2}(A_1+A_2)}. \tag{12.113}$$

This is nothing but the exponential of the total area.

For the contour in Fig. 12.9b, we obtain

$$\oint_C dx^\mu \oint_C dy^\nu D_{\mu\nu}(x-y) \;=\; A_1 + 4A_2. \tag{12.114}$$

This is easy to understand in the axial gauge where the ends of the propagator line can lie both on the exterior and interior loops, or one end at the exterior loop and the other end on the interior loop. These cases are illustrated by Fig. 12.10. The contributions of the diagrams in Figs. 12.10a–d are $A_1 + A_2$, A_2, A_2, and A_2, respectively. The result given by Eq. (12.114) is obtained by summing over all four diagrams.

For the contour in Fig. 12.9b, the Wilson loop average is

$$W(C) \;=\; e^{-\frac{\lambda}{2}(A_1+4A_2)} \tag{12.115}$$

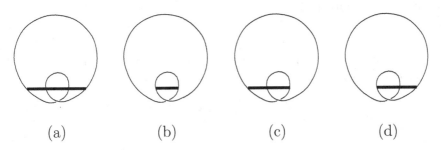

(a)   (b)   (c)   (d)

Fig. 12.10.   Three type of contribution in Eq. (12.114). The ends of the propagator line lie both on (a) exterior and (b) interior loops, or (c), (d) one end on the exterior loop and another end on the interior loop.

in the Abelian case and

$$W(C) \;=\; (1 - \lambda A_2)\, e^{-\frac{\lambda}{2}(A_1 + 2A_2)} \tag{12.116}$$

in the non-Abelian case at $N = \infty$. They coincide only to order λ as they should.

The difference to the next orders is because only the diagrams with one propagator line connecting the interior and exterior loops are planar and, therefore, contribute in the non-Abelian case. Otherwise, the diagram is nonplanar and vanishes as $N \to \infty$.

Note that the exponential of the total area $A(C) = A_1 + 2A_2$ of the surface with the folding, which is enclosed by the contour C, appears in the exponent for the non-Abelian case. The additional pre-exponential factor could be associated with the entropy of folding the surface.

The Wilson loop averages (12.113) and (12.116) in QCD$_2$ at large N as well as those for contours with arbitrary self-intersections, which have a generic form

$$W(C) \;=\; P(A_1, \ldots, A_n)\, e^{-\frac{\lambda}{2} A(C)}, \tag{12.117}$$

where P is a polynomial of the areas of individual windows and $A(C)$ is the total area of the surface with foldings, were first calculated in [KK80] by solving the two-dimensional loop equation and in [Bra80] by applying the non-Abelian Stokes theorem. The lattice version is given in [KK81].

Problem 12.10 Demonstrate that Eq. (12.104) satisfies the Abelian loop equation

$$\partial^x_\mu \frac{\delta}{\delta \sigma_{\mu\nu}(x)} W(C) \;=\; \lambda \oint_C dy_\nu\, \delta^{(d)}(x - y)\, W(C). \tag{12.118}$$

Solution   The calculation is the same as in Problem 12.8 on p. 268. In $d = 2$ one can alternatively use [OP81] the expression on the RHS of Eq. (12.112).

Problem 12.11 Obtain Eqs. (12.113) and (12.116) for the contours with one self-intersection by solving the loop equation (12.59).

Solution Let us multiply Eq. (12.59) in $d = 2$ by $\varepsilon_{\rho\nu}$ and integrate over dx^ρ along a small (open) piece C' of the contour C including the point of self-intersection. We obtain

$$\varepsilon_{\rho\nu} \int_{C'} dx_\rho\, \partial_\mu^x \frac{\delta}{\delta\sigma_{\mu\nu}(x)} W(C) = \lambda\varepsilon_{\rho\nu} \int_{C'} dx_\rho \oint_C dy_\nu\, \delta^{(2)}(x-y)\, W(C_{yx})\, W(C_{xy}).$$

(12.119)

The RHS of Eq. (12.119) can be calculated analogously to the known representation for the number of self-intersections of a loop in two dimensions. For the case of one self-intersection, we have

$$\varepsilon_{\rho\nu} \int_{C'} dx_\rho \oint_C dy_\nu\, \delta^{(2)}(x-y)\, W(C_{yx})\, W(C_{xy}) = W(C_1)\, W(C_2),$$ (12.120)

where C_1 and C_2 denote, respectively, the upper and lower loops in Fig. 12.9a or the exterior and interior loops in Fig. 12.9b.

The LHS of Eq. (12.119) can be transformed as

$$\varepsilon_{\rho\nu} \int_{C'} dx_\rho\, \partial_\mu^x \frac{\delta}{\delta\sigma_{\mu\nu}(x)} W(C) = \int_{C'} dx_\nu\, \partial_\nu^x \frac{\delta}{\delta\sigma(x)} W(C),$$ (12.121)

where $\delta/\delta\sigma(x)$ denotes the variational derivative with respect to the "scalar" area

$$\delta\sigma(x) = \frac{1}{2}\varepsilon_{\mu\nu}\delta\sigma^{\mu\nu}(x).$$ (12.122)

The integrand on the RHS of Eq. (12.121) is a total derivative and the contour integral reduces to the difference of the Ω-variations at the end points of the contour C', which would vanish if there were no self-intersections. The RHS of Eq. (12.119) also vanishes if no self-intersections are present, so $W(C)$ is determined in this case by Eq. (12.118) rather than Eq. (12.119).

For the contour in Fig. 12.9a, this gives

$$\left(-\frac{\partial}{\partial A_1} - \frac{\partial}{\partial A_2}\right) W(C).$$

(12.123)

The Ω-variation of the contour on the LHS represents the variational derivative. The minus sign in front of $\partial/\partial A_1$ on the RHS is because adding the Ω-variation in the first term on the LHS decreases the area A_1, while that in the second term increases A_2. Then, for the contour in Fig. 12.9a, Eq. (12.119) takes the form

$$\left(-\frac{\partial}{\partial A_1} - \frac{\partial}{\partial A_2}\right) W(C) = \lambda W(C_1)\, W(C_2).$$ (12.124)

For the contour in Fig. 12.9b, we obtain quite similarly

$$= \left(2\frac{\partial}{\partial A_1} - \frac{\partial}{\partial A_2}\right) W(C).$$

$$(12.125)$$

Now adding the Ω-variation in the first term on the LHS increases A_1 and decreases A_2, while that in the second term decreases A_1. Equation (12.119) takes the form

$$\left(2\frac{\partial}{\partial A_1} - \frac{\partial}{\partial A_2}\right) W(C) = \lambda W(C_1) W(C_2). \qquad (12.126)$$

The RHSs of Eqs. (12.124) and (12.126) are known since C_1 and C_2 have no self-intersections so that Eq. (12.109) holds for $W(C_1)$ and $W(C_2)$. Finally, Eqs. (12.124) and (12.126) take the explicit form [KK80]

$$\left(-\frac{\partial}{\partial A_1} - \frac{\partial}{\partial A_2}\right) W(C) = \lambda e^{-\frac{\lambda}{2}(A_1+A_2)} \qquad (12.127)$$

and

$$\left(2\frac{\partial}{\partial A_1} - \frac{\partial}{\partial A_2}\right) W(C) = \lambda e^{-\frac{\lambda}{2}(A_1+2A_2)}, \qquad (12.128)$$

respectively. Their solution is given uniquely by Eqs. (12.113) and (12.116).

It is worth noting that the linear Abelian loop equation (12.118) can be written for the contours in Figs. 12.9a and b as

$$\left(-\frac{\partial}{\partial A_1} - \frac{\partial}{\partial A_2}\right) \ln W(C) = \lambda, \qquad (12.129)$$

$$\left(2\frac{\partial}{\partial A_1} - \frac{\partial}{\partial A_2}\right) \ln W(C) = \lambda. \qquad (12.130)$$

The operators on the LHSs are always the same for the non-Abelian and Abelian loop equations, which is a general property, but the RHSs differ generically: Eqs. (12.127) and (12.129) for the contour in Fig. 12.9a coincide, while Eqs. (12.128) and (12.130) for the contour in Fig. 12.9b differ. The solution to Eq. (12.130) is given by (12.115).

Problem 12.12 Prove Eq. (12.120) for the contours with one self-intersection.

Solution Let the intersection correspond to the values s_1 and s_2 of the parameter s, i.e. $x_\mu(s_1) = x_\mu(s_2)$. Noting that only the vicinities of s_1 and s_2 contribute to the integral on the LHS of Eq. (12.120), we obtain

$$\varepsilon_{\rho\nu} \int_{C'} dx_\rho \oint_C dy_\nu\, \delta^{(2)}(x-y)\, W(C_{yx})\, W(C_{xy})$$

$$= \varepsilon_{\rho\nu}\, \dot{x}_\rho(s_1)\, \dot{x}_\nu(s_2) \int ds \int dt\, \delta^{(2)}((s-s_1)\dot{x}(s_1) - (t-s_2)\dot{x}(s_2))$$

$$\times\, W(C_{x(s_2)x(s_1)})\, W(C_{x(s_1)x(s_2)})$$

$$= \frac{\varepsilon_{\rho\nu}\, \dot{x}_\rho(s_1)\, \dot{x}_\nu(s_2)}{\sqrt{\dot{x}_\mu^2(s_1)\, \dot{x}_\nu^2(s_2) - (\dot{x}_\mu(s_1)\, \dot{x}_\mu(s_2))^2}} W(C_{x(s_2)x(s_1)})\, W(C_{x(s_1)x(s_2)})$$

$$= W(C_{x(s_2)x(s_1)})\, W(C_{x(s_1)x(s_2)}) \qquad (12.131)$$

which is precisely the RHS of Eq. (12.120).

Remark on the string representation

A nice property of QCD$_2$ at large N is that the exponential of the area enclosed by the contour C emerges* for the Wilson loop average $W(C)$. This is as it should for the Nambu–Goto string (12.99). However, the additional pre-exponential factors (such as that in Eq. (12.116)) are very difficult to interpret in string language. They may become negative for large loops, which is impossible for a bosonic string. This demonstrates explicitly in $d=2$ the statement of the previous section that the Nambu–Goto string is not a solution of the large-N loop equation. An appropriate string representation of two-dimensional large-N QCD was constructed by Gross and Taylor [GT93].

12.9 Gross–Witten transition in lattice QCD$_2$

The lattice gauge theory on a two-dimensional lattice is defined by the partition function (6.31) with $d=2$:

$$Z_{2D}(\beta) = \int \prod_x \prod_{\mu=1,2} dU_\mu(x)\, e^{-\beta S[U]}, \qquad (12.132)$$

where the action is given by Eq. (6.16).

A specific property of two dimensions is that the number of lattice sites is equal to the number of plaquettes. For this reason, we can always perform a gauge transformation such that the link variables are chosen to be equal to unity along one of the axes, say

$$U_1(x) = 1. \qquad (12.133)$$

---

* This is not true, as has already been discussed, in the Abelian case for contours with self-intersections.

Hence, the partition function (12.132) factorizes:

$$Z_{\text{2D}} = (Z_{\text{1p}})^{N_p} , \qquad (12.134)$$

where N_p denotes the number of plaquettes of the lattice and Z_{1p} is the one-matrix integral

$$Z_{\text{1p}}(\beta) = \int dU\, e^{\beta\left(\frac{1}{N}\operatorname{Re}\operatorname{tr} U - 1\right)} . \qquad (12.135)$$

In other words, the plaquette variables of the lattice gauge theory can be treated in two dimensions as being independent.

The correct interpretation of Eq. (12.135) is that it is the partition function of the one-plaquette model, i.e. the lattice gauge theory on a single plaquette. This is consistent with the gauge invariance.

The unitary one-matrix model (12.135) can be easily solved in the large-N limit using loop equations.

We first introduce the "observables" for the one-matrix model:

$$W_n = \left\langle \frac{1}{N} \operatorname{tr} U^n \right\rangle_{\text{1p}} , \qquad (12.136)$$

where the average is taken with the same weight as in Eq. (12.135). The interpretation of W_n in the language of the single-plaquette model is that these are the Wilson loop averages for contours which go along the boundary of n stacked plaquettes.

In order to derive the loop equation for the one-matrix model, we proceed quite analogous to the derivation of the loop equation in the lattice gauge theory (given in Problem 12.6 on p. 265).

Let us consider the obvious identity

$$0 = \langle \operatorname{tr} t^a U^n \rangle_{\text{1p}} , \qquad (12.137)$$

and perform the (infinitesimal) change

$$U \to U\left(1 - it^a \epsilon^a\right), \qquad U^\dagger \to \left(1 + it^a \epsilon^a\right) U^\dagger \qquad (12.138)$$

of the integration variable on the RHS of Eq. (12.137). Since the Haar measure is invariant under the change (12.138), we finally obtain

$$\left.\begin{aligned} \frac{\beta}{2N^2}\left(W_{n-1} - W_{n+1}\right) &= \sum_{k=1}^{n} W_k W_{n-k} \qquad \text{for } n \geq 1 , \\ W_0 &= 1 , \end{aligned}\right\} \qquad (12.139)$$

where[*] $\beta = N^2/\lambda$ and $\lambda \sim 1$ as $N \to \infty$.

---

[*] In contrast to the previous section, here we include the factor of a^2 in the definition of λ to make it dimensionless.

Equation (12.139) has the following exact solution:

$$W_1 = \frac{1}{2\lambda}; \qquad W_n = 0 \quad \text{for } n \geq 2, \qquad (12.140)$$

which reproduces the strong-coupling expansion. The leading order of the strong-coupling expansion turns out to be exact at $N = \infty$.

However, the solution (12.140) cannot be the desired solution at any values of the coupling constant. Since W_k are (normalized) averages of unitary matrices, they must obey

$$W_n \leq 1, \qquad (12.141)$$

which is not the case for W_1, given by Eq. (12.140), at small enough values of λ.

In order to find all solutions to Eq. (12.139), let us introduce the generating function

$$f(z) \equiv \sum_{n=0}^{\infty} W_n z^n \qquad (12.142)$$

and rewrite Eq. (12.139) as the quadratic equation

$$fz - \frac{1}{z}(f-1) + W_1 = 2\lambda\left(f^2 - f\right). \qquad (12.143)$$

A formal solution to Eq. (12.143) is

$$f(z) = -\frac{1 - 2\lambda z - z^2}{4\lambda z} + \frac{\sqrt{\left(1 + 2\lambda z + z^2\right)^2 + 4z^2\left(2\lambda W_1 - 1\right)}}{4\lambda z}, \qquad (12.144)$$

where the positive sign of the square root is chosen to satisfy $f(0) = 1$.

The RHS of Eq. (12.144) depends on an unknown function $W_1(\lambda)$, which must guarantee $f(z)$ to be a holomorphic function of the complex variable z within the unit circle $|z| < 1$. This is a consequence of the inequality (12.141) which stems from the unitarity of U.

There exist two solutions for which $f(z)$ is holomorphic inside the unit circle: the strong-coupling solution given for $\lambda \geq 1$ by Eq. (12.140) and the weak-coupling solution given for $\lambda \leq 1$ by

$$W_1 = 1 - \frac{\lambda}{2}. \qquad (12.145)$$

A comparison with Eq. (7.1) for $d = 2$ shows that the leading order of the weak-coupling expansion is now exact. Therefore, $f(z)$ is given by two *different* analytic functions for $\lambda > 1$ and $\lambda < 1$.

At the point $\lambda = 1$, a phase transition occurs as was discovered by Gross and Witten [GW80] who first solved lattice QCD_2 in the large-N limit. This phase transition is of the third order since both the first and second derivatives of the partition function are continuous at $\lambda = 1$. The discontinuity resides only in the third derivative. This phase transition is pretty unusual from the point of view of statistical mechanics where phase transitions usually occur in the limit of an infinite volume (otherwise the partition function is analytic in temperature). Now the Gross–Witten phase transition occurs even for the single-plaquette model (12.135) in the large-N limit. In other words, the number of degrees of freedom is now infinite owing to the internal symmetry group rather than an infinite volume.

Finally, we mention that since plaquette variables are independent in lattice QCD_2, the Wilson loop average for a nonintersecting lattice contour C takes the form

$$W(C) = (W_1)^A, \tag{12.146}$$

where A is the area (in the lattice units) enclosed by the contour C. W_1 in this formula is given by Eq. (12.140) in the strong coupling phase ($\lambda \geq 1$) and Eq. (12.145) in the weak-coupling phase ($\lambda \leq 1$).

The continuum formula (12.109) can be recovered for small λ from Eq. (12.146) as follows:

$$W(C) = \left(1 - \frac{\lambda a^2}{2}\right)^{A/a^2} \xrightarrow{a \to 0} e^{-\frac{\lambda}{2}A}, \tag{12.147}$$

where we have restored the a-dependence as is prescribed by the dimensional analysis.

The solution of $N = \infty$ lattice QCD_2 by the loop equations, which is described in this section, was given in [PR80, Fri81].

Problem 12.13 Calculate the density of eigenvalues for the matrix U in the one-matrix model (12.135).

Solution Let us reduce U to the diagonal form

$$U = \text{diag}\left(e^{i\alpha_1}, \ldots, e^{i\alpha_j}, \ldots, e^{i\alpha_N}\right). \tag{12.148}$$

The density of eigenvalues (or the spectral density), $\rho(\alpha)$, is then defined as a fraction of the eigenvalues which lie in the interval $[\alpha, \alpha + d\alpha]$. In other words, introducing the continuum variable $x = j/N$ ($0 \leq x \leq 1$) in the large-N limit, we have

$$\rho(\alpha) = \frac{dx}{d\alpha} \geq 0 \tag{12.149}$$

which obeys the obvious normalization

$$\int_{-\pi}^{\pi} d\alpha\, \rho(\alpha) = \int_0^1 dx = 1. \tag{12.150}$$

Given $\rho(\alpha)$, we can calculate W_n by

$$W_n = \int_{-\pi}^{\pi} d\alpha\, \rho(\alpha) \cos n\alpha. \tag{12.151}$$

It is now clear from the definition (12.136) and (12.142) that

$$f(z) = \int_{-\pi}^{\pi} d\alpha\, \rho(\alpha) \frac{1}{1 - z\, e^{-i\alpha}}. \tag{12.152}$$

Choosing $z = \exp(i\omega)$, we rewrite Eq. (12.152) as

$$f(e^{i\omega}) = \frac{1}{2} + \frac{i}{2} \int_{-\pi}^{\pi} d\alpha\, \rho(\alpha) \cot \frac{\omega - \alpha}{2}. \tag{12.153}$$

The discontinuity of this analytic function at $\omega = \alpha \pm i0$ then determines $\rho(\alpha)$. Using the explicit solution (12.144), we formally find

$$\rho(\alpha) = \frac{1}{2\lambda\pi} \sqrt{\left(\cos\alpha + \lambda + \sqrt{1 - 2\lambda W_1}\right) \left(\cos\alpha + \lambda - \sqrt{1 - 2\lambda W_1}\right)}. \tag{12.154}$$

For W_1 given by Eqs. (12.140) and (12.145) for the strong- and weak-coupling phases, we finally obtain

$$\rho(\alpha) = \frac{1}{2\pi} \left(1 + \frac{1}{\lambda} \cos\alpha\right) \qquad \text{for } \lambda \geq 1, \tag{12.155}$$

$$\rho(\alpha) = \frac{1}{\lambda\pi} \cos\frac{\alpha}{2} \sqrt{\lambda - \sin^2 \frac{\alpha}{2}} \qquad \text{for } \lambda \leq 1 \tag{12.156}$$

for the strong- and weak-coupling solutions, respectively. Note that (12.155) is nonnegative for $\lambda \geq 1$ as it should be because of the inequality (12.149). For $\lambda < 1$, the strong-coupling solution (12.155) becomes negative somewhere in the interval $[-\pi, \pi]$ which cannot happen for a dynamical system. This is the reason why the other solution (12.156) is realized for $\lambda < 1$. It has the support on the smaller interval $[-\alpha_c, \alpha_c]$, where $0 < \alpha_c < \pi$ is determined by the equation

$$\sin^2 \frac{\alpha_c}{2} = \lambda \tag{12.157}$$

which always has a solution for $\lambda < 1$. The weak-coupling spectral density (12.156) is nonnegative for $\lambda \leq 1$.

For small λ, $\alpha_c = 2\sqrt{\lambda}$ so that

$$\rho(\alpha) = \frac{1}{2\lambda\pi} \sqrt{4\lambda - \alpha^2}. \tag{12.158}$$

As $\lambda \to 0$, $\rho(\alpha) \to \delta(\alpha)$ and U freezes, modulo a gauge transformation, near a unit matrix. This guarantees the existence of the continuum limit of QCD$_2$.

The spectral densities (12.155) and (12.156) were first calculated [GW80] by a direct solution of the saddle-point equation at large N.

13

Matrix models

Matrix models first appeared in statistical mechanics and nuclear physics [Wig51, Dys62] and turned out to be very useful in the analysis of various physical systems where the energy levels of a complicated Hamiltonian can be approximated by the distribution of eigenvalues of a random matrix. The statistical averaging is then replaced by averaging over an appropriate ensemble of random matrices. This idea has been applied, in particular, in studying the low-energy chiral properties of QCD [SV93, VZ93].

Matrix models possess some features of multicolor QCD described in Chapter 11 but are simpler and can often be solved as $N \to \infty$ (i.e. in the planar limit) using the methods proposed for multicolor QCD. For the simplest case of the Hermitian one-matrix model, the genus expansion in $1/N$ can be constructed.

The Hermitian one-matrix model is related to the problem of enumeration of graphs. Its explicit solution at large N was first obtained by Brézin, Itzykson, Parisi and Zuber [BIP78] and inspired a lot of activity in this subject. Further results in this direction are linked to the method of orthogonal polynomials [Bes79, IZ80, BIZ80].

A very interesting application of the matrix models along this line is for the problem of discretization of random surfaces and two-dimensional quantum gravity [Kaz85, Dav85, ADF85, KKM85]. The continuum limits of these matrix models are associated with lower-dimensional conformal field theories and exhibit properties of integrable systems.

We shall begin this chapter by describing the original approach [BIP78] for solving the Hermitian one-matrix model at large N and then concentrate on a more general approach based on the loop equations. Our main goal is to illustrate the methods described in the two previous chapters.

13.1 Hermitian one-matrix model

The unitary one-matrix model (12.135) is generically a matrix model solvable in the large-N limit. A simplest and historically the first example of this kind is the Hermitian one-matrix model, the large-N solution of which is obtained in [BIP78].

The Hermitian one-matrix model is defined by the partition function

$$Z_{1h} = \int d\varphi\, e^{-N\,\mathrm{tr}\,V(\varphi)}, \tag{13.1}$$

where

$$d\varphi = \prod_{i=1}^{N} d\varphi_{ii} \prod_{j>i}^{N} d\,\mathrm{Re}\,\varphi_{ij}\, d\,\mathrm{Im}\,\varphi_{ij} \tag{13.2}$$

is the measure for integrating over Hermitian $N \times N$ matrices. It is invariant under the shift

$$\varphi_{ij} \;\rightarrow\; \varphi_{ij} + \epsilon_{ij} \tag{13.3}$$

by an arbitrary $N \times N$ Hermitian matrix ϵ_{ij}.

We consider the most general potential

$$V(\varphi) = \sum_k t_k\, \varphi^k, \tag{13.4}$$

where t_k are coupling constants. We shall also use another normalization

$$t_k = \frac{g_k}{k} \qquad \text{for}\quad k \geq 1, \tag{13.5}$$

which respects the cyclic symmetry of the trace. The simplest Gaussian case is associated with $g_2 = 1$ and $g_k = 0$ for $k \neq 2$.

The averages in the Hermitian one-matrix model are defined by

$$\langle F[\varphi] \rangle_{1h} = Z_{1h}^{-1} \int d\varphi\, e^{-N\,\mathrm{tr}\,V(\varphi)} F[\varphi]. \tag{13.6}$$

Performing the Gaussian integral, it is easy to calculate the propagator

$$\langle \varphi_{ij}\varphi_{kl} \rangle_{\mathrm{Gauss}} \;\stackrel{\text{def}}{=}\; \frac{\int d\varphi\, e^{-\frac{N}{2}\,\mathrm{tr}\,\varphi^2}\, \varphi_{ij}\varphi_{kl}}{\int d\varphi\, e^{-\frac{N}{2}\,\mathrm{tr}\,\varphi^2}} \;=\; \frac{1}{N}\delta_{il}\delta_{kj}. \tag{13.7}$$

Equation (13.7) can be obtained alternatively from the Schwinger–Dyson equation

$$\left\langle \frac{\partial\,\mathrm{tr}\,V(\varphi)}{\partial\varphi_{ji}} F[\varphi] \right\rangle_{1h} = \left\langle \frac{1}{N} \frac{\partial F[\varphi]}{\partial\varphi_{ji}} \right\rangle_{1h} \tag{13.8}$$

which results from the invariance of the measure under the infinitesimal shift (13.3). It is enough to choose $F[\varphi] = \varphi_{kl}$ and to calculate the derivatives of the Gaussian potential on the LHS and of φ_{kl} on the RHS by the use of

$$\frac{\partial \varphi_{kl}}{\partial \varphi_{ji}} = \delta_{il}\delta_{kj} \,. \tag{13.9}$$

Problem 13.1 Derive Eq. (13.7) by calculating the Gaussian integral.

Solution Let us substitute $\varphi_{ij} = (X_{ij} + iY_{ij})/\sqrt{2}$ with real symmetric $X_{ij} = X_{ji}$ and antisymmetric $Y_{ij} = -Y_{ji}$. The number of independent components is $N(N+1)/2$ for X and $N(N-1)/2$ for Y, i.e. N^2 in total as it should be. We then obtain

$$
\begin{aligned}
\langle \varphi_{ij}\varphi_{kl}\rangle_{\text{Gauss}} &= \frac{1}{2}\langle X_{ij}X_{kl}\rangle_{\text{Gauss}} - \frac{1}{2}\langle Y_{ij}Y_{kl}\rangle_{\text{Gauss}} \\
&= \frac{1}{2N}\left(\delta_{ik}\delta_{jl} + \delta_{il}\delta_{jk}\right) - \frac{1}{2N}\left(\delta_{ik}\delta_{jl} - \delta_{il}\delta_{jk}\right) \\
&= \frac{1}{N}\delta_{il}\delta_{jk}
\end{aligned}
\tag{13.10}
$$

as in Eq. (13.7).

The Feynman graphs of the Hermitian one-matrix model can be represented by the double index lines quite similarly to Sect. 11.1. Now there are no commutators so all vertices are symmetric in the indices.

Generically, the Hermitian one-matrix model generates graphs of a zero-dimensional field theory. Since there is no momentum variable and each propagator is $1/N$, the contribution of each graph is simply $1/N^{2\,\text{genus}}$ times a symmetry factor. Hence solving the Hermitian one-matrix model is equivalent to calculating the number of graphs with a given genus.

A very important property of the model is that $\mathrm{tr}\,V(\varphi)$ depends only on the eigenvalues of the matrix φ. Similarly, representing φ in a canonical form

$$\varphi = VPV^{\dagger} \tag{13.11}$$

with unitary $N \times N$ matrix V and diagonal

$$P = \mathrm{diag}\,\{p_1,\ldots,p_N\}\,, \tag{13.12}$$

the measure (13.2) can be written in a standard Weyl form

$$\mathrm{d}\varphi = \mathrm{d}V \prod_{i=1}^{N} \mathrm{d}p_i\, \Delta^2(P)\,, \tag{13.13}$$

where

$$\Delta(P) \;=\; \prod_{i<j} (p_i - p_j) \tag{13.14}$$

is the Vandermonde determinant.

We see that the contribution from angular degrees of freedom residing in V factorizes, so the partition function (13.1) becomes

$$Z_{1h} \;=\; \int_{-\infty}^{+\infty} \prod_{i=1}^{N} \mathrm{d}p_i \prod_{i<j} (p_i - p_j)^2 \exp\left[-N \sum_{i=1}^{N} V(p_i)\right]. \tag{13.15}$$

Problem 13.2 Derive Eq. (13.13).

Solution The representation (13.11) of φ in the canonical form reminds one of fixing a gauge where V are matrices of a gauge transformation. The measure $\mathrm{d}\varphi$ can then be represented as

$$\mathrm{d}\varphi \;=\; \mathrm{d}V \prod_{i=1}^{N} \mathrm{d}p_i \, J(P), \tag{13.16}$$

where the Jacobian $J(P)$ depends only on the eigenvalues of φ since $\mathrm{d}\varphi$ is invariant under

$$\varphi \;\to\; \Omega \varphi \Omega^\dagger. \tag{13.17}$$

To calculate the Jacobian, it is convenient [BIZ80] to apply the Faddeev–Popov method inserting

$$1 \;=\; \Delta^2(\varphi) \int \mathrm{d}\Omega \prod_{i<j} \delta^{(2)}\big([\Omega\varphi\Omega^\dagger]_{ij}\big) \tag{13.18}$$

in the measure $\mathrm{d}\varphi$. Here $\mathrm{d}\Omega$ is the Haar measure for $U(N)$ and the $N^2 - N$ distributions are only present for off-diagonal components. It is easy to see that $\Delta^2(\varphi)$ depends solely on eigenvalues of φ since the measure $\mathrm{d}\Omega$ is invariant under multiplication by a unitary matrix.

We can insert the unity (13.18) into the integral of a function $f(\varphi)$ which is invariant under (13.17) and hence depends only on the eigenvalues of φ:

$$\int \mathrm{d}\varphi f(\varphi) \;=\; \int \mathrm{d}\varphi\, \Delta^2(\varphi) \int \mathrm{d}\Omega \prod_{i<j} \delta^{(2)}\big([\Omega\varphi\Omega^\dagger]_{ij}\big) f(\varphi)$$

$$=\; \int \mathrm{d}\Omega \int \prod_{i<j} \mathrm{d}\varphi_{ij}\, \delta^{(2)}(\varphi_{ij}) \prod_{i=1}^{N} \mathrm{d}p_i\, \Delta^2(P)\, f(P)$$

$$=\; \prod_{i=1}^{N} \mathrm{d}p_i\, \Delta^2(P)\, f(P). \tag{13.19}$$

Comparing with Eq. (13.16), we conclude that $J(P) = \Delta^2(P)$.

Let us now find $\Delta^2(P)$ by evaluating the integral over Ω in Eq. (13.18). We first reduce φ to the diagonal form (13.12) by the transformation (13.17). Then Ωs which are essential in the integral of the delta-function are close to a diagonal unitary matrix Ω_0. The integral can be calculated by substituting $\Omega = (1 + ih)\Omega_0$ with an infinitesimal off-diagonal Hermitian matrix h. Since $\left[\Omega P \Omega^\dagger\right]_{ij} = ih_{ij}(p_i - p_j)$ for $i \neq j$, we obtain

$$
\begin{aligned}
\Delta^{-2}(P) &= \int d\Omega_0 \int \prod_{i<j} dh_{ij}\, \delta^{(2)}(h_{ij}(p_i - p_j)) \\
&= \prod_{i<j} (p_i - p_j)^{-2}\,.
\end{aligned}
\tag{13.20}
$$

This reproduces the Weyl measure (13.13).

We have therefore rewritten the Hermitian one-matrix model via N degrees of freedom in the spirit of Sect. 11.7. The integral on the RHS of Eq. (13.15) can be calculated as $N \to \infty$ using the saddle-point method.

To write down the saddle-point equation, let us introduce the spectral density

$$
\rho(p) = \frac{1}{N}\sum_{i=1}^{N} \delta^{(1)}(p - p_i)
\tag{13.21}
$$

which becomes a continuous function of p as $N \to \infty$. It describes the distribution of eigenvalues of the matrix φ.

The spectral density (13.21) obeys

$$
\rho(p) \geq 0,
\tag{13.22}
$$

$$
\int dp\, \rho(p) = 1
\tag{13.23}
$$

as it follows from the definition (13.21).

Given the spectral density, we have

$$
\frac{1}{N}\operatorname{tr}\varphi^k = \int dp\, \rho(p)\, p^k
\tag{13.24}
$$

and, in particular,

$$
\frac{1}{N}\operatorname{tr}V(\varphi) = \int dp\, \rho(p)\, V(p)\,.
\tag{13.25}
$$

In the large-N limit where the integral over p_i is dominated by a saddle-point configuration, we obtain

$$
W_k \overset{\text{def}}{=} \left\langle \frac{1}{N}\operatorname{tr}\varphi^k \right\rangle_{1\text{h}} \overset{N=\infty}{=} \int dp\, \rho_{\text{sp}}(p)\, p^k\,,
\tag{13.26}
$$

where $\rho_{\text{sp}}(p)$ describes the distribution of eigenvalues at the saddle point.

Extrema of the integrand on the RHS of Eq. (13.15) are reached when

$$V'(p_i) \;=\; \frac{2}{N}\sum_{j\neq i}\frac{1}{p_i - p_j}\,, \qquad\qquad (13.27)$$

where $V'(p) = \mathrm{d}V(p)/\mathrm{d}p$.

This determines the large-N saddle-point equation to be [BIP78]

$$V'(p) \;=\; 2\int \mathrm{d}\lambda\frac{\rho(\lambda)}{p - \lambda}\qquad \boxed{p \in \text{support of } \rho}\,, \qquad (13.28)$$

where the RHS involves the principal part of the integral. Equation (13.28) holds only when p belongs to the support of ρ as is clear from the derivation.

Before solving the saddle-point equation (13.28), let us mention that the support of ρ must be finite for a general potential, say, the support is to be included in an interval $[a, b]$. Otherwise, the saddle-point equation would be inconsistent as $p \to \infty$ except for $V(p)$ which behaves asymptotically as $2\ln |p|$.

Remark on discretization of random surfaces

Matrix models are associated generically [Kaz85, Dav85, ADF85, KKM85] with discretization of random surfaces. The simplest Hermitian one-matrix model corresponds to a zero-dimensional embedding space, i.e. to two-dimensional Euclidean quantum gravity described by the partition function

$$Z_{\mathrm{2DG}} \;=\; \int \mathcal{D}g\, e^{-\int \mathrm{d}^2 x \sqrt{g}(\Lambda - R/4\pi\mathcal{G})}$$

$$=\; \int \mathcal{D}g\, e^{-\int \mathrm{d}^2 x \sqrt{g}\Lambda + \chi/\mathcal{G}}\,. \qquad (13.29)$$

Here Λ denotes the cosmological constant, R is the scalar curvature, and χ is the Euler characteristic of the two-dimensional world, while the coupling $\mathcal{G}$ weights topologies. The path integral in Eq. (13.29) is over all metrics $g_{\mu\nu}(x)$.

The idea of dynamical triangulation of random surfaces is to approximate a surface by a set of equilateral triangles. The coordination number (the number of triangles meeting at a vertex) does not necessarily equal six, which represents internal curvature of the surface.

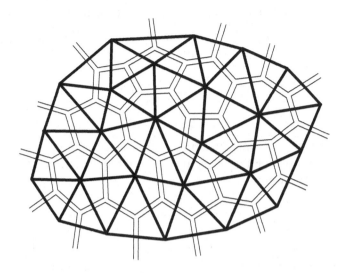

Fig. 13.1. Generic graph constructed from equilateral triangles (depicted by bold lines) and associated with dynamical triangulation of random surfaces. Its dual graph (depicted by double lines) coincides with that in the Hermitian one-matrix model with a cubic interaction potential.

The partition function (13.29) is approximated by

$$Z_{\text{DT}} \; = \; \sum_h e^{2(1-h)/\mathcal{G}} \sum_{T_h} e^{-\sigma n_t} , \qquad (13.30)$$

where we split the sum over triangles into the sum over genus h and the sum over all possible triangulations T_h at fixed h. In (13.30) n_t denotes the number of triangles which is not fixed at the outset, but rather is a dynamical variable similar to that in Problem 1.12 on p. 27 for random paths.

The partition function (13.30) can be represented as a matrix model. A graph dual to a generic set of equilateral triangles coincides with a graph in the Hermitian one-matrix model with a cubic interaction as is depicted in Fig. 13.1. The precise statement is that Z_{DT} equals the (logarithm of the) partition function (13.1) with $N = \exp\left(1/\mathcal{G}\right)$ and the cubic coupling constant $g_3 = \exp\left(-\sigma\right)$. This can be easily shown by comparing the graphs. The logarithm is needed to pick up connected graphs in the matrix model.

Analogously, the interaction $\text{tr}\,\varphi^k$ in the matrix model is associated with discretization of random surfaces by regular k-gons, the area of which is $k - 2$ times the area of the equilateral triangle.

13.2 Hermitian one-matrix model (solution at $N = \infty$)

The saddle-point equation (13.28) can be solved by the Riemann–Hilbert method introducing an analytic function

$$W(\omega) \;=\; \int\limits_a^b d\lambda \frac{\rho(\lambda)}{\omega - \lambda}.$$

(13.31)

It has cuts (or simply one cut) in the complex ω-plane at the real axis where ρ has support. These cuts are included in the interval $[a, b]$.

Asymptotically, we have

$$W(\omega) \;\to\; \frac{1}{\omega}$$

(13.32)

as $\omega \to \infty$, as a consequence of the normalization (13.23) of ρ.

The idea is now to have $\operatorname{Re} W = V'/2$ at the support of ρ, i.e. where $\operatorname{Im} W \neq 0$, to satisfy Eq. (13.28). This is equivalent to the equation

$$\operatorname{Im} \left(V'W - W^2\right) \;=\; \left(V' - 2\operatorname{Re} W\right) \operatorname{Im} W \;=\; 0$$

(13.33)

which holds for the whole real axis: at the support owing to Eq. (13.28) and outside of the support since there $\operatorname{Im} W = 0$.

Equation (13.33) tells us that

$$V'W - W^2 \;=\; Q,$$

(13.34)

where an analytic function $Q(\omega)$ should have no singularities at the real axis. For a polynomial $V(p)$ it must be a polynomial of the same degree as $V'(p)/p$ to satisfy asymptotically Eq. (13.32).

We therefore find

$$W \;=\; \frac{V'}{2} - \frac{1}{2}\sqrt{(V')^2 - 4Q},$$

(13.35)

where the minus sign is chosen to again provide the asymptotic behavior (13.32). Then ρ is given by the discontinuity of this W at the cuts (cut):

$$W(p \pm i0) \;=\; \frac{V'(p)}{2} \mp i\pi\rho(p).$$

(13.36)

The simplest example is the Hermitian one-matrix model with the Gaussian potential when $V'(p) = \mu p$ ($\mu \equiv g_2$). The asymptotic behavior of Eq. (13.34) fixes $Q(p) = \mu$. Then Eq. (13.35) simplifies to

$$W(\omega) \;=\; \frac{\mu\omega}{2} - \frac{\mu}{2}\sqrt{\omega^2 - \frac{4}{\mu}}$$

(13.37)

for which the discontinuity determines the spectral density

$$\rho(p) = \frac{\mu}{2\pi}\sqrt{\frac{4}{\mu} - p^2}. \tag{13.38}$$

Note that this spectral density is nonnegative and has support on a finite interval $[-2/\sqrt{\mu}, 2/\sqrt{\mu}]$. The spectral density (13.38) was first calculated by Wigner [Wig51] and is called Wigner's *semicircle law*.

Problem 13.3 Calculate the density of eigenvalues for the Gaussian Hermitian one-matrix model using the Schwinger–Dyson equations.

Solution The calculation is similar to that in Problem 12.13 on p. 285. The difference is that now φ_{ij} is a Hermitian matrix with eigenvalues p_i which can take on values along the whole real axis.

The Schwinger–Dyson equations for W_n, defined in the Hermitian one-matrix model by Eq. (13.26), can be obtained from Eq. (13.8) by choosing $F[\varphi] = (\varphi^n)_{ij}$. Proceeding as before and using the large-N factorization, we obtain the set of equations [Wad81]

$$\left. \begin{array}{rcl} \mu\, W_{n+1} & = & \displaystyle\sum_{k=0}^{n-1} W_k W_{n-k} \quad \text{for } n \ge 0, \\[2mm] W_0 & = & 1. \end{array} \right\} \tag{13.39}$$

Introducing the generating function

$$W(p) \equiv \left\langle \frac{1}{N} \operatorname{tr} \frac{1}{p-\varphi} \right\rangle_{1h} = \sum_{n=0}^{\infty} \frac{W_n}{p^{n+1}}, \tag{13.40}$$

we rewrite Eq. (13.39) as the quadratic equation

$$\mu\, p\, W(p) - \mu = W^2(p), \tag{13.41}$$

the solution of which is given by Eq. (13.37) determining the spectral density (13.38) which has support on a finite interval $[-2/\sqrt{\mu}, 2/\sqrt{\mu}]$ in analogy with the unitary one-matrix model it the weak-coupling regime.

We have already met the semicircle distribution in Problem 12.13 for the spectral density of the unitary one-matrix model at small λ (see Eq. (12.158)). This is because we can always substitute $U = \exp{(i\varphi)}$ where U is unitary and φ is Hermitian and expand for small λ in φ up to the quadratic term. We then obtain the Hermitian model (13.1) with $\mu = 1/\lambda$ from the unitary model (12.135).

For a general polynomial potential, we are looking for a one-cut solution at small couplings $g_3, g_4, \dots$ bearing in mind that it should look similar to the Gaussian case which is perturbed by the interactions. The expression (13.35) then takes the form

$$W(\omega) = \frac{V'(\omega)}{2} - \frac{M(\omega)}{2}\sqrt{(\omega - a)(\omega - b)}, \tag{13.42}$$

where a and b are the ends of the cut and $M(\omega)$ is a polynomial of degree $K - 2$ if $V(\omega)$ is a polynomial of degree K.

The coefficients of M are determined together with a and b from the asymptotic condition

$$\frac{V'(\omega)}{\sqrt{(\omega - a)(\omega - b)}} - M(\omega) \;\to\; \frac{2}{\omega^2} \tag{13.43}$$

as $\omega \to \infty$. There are precisely K conditions in Eq. (13.43) to unambiguously determine these K numbers.

A solution is acceptable if $M(p)$ is not negative in the interval $[a, b]$. Then the spectral density equals

$$\rho(p) \;=\; \frac{M(p)}{2\pi} \sqrt{(p - a)(b - p)} \tag{13.44}$$

which solves the problem for a general polynomial potential. This solution was first obtained in [BIP78] for cubic and quartic potentials.

For small values of the couplings $g_3, g_4, \ldots$, the one-cut solution is always realized. With increasing coupling, a third-order phase transition of the Gross–Witten type (see Sect. 12.9) may occur after which a more complicated multicut solution is realized.

An example of when such a phase transition happens is the quartic potential

$$V(p) \;=\; \frac{\mu}{2} p^2 + \frac{g_4}{4} p^4 \tag{13.45}$$

when the one-cut solution exists only for $-g_4 \le \mu^2/12$.

Problem 13.4 Elaborate the solution (13.44) for the quartic potential (13.45).

Solution Substituting the quartic potential (13.45) into Eq. (13.42), we obtain

$$W(p) \;=\; \frac{\mu p + g_4 p^3}{2} - \left(\frac{\mu + g_4 p^2 + g_4 a^2/2}{2} \right) \sqrt{p^2 - a^2}, \tag{13.46}$$

where

$$a^2 \;=\; \frac{2\mu}{3 g_4} \left(-1 + \sqrt{1 + \frac{12 g_4}{\mu^2}} \right) \tag{13.47}$$

reproducing $a^2 \to 4/\mu$ as $g_4 \to 0$.

The RHS of Eq. (13.47) is well-defined only for $-g_4 \le \mu^2/12$ which determines the critical value $(g_4)_* = -\mu^2/12$. At $-g_4 \to \mu^2/12$ from below, two zeros of $M(p)$ approach two ends of the cut so that

$$\rho(p) \;\to\; \frac{\mu^2}{24\pi} \left(\frac{8}{\mu} - p^2 \right)^{3/2}. \tag{13.48}$$

The one-cut solution is no longer realized for $-g_4 > \mu^2/12$.

The reason why we are interested in the one-cut solution is simple. This solution sums planar graphs of the Hermitian one-matrix model.

Remember that an effective expansion parameter associated with each quartic vertex in graphs is $-g_4/\mu^2$. Therefore, g_4 must be negative for the weight of each graph to be positive. At the critical value $-\mu^2/(g_4)_* = 12$, the sum of the planar graphs diverges, which determines the constant in Eq. (11.23) for the number of planar graphs with quartic vertices. Analogously, for a cubic potential, the solution (13.42) gives $\mu^3(g_3)_*^{-2} = 12\sqrt{3}$, which results in Eq. (11.43) for the number of trivalent planar graphs.

13.3 The loop equation

In the previous section, we have solved the Hermitian one-matrix model at $N = \infty$ using the saddle-point equation for the spectral density. This method of solution was historically the first one but cannot be extended to higher orders in $1/N^2$. In this section we present a very closely related method of solving the matrix model using loop equations which allows us to find a solution systematically order by order in $1/N^2$.

Choosing $F[\varphi] = (p - \varphi)_{ij}^{-1}$ in Eq. (13.8), we obtain the Schwinger–Dyson equation

$$\left\langle \frac{1}{N} \operatorname{tr} \frac{V'(\varphi)}{p - \varphi} \right\rangle_{1h} = \left\langle \frac{1}{N^2} \operatorname{tr} \frac{1}{p - \varphi} \operatorname{tr} \frac{1}{p - \varphi} \right\rangle_{1h}. \tag{13.49}$$

Equation (13.49) can be expressed entirely via the resolvent

$$W(p) = \left\langle \frac{1}{N} \operatorname{tr} \frac{1}{p - \varphi} \right\rangle_{1h} \tag{13.50}$$

which is a Laplace transform of the "Wilson loop":

$$W(p) = \int_0^\infty \mathrm{d}l \, e^{-pl} \left\langle \frac{1}{N} \operatorname{tr} e^{l\varphi} \right\rangle_{1h}. \tag{13.51}$$

The resulting loop equation reads

$$\int_{C_1} \frac{\mathrm{d}\omega}{2\pi i} \frac{V'(\omega)}{(p - \omega)} W(\omega) = W^2(p) + \frac{1}{N^2} \frac{\delta}{\delta V(p)} W(p), \tag{13.52}$$

where the contour C_1 encloses counterclockwise singularities of $W(\omega)$ leaving outside the pole at $\omega = p$ as depicted in Fig. 13.2. The contour integral on the LHS simply acts as a projector picking up negative powers of p.

At $N = \infty$, when the second term on the RHS can be omitted, Eq. (13.52) coincides for polynomial V with Eq. (13.34) derived above by

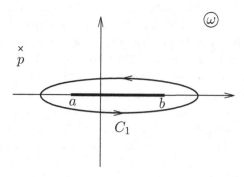

Fig. 13.2.   Contour C_1 in the ω-plane for integration on the LHS of Eq. (13.52).

the other method. The polynomial Q can then be calculated by deforming the contour to infinity and taking the residue at $\omega = \infty$. The residue at $\omega = p$ simply yields $V'(p)W(p)$ which enters the LHS of Eq. (13.34).

The first term on the RHS of Eq. (13.52) is associated with the factorized part of the correlator, while the second term represents the connected part of the two-loop correlator which is $\sim 1/N^2$ as $N \to \infty$. It involves the variational derivative

$$\frac{\delta}{\delta V(p)} = -\sum_{k=0}^{\infty} p^{-k-1}\frac{\partial}{\partial t_k} \tag{13.53}$$

acting on $W(p)$. For this reason the operator (13.53) is often called the *loop insertion operator*.

Consequently, Eq. (13.52) is closed and determines $W(p)$ unambiguously, providing the boundary condition $W(p) \to 1/p$ is imposed as $p \to \infty$.

Note that we obtained a *single* (functional) equation for $W(p)$. This is due to the fact that $\mathrm{tr}\,V(\varphi)$ contains a complete set of traces $\mathrm{tr}\,\varphi^k$. They become independent as $N \to \infty$.

Problem 13.5 Obtain Eq. (13.52) from Eq. (13.49).

Solution The coupling t_k plays the role of a source for $\mathrm{tr}\,\varphi^k$:

$$\left\langle \frac{1}{N}\,\mathrm{tr}\,\varphi^k \right\rangle_{1\mathrm{h}} = -\frac{\partial}{\partial t_k}F\,, \tag{13.54}$$

where the free energy is

$$F = \frac{1}{N^2}\ln Z_{1\mathrm{h}}\,. \tag{13.55}$$

Analogously, using the definition (13.53) and Eq. (13.54), we find

$$\left\langle \frac{1}{N}\,\mathrm{tr}\,\frac{1}{p-\varphi} \right\rangle_{1\mathrm{h}} = \frac{\delta}{\delta V(p)}F\,. \tag{13.56}$$

Applying the operator (13.53) one more time, we obtain [AM90] the connected correlator of two Wilson loops:

$$\frac{\delta}{\delta V(p_2)} W(p_1) = \left\langle \operatorname{tr} \frac{1}{(p_1 - \varphi)} \operatorname{tr} \frac{1}{(p_2 - \varphi)} \right\rangle^{\mathrm{conn}}_{\mathrm{1h}}, \tag{13.57}$$

which enters the RHS of Eq. (13.52). The higher-loop correlators can be obtained by further applying $\delta/\delta V(p_i)$.

Instead of introducing the sources t_k, we can consider $V(p)$ as a source for the Wilson loop $\operatorname{tr}[1/(p - \varphi)]$ from the very beginning by writing

$$\operatorname{tr} V(\varphi) = \int\limits_{-i\infty+0}^{+i\infty+0} \frac{d\omega}{2\pi i} V(\omega) \operatorname{tr} \frac{1}{\omega - \varphi}. \tag{13.58}$$

According to this definition $\delta V(p)/\delta V(q) = 1/(p - q)$ which plays the role of a delta-function when integrated along the imaginary axis.

The LHS of Eq. (13.49) is transformed into the LHS of Eq. (13.52) using

$$\frac{1}{N} \operatorname{tr} \frac{V'(\varphi)}{p - \varphi} = \int\limits_{-i\infty+0}^{+i\infty+0} \frac{d\omega}{2\pi i} \frac{V'(\omega)}{(p - \omega)} \frac{1}{N} \operatorname{tr} \frac{1}{\omega - \varphi}, \tag{13.59}$$

taking the average and deforming the contour to encircle singularities of $W(\omega)$.

Remark on the Virasoro constraints

The loop equation (13.52) can be represented as a set of Virasoro constraints imposed on the partition function.

We first rewrite Eq. (13.52) using the definitions (13.1) and (13.4) as

$$\frac{1}{Z_{1h}} \sum_{n=-1}^{\infty} \frac{1}{p^{n+2}} L_n Z_{1h} = 0, \tag{13.60}$$

where the operators

$$L_n = \sum_{k=0}^{\infty} k t_k \frac{\partial}{\partial t_{k+n}} + \frac{1}{N^2} \sum_{0 \leq k \leq n} \frac{\partial^2}{\partial t_k \partial t_{n-k}} \tag{13.61}$$

satisfy the Virasoro algebra

$$[L_n, L_m] = (n - m) L_{n+m}. \tag{13.62}$$

Equation (13.52) is therefore represented as the Virasoro constraints

$$L_n Z_{1h} = 0 \qquad \text{for } n \geq -1. \tag{13.63}$$

It is enough to consider the constraints (13.63) only with $n = 2$ and $n = -1$. Then all the others are satisfied because of Eq. (13.62).

13.4 Solution in $1/N$

Equation (13.52) can be solved order by order of the genus expansion in $1/N^2$.

The genus zero one-cut solution to Eq. (13.52) can be written as [Mig83]

$$W_0(p) = \int\limits_{C_1} \frac{d\omega}{4\pi i} \frac{V'(\omega)}{(p-\omega)} \sqrt{\frac{(p-a)(p-b)}{(\omega-a)(\omega-b)}}, \qquad (13.64)$$

where a and b are determined by

$$\int\limits_{C_1} \frac{d\omega}{2\pi i} \frac{V'(\omega)}{\sqrt{(\omega-a)(\omega-b)}} = 0, \qquad \int\limits_{C_1} \frac{d\omega}{2\pi i} \frac{\omega V'(\omega)}{\sqrt{(\omega-a)(\omega-b)}} = 2. \qquad (13.65)$$

Performing the contour integral in (13.64) by taking the residues at $\omega = p$ and $\omega = \infty$, we reproduce Eq. (13.42) for polynomial V. However, Eq. (13.64) remains valid in the more general case of nonpolynomial V, e.g. having logarithmic singularities. The position of the cut is always such as to avoid these singularities of V.

Problem 13.6 Reproduce Eqs. (13.37) and (13.46) for the Gaussian and quartic potentials from the general one-cut solution (13.64) and (13.65).

Solution We substitute $a = -b$ for even potentials, then the first equation in (13.65) is always satisfied, while the second one yields

$$\sum_{j=1}^{J} g_{2j} \frac{(2j)!}{(j!)^2} \left(\frac{a}{2}\right)^{2j} = 2 \qquad (13.66)$$

for an even polynomial potential of degree $K = 2J$. This equation is derived by expanding the square root in a^2/ω^2 and taking the residue at infinity. Analogously, Eq. (13.64) yields

$$M(p) = \sum_{j=1}^{J} p^{2j-2} \sum_{k=0}^{J-j} g_{2k+2j} \frac{(2k)!}{(k!)^2} \left(\frac{a}{2}\right)^{2k} \qquad (13.67)$$

for the polynomial $M(p)$ in Eq. (13.42).

A solution to Eq. (13.66) reproduces the above explicit calculation for the Gaussian and quartic potentials. Analogously, Eqs. (13.37) and (13.46) are reproduced by substituting Eq. (13.67) into Eq. (13.42).

Problem 13.7 Elaborate the solution (13.64) and (13.65) for the Penner model where the potential is logarithmic:

$$V(\varphi) = \varphi - \lambda \ln \varphi. \qquad (13.68)$$

Solution The calculation is similar to the previous Problem, while now the residue is to be taken at $\omega = 0$ since

$$V'(\omega) = 1 - \frac{\lambda}{\omega} \tag{13.69}$$

has a pole there. For $\lambda > 0$ we find

$$a = 2 + \lambda - 2\sqrt{1 + \lambda}, \qquad b = 2 + \lambda + 2\sqrt{1 + \lambda} \tag{13.70}$$

and

$$\rho(p) = \frac{\sqrt{(b - p)(p - a)}}{2\pi p} \tag{13.71}$$

so that $W(p)$ is analytic at $p = 0$. The Gaussian formula (13.38) is reproduced as $\lambda \to \infty$ substituting $p \to \lambda + p$.

Note that both a and b are positive so the support is located for $\lambda > 0$ at the positive real axis where $\rho(p) > 0$. This is a manifestation of the general property already mentioned that the cut always avoids possible singularities of the potential. The location of the support of eigenvalues in the complex ω-plane for $\lambda < 0$ is studied in [AKM94].

The multiloop correlators in genus zero can be obtained from $W_0(p)$ given by Eq. (13.64) applying the loop insertion operator (13.53). For example, the two-loop correlator [AJM90]

$$W_0(p, q) = \frac{1}{4(p - q)^2} \left\{ \frac{2pq - (p + q)(a + b) + 2ab}{\sqrt{(p - a)(p - b)}\sqrt{(q - a)(q - b)}} - 2 \right\} \tag{13.72}$$

depends on the potential V only via a and b but not explicitly. This property is called *universality*.[*] It does not hold for higher multiloop correlators.

To calculate the $1/N^2$ correction to the genus-zero result (13.64), we substitute

$$W_0(p, p) = \frac{(a - b)^2}{16(p - a)^2(p - b)^2} \tag{13.73}$$

extracted from Eq. (13.72) into the RHS of Eq. (13.52). We can now obtain $W_1(p)$ by solving a linear equation which, in turn, determines F_1.

An advantage of this method of solving the Hermitian one-matrix model using the loop equation over the orthogonal polynomial technique, used originally [Bes79, IZ80, BIZ80] in calculating the higher genera for polynomial potentials, is that now the free energy generates all multiloop correlators at a given genus.

---

[*] An analog of this correlator in condensed-matter physics is the correlator of two densities of energy eigenvalues which is universal [BZ93].

Remark on the iterative solution

The iterative procedure [ACK93] of solving the loop equation is based on the genus-zero solution (13.64). Inserting the genus expansion of $W(p)$ and F:

$$W(p) = \sum_{h=0}^{\infty} \frac{1}{N^{2h}} W_h(p), \qquad F = \sum_{h=0}^{\infty} \frac{1}{N^{2h}} F_h \quad \text{with} \quad W_h(p) = \frac{\delta F_h}{\delta V(p)},$$

$$(13.74)$$

into Eq. (13.52), we obtain the following equation for $W_h(p)$ at $h \geq 1$:

$$\int_{C_1} \frac{d\omega}{2\pi i} \frac{V'(\omega)}{(p - \omega)} W_h(\omega) - 2 W_0(p) W_h(p)$$

$$= \sum_{h'=1}^{h-1} W_{h'}(p) W_{h-h'}(p) + \frac{\delta}{\delta V(p)} W_{h-1}(p). \qquad (13.75)$$

It expresses $W_h(p)$ entirely in terms of $W_{h'}(p)$ with $h' < h$. This makes it possible to solve Eq. (13.75) iteratively genus by genus.

The iterative procedure simplifies if we introduce, instead of the coupling constants t_j, the moments M_k and J_k defined for $k \geq 1$ by

$$\left. \begin{aligned} M_k &= \int_{C_1} \frac{d\omega}{2\pi i} \frac{V'(\omega)}{(\omega - a)^{k+1/2} (\omega - b)^{1/2}}, \\[2mm] J_k &= \int_{C_1} \frac{d\omega}{2\pi i} \frac{V'(\omega)}{(\omega - a)^{1/2} (\omega - b)^{k+1/2}}. \end{aligned} \right\} \qquad (13.76)$$

These moments depend on the couplings t_j both explicitly and via a and b which are determined by Eq. (13.65). Note that M_k and J_k depend explicitly only on t_j with $j \geq k + 1$.

Problem 13.8 Elaborate the moments (13.76) for the quartic potential (13.45).

Solution   Given Eq. (13.47) for a, we simply calculate the moments (13.76) for the quartic potential (13.45), taking the residue at infinity. This results in an explicit representation of the moments via μ and g_4 which are given by the algebraic formulas

$$M_1 = J_1 = \mu + \frac{3}{2} g_4 a^2, \qquad M_2 = -J_2 = 2 g_4 a, \qquad M_3 = J_3 = g_4,$$

$$(13.77)$$

and $M_k = J_k = 0$ for $k \geq 4$.

The main motivation for introducing the moments (13.76) is that $W_h(\lambda)$ depends only on $2 \times (3h - 1)$ lower moments ($2 \times (3h - 2)$ for F_h). This is in contrast to the t-dependence of W_h and F_h which always depend on the infinite set of t_j ($1 \le j < \infty$).

To find F_h, we first solve Eq. (13.75) for $W_h(\lambda)$ and then use the last equation in (13.74). The result in genus one reads [ACM92]

$$F_1 = -\frac{1}{24} \ln (M_1 J_1) - \frac{1}{6} \ln (b - a). \tag{13.78}$$

The genus-two results are obtained in [ACK93]. More details on this subject can be found in Section 4.3 of the book [ADJ97].

The results for F_1 and F_2 for the quartic potential were first obtained in [Bes79] using the method of orthogonal polynomials.

13.5 Continuum limit

Continuum limits of the Hermitian one-matrix model are reached at the points of phase transitions. While no phase transition is possible at finite N since the system has a finite number of degrees of freedom, it may occur as $N \to \infty$ which plays the role of a statistical limit as has already been pointed out in the Remark in Sect. 11.8.

This third-order phase transition is of Gross–Witten type (see Sect. 12.9). It is associated with divergence of the sum over graphs at each fixed genus rather than with divergence of the sum over all graphs. The contribution of a graph with n_0 trivalent vertices is $\sim (-g_3)^{n_0}$ but an entropy (= the number) of such graphs at fixed genus is given by Eq. (11.23) so the sum can diverge at a certain critical value of g_3 calculated in Sect. 13.2.

This divergence has nothing to do with the divergence of the sum over all graphs which always occurs owing to a factorial growth of the total number of diagrams. The latter divergence is simply associated with the divergence of the integral over φ. For an even potential V, the couplings g_k are negative for $k > 2$ so the potential V is upside-down.

The phase structure of the Hermitian one-matrix model can be determined from the spectral density $\rho(p)$ given by (13.44) which vanishes under normal circumstances as a square root at both ends of its support. The critical behavior emerges when one or more roots of $M(p)$ approaches the end points a or b.

For example, the even potential

$$V(\varphi) = \frac{1}{\beta} \sum_{j=1}^{J} (-1)^{j-1} \frac{J!(j-1)!}{(J-j)!(2j)!} \left(\frac{\mu \varphi^2}{2} \right)^j \tag{13.79}$$

becomes critical at $\beta = 1$ when $(J - 1)$ zeros of $M(p)$ approach each of the two end points of the cut. They are determined by the equation

$$\beta = \sum_{j=1}^{J} (-1)^{j-1} \frac{J!}{(J-j)!j!} \left(\frac{\mu a^2}{8}\right)^j = 1 - \left(1 - \frac{\mu a^2}{8}\right)^J \qquad (13.80)$$

which results from the substitution of Eq. (13.79) into Eq. (13.66).

The critical potential (13.79) with $\beta = 1$ is associated with the Jth *multicritical* point [Kaz89]. The case of $J = 2$ describes two-dimensional quantum gravity. The resulting continuum theory is unitary only at this critical point.

The continuum limit can be obtained near the critical point:

$$p^2 \rightarrow a_c^2 + \epsilon \pi, \qquad a^2 \rightarrow a_c^2 - \epsilon \sqrt{\Lambda}, \qquad (13.81)$$

so that π plays the role of the continuum momentum and Λ is the cosmological constant.

The susceptibility near the critical point can be represented by the genus expansion

$$f(\beta) \stackrel{\text{def}}{=} \frac{1}{N^2} \left(\frac{d}{d\,1/\beta}\right)^2 \ln Z_{1h}$$

$$= \text{const} + \sum_h N^{-2h} (1 - \beta)^{-\gamma_h} f_h \qquad (13.82)$$

with the indices

$$\gamma_h = -\frac{1}{J} + \frac{2J+1}{J} h. \qquad (13.83)$$

The genus-zero contribution to the susceptibility (13.82) does not diverge but rather exhibits a root singularity. This can be easily deduced by noting that f to genus zero is analytic in a^2 and contains a term $\sim \epsilon$, when the expansion (13.81) is substituted. According to Eq. (13.80), we have

$$\epsilon = \frac{a_c^2}{\sqrt{\Lambda}} (1 - \beta)^{1/J} \qquad (13.84)$$

for the Jth multicritical point which explains Eqs. (13.82) and (13.83) to genus zero.

The dimensional cutoff ϵ should depend on N in such a way for the parameter $\mathcal{G} = N^{-2} \epsilon^{-2J-1}$ of the genus expansion to remain finite as $N \rightarrow \infty$. Then all terms of the genus expansion contribute in the continuum limit. This continuum limit was obtained in [BK90, DS90, GM90a] and is called the *double scaling limit*.

The double scaling limit of the partition function (13.1) determines the genus expansion of the continuum partition function:

$$
\ln Z_{1h} \quad \rightarrow \quad \text{const} + \sum_h N^{2-2h} \frac{(1-\beta)^{2-\gamma_h}}{(2-\gamma_h)(1-\gamma_h)} f_h
$$

$$
= \quad \text{const} + \sum_h \left(\frac{\mathcal{G}}{\Lambda^{J+1/2}} \right)^{h-1} \frac{a_c^{2(2J+1)(h-1)}}{(2-\gamma_h)(1-\gamma_h)} f_h . \qquad (13.85)
$$

At $J = 2$ this determines the partition function (13.29) of two-dimensional quantum gravity.

It is possible to construct explicitly a continuum theory which interpolates between multicritical points. We associate with Jth multicritical behavior a conformal operator of a certain scale dimension and introduce a proper source T_k.

The relation between the set of sources t_{2k} for the Hermitian one-matrix model with an even potential when $t_{2k+1} = 0$ and their continuum counterparts T_k can be obtained* from the equation

$$
W(p) - \frac{1}{2} V'(p) = \frac{1}{\epsilon \sqrt{\mathcal{G}}} \left[2 W_{\text{cont}}(\pi) - \mathcal{V}'(\pi) \right] \qquad (13.86)
$$

describing [Dav90, AM90, FKN91] a multiplicative renormalization of the Wilson loops.

Then a source for a continuum Wilson loop is

$$
\mathcal{V}(\pi) = \sum_{n=0}^{\infty} T_n \pi^{n+1/2} \qquad (13.87)
$$

and

$$
\frac{\delta}{\delta \mathcal{V}(\pi)} = -\sum_{n=0}^{\infty} \pi^{-n-3/2} \frac{\partial}{\partial T_n} \qquad (13.88)
$$

is the continuum loop insertion operator:

$$
W_{\text{cont}}(\pi_1, \ldots, \pi_m) = \mathcal{G} \frac{\delta}{\delta \mathcal{V}(\pi_1)} \cdots \frac{\delta}{\delta \mathcal{V}(\pi_m)} \ln Z_{\text{cont}} , \qquad (13.89)
$$

where

$$
Z_{\text{cont}} \propto \sqrt{Z_{1h}} \qquad \boxed{\text{even } V(\varphi)} \qquad (13.90)
$$

up to an infinite constant which is determined only by genus zero. The appearance of the square root is associated with a "doubling" of degrees of freedom for the even potential.

---

* These (linear algebraic) relations are obtained [MMM91] equating positive powers of p or π in Eq. (13.86) and using Eq. (13.81).

The continuum loop equation can be obtained from Eq. (13.52) by substituting Eq. (13.86) and using Eqs. (13.87) and (13.88):

$$\int_{C_1} \frac{d\Omega}{2\pi i} \frac{\mathcal{V}'(\Omega)}{(\pi - \Omega)} W_{\text{cont}}(\Omega) = W_{\text{cont}}^2(\pi) + \mathcal{G} \frac{\delta W_{\text{cont}}(\pi)}{\delta \mathcal{V}(\pi)} + \frac{\mathcal{G}}{16\pi^2} + \frac{T_0^2}{16\pi}.$$

$$(13.91)$$

This equation describes a model which interpolates between different multicritical points. The Jth multicritical point corresponds to $T_k = 0$ except for $k = 0$ and $k = J$ while

$$\sqrt{\Lambda} = \left(\frac{(-1)^{J-1} 2^J J! \, T_0}{(2J+1)!!} \frac{T_0}{T_J} \right)^{1/J}.$$

$$(13.92)$$

The continuum loop equation (13.91) can be solved order by order in $\mathcal{G}$ (genus expansion) analogously to that of Sect. 13.4. If $\mathcal{V}(\pi)$ is a polynomial ($T_k = 0$ for $k > K$), $K - 1$ lower coefficients of the asymptotic expansion of $W_{\text{cont}}(\pi)$ are not fixed and should be determined by requiring the one-cut analytic structure in π.

The continuum analog of Eq. (13.64) is given by

$$W_{\text{cont}}^{(0)}(\pi) = \int_{C_1} \frac{d\Omega}{4\pi i} \frac{\mathcal{V}'(\Omega)}{(\pi - \Omega)} \frac{\sqrt{\pi - u}}{\sqrt{\Omega - u}},$$

$$(13.93)$$

where u coincides to genus zero with $-\sqrt{\Lambda}$ at a given multicritical point. Then the vanishing of the $1/\sqrt{\pi}$ term is equivalent to Eq. (13.92). The cut of $W_{\text{cont}}^{(0)}(\pi)$ is from u to ∞. This is because we are magnifying the region near the end a of the cut in the one-matrix model.

The function u versus T_k is determined to all genera from the asymptotic behavior. This dependence can be obtained by comparing $1/\pi$ terms in Eq. (13.91). Denoting the derivative with respect to $x = -T_0/2$ by D, this relation can be represented conveniently as

$$\int_{C_1} \frac{d\Omega}{2\pi i} \mathcal{V}'(\Omega) \left(D W_{\text{cont}}(\Omega) + \frac{1}{2\sqrt{\Omega}} \right) = 0,$$

$$(13.94)$$

where

$$D W_{\text{cont}}^{(0)}(\pi) + \frac{1}{2\sqrt{\pi}} = \frac{1}{2\sqrt{\pi - u}}$$

$$(13.95)$$

for the genus-zero solution (13.93).

Remark on the KdV hierarchy

Equation (13.95) can be extended to all genera using the representation

$$DW_{\text{cont}}(\pi) + \frac{1}{2\sqrt{\pi}} = \left\langle x \left| \left(-\mathcal{G}D^2 - u(x) + \pi \right)^{-1} \right| x \right\rangle$$

$$= \sum_{n=0}^{\infty} \frac{R_n[u]}{\pi^{n+1/2}} \equiv R(\pi), \qquad (13.96)$$

where the diagonal resolvent of the Sturm–Liouville operator is expressed via the Gel'fand–Dikii differential polynomials [GD75]

$$R_n[u] = 2^{-n-1} \left(\frac{\mathcal{G}}{2} D^2 + u + D^{-1} u D \right)^n \cdot 1. \qquad (13.97)$$

We have explicitly

$$R_0 = \frac{1}{2}, \quad R_1 = \frac{u}{4}, \quad R_2 = \frac{\mathcal{G}}{16} D^2 u + \frac{3}{16} u^2, \quad \ldots \qquad (13.98)$$

for the lower polynomials. Equation (13.97) can be easily obtained from Eq. (1.127) derived in Problem 1.11 on p. 25.

Substituting the RHS of Eq. (13.96) into Eq. (13.94), we obtain the *string equation* [GM90b, BDS90]

$$\sum_{k=0}^{\infty} \left(k + \tfrac{1}{2} \right) T_k R_k[u] = 0 \qquad (13.99)$$

which determines u versus T_k.

The meaning of u is clear from Eqs. (13.96) and (13.89):

$$u = 4R_1 = 2\mathcal{G}D^2 \ln Z_{\text{cont}}, \qquad (13.100)$$

i.e. u is the continuum susceptibility. It is negative to genus zero because of the performed "renormalization".

Problem 13.9 Elaborate Eq. (13.99) for two-dimensional quantum gravity.

Solution Choosing $T_2 = 16/15$ and using Eq. (13.92), which gives $\Lambda = -T_0/2$, and Eq. (13.98), we represent Eq. (13.99) as the Painlevé equation

$$\Lambda = u^2 + \frac{\mathcal{G}}{3} D^2 u, \qquad D = \frac{d}{d\Lambda}, \qquad (13.101)$$

the solution of which is given by a Painlevé transcendental. It can be found by solving Eq. (13.101) iteratively in $\mathcal{G}$:

$$u = \sqrt{\Lambda} \left[-1 + \sum_{h=1}^{\infty} \left(\frac{\mathcal{G}}{\Lambda^{5/2}} \right)^h \chi_h \right], \qquad (13.102)$$

where the numerical coefficients $\chi_h > 0$ are determined by a recursion relation. This reproduces the indices (13.83) for $J = 2$. Substituting into Eq. (13.100) and integrating, one can obtain the genus expansion of the partition function of two-dimensional Euclidean quantum gravity introduced in Eq. (13.85). The series is asymptotic, since $\chi_h \sim (2h)!$ for large h.

These results were first obtained in [BK90, DS90, GM90a].

Problem 13.10 Show that the ansatz (13.96) satisfies Eq. (13.91).

Solution It is convenient to introduce

$$\widetilde{W}(\pi) = W_{\text{cont}}(\pi) - \frac{T_0}{4\sqrt{\pi}}, \tag{13.103}$$

$$\widetilde{\mathcal{V}}(\pi) = \mathcal{V}(\pi) - T_0\sqrt{\pi} = \sum_{n=1}^{\infty} T_n \pi^{n+1/2}. \tag{13.104}$$

In the new variables, the last term on the RHS of the loop equation (13.91) disappears and it can be written as

$$\int_{C_1} \frac{d\Omega}{2\pi i} \frac{\widetilde{\mathcal{V}}'(\Omega)}{(\pi - \Omega)} \widetilde{W}(\Omega) = \widetilde{W}^2(\pi) + \mathcal{G}\frac{\delta \widetilde{W}(\pi)}{\delta \mathcal{V}(\pi)} - \frac{3\mathcal{G}}{16\pi^2}. \tag{13.105}$$

We then apply the operator

$$\Delta_\pi = -\left(\frac{\mathcal{G}}{2}D^3 + uD + Du - 2\pi D\right)D, \tag{13.106}$$

which annihilates $\widetilde{W}(\pi)$ given by the Gel'fand–Dikii ansatz (13.96) (cf. Eq. (1.127)), to both sides of Eq. (13.105).

The following terms emerge:

$$\Delta_\pi \int_{C_1} \frac{d\Omega}{2\pi i} \frac{\widetilde{\mathcal{V}}'(\Omega)}{(\pi - \Omega)} \widetilde{W}(\Omega) = \int_{C_1} \frac{d\Omega}{2\pi i} \frac{\widetilde{\mathcal{V}}'(\Omega)}{(\pi - \Omega)} \Delta_\Omega \widetilde{W}(\Omega)$$

$$+2D \int_{C_1} \frac{d\Omega}{2\pi i} \widetilde{\mathcal{V}}'(\Omega) D\widetilde{W}(\Omega) = 2D \sum_{k=1}^{\infty} (k + \tfrac{1}{2})T_k R_k[u] = 1, \tag{13.107}$$

$$\Delta_\pi \widetilde{W}^2(\pi) = 2\widetilde{W}\Delta\widetilde{W} - 4\mathcal{G}D\widetilde{W}D^3\widetilde{W} - 3\mathcal{G}(D^2\widetilde{W})^2 + 4(\pi - u)(D\widetilde{W})^2$$

$$= -4\mathcal{G}RD^2R - 3\mathcal{G}(DR)^2 + 4(\pi - u)R^2, \tag{13.108}$$

$$\Delta_\pi \frac{\delta}{\delta\mathcal{V}(\pi)}\widetilde{W}(\pi) = \frac{\delta}{\delta\mathcal{V}(\pi)}\Delta_\pi\widetilde{W}(\pi) + 2\frac{\delta u}{\delta\mathcal{V}(\pi)}D^2\widetilde{W}(\pi) + \frac{\delta(Du)}{\delta\mathcal{V}(\pi)}D\widetilde{W}(\pi)$$

$$= 4(DR)^2 + 2RD^2R. \tag{13.109}$$

We have used Eq. (13.99) in deriving Eq. (13.107), the equation

$$\frac{\delta}{\delta\mathcal{V}(\pi)}u = 2DR(\pi), \tag{13.110}$$

which arises from acting by the loop insertion operator on Eq. (13.100) and the expansion of which in $1/\pi$ reproduces the Korteweg–de Vries (KdV) hierarchy,

$$-\frac{\partial u}{\partial T_n} = 2DR_{n+1}[u], \tag{13.111}$$

in deriving Eq. (13.109), and the fact that $\Delta\widetilde{W} = 0$ for the Gel'fand–Dikii ansatz owing to Eq. (1.127).

Combining the RHSs of Eqs. (13.107), (13.108), and (13.109), we obtain

$$1 = -2\mathcal{G}RD^2R + \mathcal{G}(DR)^2 + 4(\pi - u)R^2 \tag{13.112}$$

which is the same as Eq. (1.134) satisfied by the Gel'fand–Dikii resolvent. Its solution is unambiguous (at least perturbatively in $\mathcal{G}$).

Thus, the Gel'fand–Dikii ansatz is obtained [DVV91] as a solution of the continuum loop equation (13.91).

Remark on the continuum Wilson loop

The continuum Wilson loops are related [Kaz89] to boundaries of surfaces in two-dimensional quantum gravity. Given the path integral (13.29) over surfaces with fixed boundary, we integrate over metrics (including the metric at the boundary). The result can depend solely on the length of the boundary which is the only invariant. Then $W_{\text{cont}}(\pi)$ is simply the Laplace transform of this object.

Remark on the continuum Virasoro constraints

The continuum loop equation (13.91) can be represented as a set of the continuum Virasoro constraints [FKN91, DVV91]

$$\mathcal{L}_n^{\text{cont}} Z_{\text{cont}} = 0 \qquad \text{for } n \geq -1, \tag{13.113}$$

where

$$\mathcal{L}_n^{\text{cont}} = \sum_{k=0}^{\infty} (k + \tfrac{1}{2}) T_k \frac{\partial}{\partial T_{k+n}} + \mathcal{G} \sum_{0 \leq k \leq n-1} \frac{\partial^2}{\partial T_k \partial T_{n-k-1}} + \frac{\delta_{0,n}}{16} + \frac{\delta_{-1,n} T_0^2}{16\mathcal{G}} \tag{13.114}$$

obey the Virasoro algebra (13.62). This is a consequence of conformal invariance of the continuum theory.

The Virasoro constraints (13.113) and (13.114) can be obtained [MMM91] from their matrix-model counterparts (13.63), (13.61) by passing from the variables t_{2k} to the variables T_k.

Remark on the Kontsevich matrix model

The above continuum model interpolating between multicritical points can be formulated as a matrix model [Kon91]:

$$Z_{\text{Kont}}[M] \;=\; \frac{\displaystyle\int dX\, e^{\,\text{tr}\,(\frac{\sqrt[4]{\mathcal{G}}}{6}X^3 - \frac{1}{2}MX^2)}}{\displaystyle\int dX\, e^{-\frac{1}{2}\,\text{tr}\,MX^2}}\,, \qquad (13.115)$$

where the integral goes over the Hermitian $N \times N$ matrix X. The RHS of Eq. (13.115) is well-defined perturbatively in $\mathcal{G}$.

The couplings T_k are expressed via the positive-definite Hermitian matrix M by

$$T_k \;=\; \frac{\sqrt{\mathcal{G}}}{k + \frac{1}{2}}\,\text{tr}\,(M^{-2k-1}) - \frac{2}{3}\delta_{1k}\,. \qquad (13.116)$$

The identification (13.116) makes sense as $N \to \infty$ when all $\text{tr}\,(M^{-2k-1})$ become independent but M is chosen such that they are finite. Alternatively, the standard topological expansion of the Kontsevich model in $1/N^2$ is associated with $\mathcal{G} \sim 1/N^2$.

The partition function of continuum two-dimensional quantum gravity coincides with the partition function of the Kontsevich model:

$$Z_{\text{cont}}[T] \;=\; Z_{\text{Kont}}[M]\,. \qquad (13.117)$$

This equality is valid in the sense of an asymptotic expansion at large M, each term of which is finite providing M is positive definite.

Remark on 2D topological gravity

Equation (13.117) represents the fact that quantum gravity is, in fact, a *topological* theory in two dimensions:

$$\boxed{\text{2D quantum gravity}} \;=\; \boxed{\text{2D topological gravity}}\,. \qquad (13.118)$$

A crucial property of topological theories is that correlators of operators $\sigma_{n_i}(x_i)$ with definite (nonnegative integer) scale dimension n_i, located at the point x_i of a two-dimensional Riemann surface of genus h, depend only on the dimensions n_i and genus h but not on the metric on the surface and, therefore, not on the positions of the punctures x_i. The Kontsevich matrix model appeared [Kon91] as an explicit realization of the Witten geometric formulation [Wit90] of two-dimensional topological gravity.

13.6 Hermitian multimatrix models

An obvious extension of the Hermitian one-matrix model is the model of two Hermitian matrices φ_1 and φ_2. The partition function of the Hermitian two-matrix model is

$$Z_{2h} = \int d\varphi_1 \, d\varphi_2 \, e^{N \operatorname{tr}\left[-V(\varphi_1)-V(\varphi_2)+\varphi_1\varphi_2\right]}, \qquad (13.119)$$

where for simplicity we take the same potentials for self-interactions of each matrix.

The presence of two matrices adds matter to two-dimensional gravity. The Hermitian two-matrix model is precisely associated with the Ising model on a random two-dimensional lattice.

There is a vast literature on the Hermitian two-matrix model starting from the work by Itzykson and Zuber [IZ80], who showed how to reduce it to an eigenvalue problem. We shall rather briefly review the loop equations for the Hermitian two-matrix model.

Let us define the Wilson loop average and the one-link correlator in the Hermitian two-matrix model (13.119), respectively, by

$$W(\lambda) = \left\langle \frac{1}{N} \operatorname{tr} \frac{1}{\lambda - \varphi_1} \right\rangle_{2h}, \qquad (13.120)$$

$$G(\nu, \lambda) = \left\langle \frac{1}{N} \operatorname{tr} \left(\frac{1}{(\nu - \varphi_1)} \frac{1}{(\lambda - \varphi_2)} \right) \right\rangle_{2h}. \qquad (13.121)$$

The definition of $W(\lambda)$ is similar to Eq. (13.50) while $G(\nu, \lambda)$, which is symmetric in ν and λ since the potentials of self-interaction are the same for both matrices, is absent in the one-matrix model. Expanding $G(\nu, \lambda)$ in $1/\nu$, we obtain

$$\left.\begin{aligned}
G(\nu, \lambda) &= \frac{W(\lambda)}{\nu} + \sum_{n=1}^{\infty} \frac{G_n(\lambda)}{\nu^{n+1}}, \\[2mm]
G_n(\lambda) &= \left\langle \frac{1}{N} \operatorname{tr} \left(\varphi_1^n \frac{1}{\lambda - \varphi_2} \right) \right\rangle_{2h}.
\end{aligned}\right\} \qquad (13.122)$$

In the large-N limit, the correlator $G(\nu, \lambda)$ obeys the following loop equation:

$$\int_{C_1} \frac{d\omega}{2\pi i} \frac{V'(\omega)}{(\nu - \omega)} G(\omega, \lambda) = W(\nu) G(\nu, \lambda) + \lambda G(\nu, \lambda) - W(\nu), \qquad (13.123)$$

where the contour C_1 encircles counterclockwise the cut (or cuts) of the function $G(\omega, \lambda)$ as depicted in Fig. 13.2.

To analyze Eq. (13.123), let us consider the Hermitian two-matrix model with the general potential (13.4). The solution for $W(\lambda)$ versus $V(\lambda)$ is determined by the equation

$$\sum_{k \geq 1} g_k G_{k-1}(\lambda) \;=\; \lambda W(\lambda) - 1 \qquad (13.124)$$

which is just the $1/\nu$ term of the expansion of Eq. (13.123) in $1/\nu$.

The functions $G_n(\lambda)$ are expressed via $W(\lambda)$ using the recurrence relation

$$\left. \begin{aligned} G_{n+1}(\lambda) &= \int_{C_1} \frac{d\omega}{2\pi i} \frac{V'(\omega)}{(\lambda - \omega)} G_n(\omega) - W(\lambda) G_n(\lambda), \\ G_0(\lambda) &= W(\lambda) \end{aligned} \right\} \qquad (13.125)$$

which is obtained by expanding Eq. (13.123) in $1/\lambda$. If $V(\lambda)$ is a polynomial of degree K, Eq. (13.124) contains $W(\lambda)$ up to degree K and the solution is algebraic [GN91, Alf93, Sta93].

For a cubic potential, this equation for $W(\lambda)$ is cubic and determines the critical index of the susceptibility $\gamma_0 = -1/3$. This is in contrast to the Hermitian one-matrix model where the loop equation is quadratic in $W(\lambda)$. We see that matter changes [Kaz86] the critical behavior of pure quantum gravity. The continuum theory associated with the $\gamma_0 = -1/3$ critical point of the Hermitian two-matrix model is unitary.

The correlator $G(\nu, \lambda)$ is symmetric in ν and λ for any solution of Eq. (13.124). This symmetry requirement can be used directly to determine $W(\lambda)$ alternatively to Eq. (13.124).

It is possible to further extend the Hermitian two-matrix model by considering a chain of matrices with the nearest-neighbor interaction:

$$Z_{qh} \;=\; \prod_{i=1}^{q} d\varphi_i \, \exp \left\{ N \operatorname{tr} \left[- \sum_{i=1}^{q} V(\varphi_i) + \sum_{i=1}^{q-1} \varphi_i \varphi_{i+1} \right] \right\}. \qquad (13.126)$$

In the limit of $q \to \infty$, we obtain an infinite chain associated with discretization of a one-dimensional theory.

The Hermitian q-matrix model possesses unitary continuum limits with $\gamma_0 = -1/(q+1)$. In the $q \to \infty$ limit, this gives $\gamma_0 \to 0$.

Remark on the $d = 1$ barrier

The $d = 1$ barrier is associated with the formula [GN84, OW85, KPZ88]

$$\gamma_0 \;=\; \frac{d - 1 - \sqrt{(1-d)(25-d)}}{12} \qquad (13.127)$$

for the critical index of string susceptibility of the bosonic string in a d-dimensional embedding space. Alternatively, it describes two-dimensional quantum gravity interacting with conformal matter of central charge $c = d$.

The RHS of Eq. (13.127) is well-defined for $d \leq 1$, where it is associated with topological theories of gravity (with matter). They can also be described by the Hermitian (multi)matrix models.

The RHS of Eq. (13.127) becomes complex for $d > 1$ which is physically unacceptable. This is termed the $d = 1$ *barrier*.

Remark on the Kazakov–Migdal model

A natural multidimensional extension of the matrix chain (13.126) is the Kazakov–Migdal model [KM92], which is defined by the partition function

$$Z_{\mathrm{KM}} = \int \prod_{x,\mu} dU_\mu(x) \prod_x d\varphi_x \, e^{-S_{\mathrm{KM}}[U,\varphi]} \tag{13.128}$$

with the action

$$S_{\mathrm{KM}}[U,\varphi] = N \operatorname{tr}\left[-\sum_{x,\mu} \varphi_{x+a\hat{\mu}} U_\mu(x) \varphi_x U_\mu^\dagger(x) + \sum_x V(\varphi_x) \right]. \tag{13.129}$$

Here φ_x and $U_\mu(x)$ are $N \times N$ Hermitian and unitary matrices, respectively, with x labeling lattice sites on a d-dimensional hypercubic lattice. The integration over the gauge field $U_\mu(x)$ is over the Haar measure on $SU(N)$ at each link of the lattice.

The Kazakov–Migdal model is of the same type as Wilson's lattice gauge theory with adjoint matter but without the action for the gauge field, i.e. at $\beta = 0$ in front of the plaquette term. When integrated over φ_x, it induces an action for the gauge field $U_\mu(x)$ of the type discussed in Problem 8.6 on p. 155.

The model (13.128) obviously recovers the open matrix chain (13.126) if the lattice is just a one-dimensional sequence of points for which the gauge field can be absorbed by a unitary transformation of φ_x.

The Kazakov–Migdal model is described at $N = \infty$ by the loop equation which coincides with Eq. (13.123) for the two-matrix model with the potential [DMS93]

$$V'(\omega) \quad \rightarrow \quad \mathcal{V}'(\omega) \equiv V'(\omega) - (2d-1)F(\omega). \tag{13.130}$$

The function

$$F(\omega) = \sum_n F_n \omega^n \tag{13.131}$$

is defined by the pair correlator of the gauge fields

$$\frac{\int dU\, e^{N\,\mathrm{tr}\,(\Phi U \Psi U^\dagger)} \frac{1}{N}\,\mathrm{tr}\left(t^a U \Psi U^\dagger\right)}{\int dU\, e^{N\,\mathrm{tr}\,(\Phi U \Psi U^\dagger)}} = \sum_{n=1}^{\infty} F_n \frac{1}{N}\,\mathrm{tr}\left(t^a \Phi^n\right), \qquad (13.132)$$

where Φ and Ψ play the role of external fields and t^a ($a = 1, \ldots, N^2 - 1$) denote the generators of $SU(N)$. Eq. (13.132) holds [Mig92] at $N = \infty$. The function $F(\omega)$ is determined by the loop equation itself.

The loop equation of the Hermitian two-matrix model emerges because the last term on the RHS of Eq. (13.130) disappears at $d = 1/2$, which is associated with the Hermitian two-matrix model, and we simply have $\mathcal{V}(\omega) = V(\omega)$.

An exact solution of the Kazakov–Migdal model was found for the quadratic potential [Gro92] and the logarithmic potential [Mak93].

Continuum limits of the Kazakov–Migdal model are associated again with lower-dimensional theories. It does not allow us to go beyond the $d = 1$ barrier.

Bibliography to Part 3

Reference guide

The large-N methods are briefly described in the book by Polyakov [Pol87]. The book edited by Brézin and Wadia [BW93] contains reprints of original papers on this subject.

The large-N limit of the four-Fermi and φ^4 theories was obtained in the paper by Wilson [Wil73]. The renormalizability of the $1/N$-expansion of four-Fermi theory in $d < 4$ dimensions was demonstrated by Parisi [Par75]. The appearance of conformal invariance in the $1/N$-expansion of four-Fermi theory in three dimensions is discussed in [CMS93]. The scale and conformal symmetries are described in the lectures by Jackiw [Jac72].

The $1/N$-expansion of $SU(N)$ Yang–Mills theory and its relation to the topology of Riemann surfaces was introduced by 't Hooft [Hoo74a]. The incorporation of quarks into this picture was accomplished by Veneziano [Ven76]. The geometric growth of the number of planar graphs was demonstrated by Koplik, Neveu and Nussinov [KNN77]. The large-N factorization was observed by A.A. Migdal in the late 1970s (first published in [MM79]). Its consequences for the semiclassical nature of the large-N limit are discussed in the lectures by Witten [Wit79] and Coleman [Col79].

The loop equation of multicolor QCD was derived in [MM79]. The program of reformulating QCD entirely in loop space was realized in [MM81]. The renormalization of the Wilson loops was investigated in [GN80, Pol80, DV80, BNS81]. A solution of the loop equation in two dimensions was found by Kazakov and Kostov [KK80]. A string representation of large-N QCD$_2$ was constructed by Gross and Taylor [GT93].

For a canonical book on the matrix models see the one by Mehta [Meh67]. The solution of the Hermitian one-matrix model at large N

was found by Brézin, Itzykson, Parisi and Zuber [BIP78]. The large-N phase transition in lattice QCD$_2$ was first observed by Gross and Witten [GW80]. The application of matrix models to discretization of random surfaces is described in the review by Di Francesco, Ginsparg, and Zinn-Justin [DGZ95] and in the book by Ambjørn, Durhuus, and Jonsson [ADJ97] which contain extensive references.

References

[ACK93]   AMBJØRN J., CHEKHOV L., KRISTJANSEN C.F., AND MAKEENKO YU. 'Matrix model calculations beyond the spherical limit'. *Nucl. Phys.* **B404** (1993) 127.

[ACM92]   AMBJØRN J., CHEKHOV L., AND MAKEENKO YU. 'Higher genus correlators from the Hermitean one-matrix model'. *Phys. Lett.* **B282** (1992) 341.

[ADF85]   AMBJØRN J., DURHUUS B., AND FRÖLICH J. 'Diseases of triangulated random surface models, and possible cures'. *Nucl. Phys.* **B257[FS14]** (1985) 433.

[ADJ97]   AMBJØRN J., DURHUUS B., AND JONSSON T. *Quantum geometry: a statistical field theory approach.* (Cambridge Univ. Press, 1997).

[AJM90]   AMBJØRN J., JURKIEWICZ J., AND MAKEENKO YU. 'Multiloop correlators for two-dimensional quantum gravity'. *Phys. Lett.*, **B251** (1990) 517.

[AKM94]   AMBJØRN J., KRISTJANSEN C.F., AND MAKEENKO YU. 'Generalized Penner models beyond the spherical limit'. *Phys. Rev.* **D50** (1994) 5193.

[Alf93]   ALFARO J. 'The large-N limit of the two-Hermitian-matrix model by the hidden BRST method'. *Phys. Rev.* **D47** (1993) 4714.

[Ans59]   ANSELM A.A. 'A model of a field theory with nonvanishing renormalized charge'. *Sov. Phys. JETP* **9** (1959) 608.

[AM90]   AMBJØRN J. AND MAKEENKO YU. 'Properties of loop equations for the Hermitean matrix model and for two-dimensional quantum gravity'. *Mod. Phys. Lett.* **A5** (1990) 1753.

[BDS90]   BANKS T., DOUGLAS M., SEIBERG N., AND SHENKER S. 'Microscopic and macroscopic loops in nonperturbative two-dimensional gravity'. *Phys. Lett.*, **B238** (1990) 279.

[Bes79]   BESSIS D. 'A new method in the combinatorics of the topological expansion'. *Commun. Math. Phys.* **69** (1979) 147.

[BGS82]   BRANDT R.A., GOCKSCH A., SATO M., AND NERI F. 'Loop space'. *Phys. Rev.* **D26** (1982) 3611.

[BIP78]   BRÉZIN E., ITZYKSON C., PARISI G., AND ZUBER J.B. 'Planar diagrams'. *Commun. Math. Phys.* **59** (1978) 35.

[BIZ80]   BESSIS D., ITZYKSON C., AND ZUBER J.B. 'Quantum field theory techniques in graphical enumeration'. *Adv. Appl. Math.* **1** (1980) 109.

[BK90]   BRÉZIN E. AND KAZAKOV V.A. 'Exactly solvable field theories of closed strings'. *Phys. Lett.* **B236** (1990) 144.

[BNS81]   BRANDT R.A., NERI F., AND SATO M. 'Renormalization of loop functions for all loops'. *Phys. Rev.* **D24** (1981) 879.

[BNZ79]   BRANDT R.A., NERI F., AND ZWANZIGER D. 'Lorentz invariance from classical particle paths in quantum field theory of electric and magnetic charge'. *Phys. Rev.* **D19** (1979) 1153.

[BP75]   BELAVIN A.A. AND POLYAKOV A.M. 'Metastable states of two-dimensional isotropic ferromagnet'. *JETP Lett.* **22** (1975) 245.

[Bra80]   BRALIĆ N. 'Exact computation of loop averages in two-dimensional Yang–Mills theory'. *Phys. Rev.* **D22** (1980) 3090.

[BW93]   BRÉZIN E. AND WADIA S. *The large-N expansion in quantum field theory and statistical physics: from spin systems to two-dimensional gravity* (World Scientific, Singapore, 1993).

[BZ93]   BRÉZIN E. AND ZEE A. 'Universality of the correlations between eigenvalues of large random matrices'. *Nucl. Phys.* **B402** (1993) 613.

[CJP74]   COLEMAN S., JACKIW R., AND POLITZER H.D. 'Spontaneous symmetry breaking in the $O(N)$ model at large N'. *Phys. Rev.* **D10** (1974) 2491.

[CLS82]   CVITANOVIĆ P., LAUWERS P.G., AND SCHARBACH P.N. 'The planar sector of field theories'. *Nucl. Phys.* **B203** (1982) 385.

[CMS93]   CHEN W., MAKEENKO Y., AND SEMENOFF G.W. 'Four-fermion theory and the conformal bootstrap'. *Ann. Phys.* **228** (1993) 341.

[Col79]   COLEMAN S. '1/N', in Proc. of Erice Int. School of subnuclear physics 1979 (Plenum, New York, 1982) p. 805. Reprinted in COLEMAN S. *Aspects of symmetry* (Cambridge Univ. Press, 1985) p. 351.

[Cvi81]   CVITANOVIĆ P. 'Planar perturbation expansion'. *Phys. Lett.* **B99** (1981) 49.

[Dav85]   DAVID F. 'Planar diagrams, two-dimensional lattice gravity and surface models'. *Nucl. Phys.* **B257[FS14]** (1985) 45.

[Dav90]   DAVID F. 'Loop equations and nonperturbative effects in two-dimensional quantum gravity'. *Mod. Phys. Lett* **A5** (1990) 1019.

[DGZ95]   DI FRANCESCO P., GINSPARG P., AND ZINN-JUSTIN J. '2D gravity and random matrices'. *Phys. Rep.* **254** (1995) 1.

[DKS93]   DERKACHOV S.E., KIVEL N.A., STEPANENKO A.S., AND VASIL'EV A.N. 'On calculation of $1/N$ expansions of critical exponents in the Gross–Neveu model with the conformal technique'. `hep-th/9302034`.

[Dou95]   DOUGLAS M.R. 'Stochastic master fields'. *Phys. Lett.* **B344** (1995) 117.

[DMS93] DOBROLIUBOV M.I., MAKEENKO YU., AND SEMENOFF G.W. 'Correlators of the Kazakov–Migdal model'. *Mod. Phys. Lett.* **A8** (1993) 2387.

[DS90]   DOUGLAS M. AND SHENKER S. 'Strings in less than one dimensions'. *Nucl. Phys.* **B335** (1990) 635.

[DV80]   DOTSENKO V.S. AND VERGELES S.N. 'Renormalizability of phase factors in non-Abelian gauge theory'. *Nucl. Phys.* **B169** (1980) 527.

[DVV91] DIJKGRAAF R., VERLINDE H., AND VERLINDE E. 'Loop equations and Virasoro constraints in nonperturbative 2D quantum gravity'. *Nucl. Phys.* **B348** (1991) 435.

[Dys62] DYSON F.J. 'Statistical theory of the energy levels of complex systems. I'. *J. Math. Phys.* **3** (1962) 140.

[Egu79] EGUCHI T. 'Strings in $U(N)$ lattice gauge theory'. *Phys. Lett.* **B87** (1979) 91.

[EK82]   EGUCHI T. AND KAWAI H. 'Reduction of dynamical degrees of freedom in the large-N gauge theory'. *Phys. Rev. Lett.* **48** (1982) 1063.

[Fel86] FELLER M.N. 'Infinite-dimensional elliptic equations and operators of the type by P. Lévy'. *Russ. Math. Surv.* **41** (1986) 97.

[FKN91] FUKUMA M., KAWAI H., AND NAKAYAMA R. 'Continuum Schwinger–Dyson equations and universal structures in 2D quantum gravity'. *Int. J. Mod. Phys.* **A6** (1991) 1385.

[Foe79] FOERSTER D. 'Yang–Mills theory – a string theory in disguise'. *Phys. Lett.* **B87** (1979) 87.

[Fri81] FRIEDAN D. 'Some non-Abelian toy models in the large-N limit'. *Commun. Math. Phys.* **78** (1981) 353.

[GD75]   GEL'FAND I.M. AND DIKII L.A. 'Asymptotic behavior of the resolvent of Sturm–Liouville equations and the algebra of the Korteweg–de Vries equations'. *Russ. Math. Surv.* **30** (1975) v. 5, p. 77.

[GG95]   GOPAKUMAR R. AND GROSS G.J. 'Mastering the master field'. *Nucl. Phys.* **B451** (1995) 379.

[GM90a] GROSS D. AND MIGDAL A.A. 'Nonperturbative two-dimensional quantum gravity'. *Phys. Rev. Lett.* **64** (1990) 127.

[GM90b] GROSS D. AND MIGDAL A.A. 'A nonperturbative treatment of two-dimensional quantum gravity'. *Nucl. Phys.* **B340** (1990) 333.

[GN74]   GROSS D.J. AND NEVEU A. 'Dynamical symmetry breaking in asymptotically free field theories'. *Phys. Rev.* **D10** (1974) 3235.

[GN79a] GERVAIS J.L. AND NEVEU A. 'The quantum dual string wave functional in Yang–Mills theories'. *Phys. Lett.* **B80** (1979) 255.

[GN79b] GERVAIS J.L. AND NEVEU A. 'Local harmonicity of the Wilson loop integral in classical Yang–Mills theory'. *Nucl. Phys.* **B153** (1979) 445.

[GN80]   GERVAIS J.L. AND NEVEU A. 'The slope of the leading Regge trajectory in quantum chromodynamics'. *Nucl. Phys.* **B163** (1980) 189.

[GN84]   GERVAIS J.L. AND NEVEU A. 'Novel triangle relation and absence of tachyons in Liouville string field theory'. *Nucl. Phys.* **B238** (1984) 125.

[GN91]   GAVA E. AND NARAIN K.S. 'Schwinger–Dyson equations for the two-matrix model and $W(3)$ algebra'. *Phys. Lett.* **B263** (1991) 213.

[Gra91]   GRACEY J.A. 'Calculation of exponent η to $O(1/N^2)$ in the $O(N)$ Gross–Neveu model'. *Int. J. Mod. Phys.* **A6** (1991) 395, 2755(E).

[Gra93]   GRACEY J.A. 'Computation of critical exponent η at $O(1/N^3)$ in the four-Fermi model in arbitrary dimensions'. *Int. J. Mod. Phys.* **A9** (1994) 727.

[Gro92]   GROSS D. 'Some remarks about induced QCD'. *Phys. Lett.* **B293** (1992) 181.

[GT93]   GROSS D.J. AND TAYLOR W.I. 'Two-dimensional QCD is a string theory'. *Nucl. Phys.* **B400** (1993) 181.

[GW70]   GROSS D. AND WESS J. 'Scale invariance, conformal invariance, and the high-energy behavior of scattering amplitudes'. *Phys. Rev.* **D2** (1970) 753.

[GW80]   GROSS D. AND WITTEN E. 'Possible third-order phase transition in the large-N lattice gauge theory'. *Phys. Rev.* **D21** (1980) 446.

[Haa81]   HAAN O. 'Large N as a thermodynamic limit'. *Phys. Lett.* **106B** (1981) 207.

[HM89]   HALPERN M.B. AND MAKEENKO YU.M. 'Continuum-regularized loop-space equation'. *Phys. Lett.* **B218** (1989) 230.

[Hoo74a] 'T HOOFT G. 'A planar diagram theory for strong interactions'. *Nucl. Phys.* **B72** (1974) 461.

[Hoo74b] 'T HOOFT G. 'A two-dimensional model for mesons'. *Nucl. Phys.* **B75** (1974) 461.

[IZ80]   ITZYKSON C. AND ZUBER J.-B. 'The planar approximation. II'. *J. Math. Phys.* **21** (1980) 411.

[Jac72]   JACKIW R. 'Field theoretic investigations in current algebra', in *Lectures on current algebra and its applications* by S.B. Treiman, R. Jackiw, and D.J. Gross (Princeton Univ. Press, 1972) p. 97; also in *Current algebra and anomalies* by S.B. Treiman, R. Jackiw, B. Zumino, and E. Witten (World Scientific, Singapore, 1985) p. 81.

[Joh61]   JOHNSON K. 'Solution of the equations for the Green's functions of a two-dimensional relativistic field theory'. *Nuovo Cim.* **20** (1961) 773.

[Kaz85]   KAZAKOV V.A. 'Bilocal regularization of models of random surfaces'. *Phys. Lett.* **B150** (1985) 282.

[Kaz86]   KAZAKOV V.A. 'Ising model on a dynamical planar random lattice: exact solution'. *Phys. Lett.* **A119** (1986) 140.

[Kaz89]   KAZAKOV V.A. 'The appearance of matter fields from quantum fluctuations of 2D gravity'. *Mod. Phys. Lett.* **A4** (1989) 2125.

[KK80]   KAZAKOV V.A. AND KOSTOV I.K. 'Nonlinear strings in two-dimensional $U(\infty)$ theory'. *Nucl. Phys.* **B176** (1980) 199.

[KK81]   KAZAKOV V.A. AND KOSTOV I.K. 'Computation of the Wilson loop functional in two-dimensional $U(\infty)$ lattice gauge theory'. *Phys. Lett.* **B105** (1981) 453.

[KKM85]   KAZAKOV V.A., KOSTOV I.K., AND MIGDAL A.A. 'Critical properties of randomly triangulated planar random surfaces'. *Phys. Lett.* **B157** (1985) 295.

[KM81]   KHOKHLACHEV S.B. AND MAKEENKO YU.M. 'Phase transition over the gauge group center and quark confinement in QCD'. *Phys. Lett.* **B101** (1981) 403.

[KM92]   KAZAKOV V.A. AND MIGDAL A.A. 'Induced lattice gauge theory at large N'. *Nucl. Phys.* **B397** (1993) 214.

[KNN77]   KOPLIK J., NEVEU A., AND NUSSINOV S. 'Some aspects of the planar perturbation series'. *Nucl. Phys.* **B123** (1977) 109.

[Kon91]   KONTSEVICH M.L. 'Theory of intersections on moduli space of curves'. *Funk. Anal. Prilozh.* **25** (1991) v. 2, p. 50 (in Russian).

[KPZ88]   KNIZHNIK V., POLYAKOV A., AND ZAMOLODCHIKOV A. 'Fractal structure of 2D quantum gravity'. *Mod. Phys. Lett.* **A3** (1988) 819.

[Lev51]   LÉVY P. *Problèmes concrets d'analyse fonctionnelle* (Gauthier-Villars, Paris, 1951).

[Mak88]   MAKEENKO YU.M. 'Polygon discretization of the loop-space equation'. *Phys. Lett.* **B212** (1988) 221.

[Mak93]   MAKEENKO YU. 'Some remarks about the two-matrix Penner model and the Kazakov–Migdal model'. *Phys. Lett.* **B314** (1993) 197.

[Mak94]   MAKEENKO Y. 'Exact multiparticle amplitudes at threshold in large-N component ϕ^4 theory'. *Phys. Rev.* **50** (1994) 4137.

[Man79]   MANDELSTAM S. 'Charge-monopole duality and the phases of non-Abelian gauge theories'. *Phys. Rev.* **D19** (1979) 2391.

[Meh67]   MEHTA M.L. *Random matrices* (Academic Press, New York, 1967).

[Mig71]   MIGDAL A.A. 'On hadronic interactions at small distances'. *Phys. Lett.* **B37** (1971) 98.

[Mig80]   MIGDAL A.A. 'Properties of the loop average in QCD'. *Ann. Phys.* **126** (1980) 279.

[Mig81]   MIGDAL A.A. 'QCD = Fermi string theory'. *Nucl. Phys.* **189** (1981) 253.

[Mig83]   MIGDAL A.A. 'Loop equations and $1/N$ expansion'. *Phys. Rep.* **102** (1983) 199.

[Mig92]   MIGDAL A.A. 'Exact solution of induced lattice gauge theory at large N'. *Mod. Phys. Lett.* **A8** (1993) 359.

[MM79]   MAKEENKO YU.M. AND MIGDAL A.A. 'Exact equation for the loop average in multicolor QCD'. *Phys. Lett.* **B88** (1979) 135.

[MM80]   MAKEENKO YU.M. AND MIGDAL A.A. 'Self-consistent area law in QCD'. *Phys. Lett.* **B97** (1980) 235.

[MM81]   MAKEENKO YU.M. AND MIGDAL A.A. 'Quantum chromodynamics as dynamics of loops'. *Nucl. Phys.* **B188** (1981) 269.

[MMM91]   MAKEENKO YU., MARSHAKOV A., MIRONOV A., AND MOROZOV A. 'Continuum versus discrete Virasoro in one-matrix models'. *Nucl. Phys.* **B356** (1991) 574.

[MS69]   MACK G. AND SALAM A. 'Finite-component field representation of the conformal group'. *Ann. Phys.* **53** (1969) 144.

[Nam79]   NAMBU Y. 'QCD and the string model'. *Phys. Lett.* **B80** (1979) 372.

[OP81]   OLESEN P. AND PETERSEN J.L. 'The Makeenko–Migdal equation in a domained QCD vacuum'. *Nucl. Phys.* **181** (1981) 157.

[OW85]   OTTO H.J. AND WEIGT G. 'Exponential Liouville field operators for regular periodic fields'. *Phys. Lett.* **B159** (1985) 341.

[Par75]   PARISI G. 'The theory of nonrenormalizable interaction'. *Nucl. Phys.* **B100** (1975) 368.

[Pol70]   POLYAKOV A.M. 'Conformal symmetry of critical fluctuations'. *JETP Lett.* **12** (1970) 381.

[Pol75]   POLYAKOV A.M. 'Interaction of Goldstone particles in two dimensions. Applications to ferromagnets and massive Yang–Mills fields'. *Phys. Lett.* **B59** (1975) 79.

[Pol79]   POLYAKOV A.M. 'String representations and hidden symmetries for gauge fields'. *Phys. Lett.* **B82** (1979) 247.

[Pol80]   POLYAKOV A.M. 'Gauge fields as rings of glue'. *Nucl. Phys.* **B164** (1980) 171.

[Pol87]   POLYAKOV A.M. *Gauge fields and strings* (Harwood Academic Pub., Chur, 1987).

[PR80]   PAFFUTI G. AND ROSSI P. 'A solution of Wilson's loop equation in lattice QCD$_2$'. *Phys. Lett.* **B92** (1980) 321.

[Sch74]   SCHNITZER H.J. 'Nonperturbative effective potential for $\lambda\varphi^4$ theory in the many-particle limit'. *Phys. Rev.* **D10** (1974) 1800.

[Sta68]   STANLEY H.E. 'Spherical model as the limit of infinite spin dimensionality'. *Phys. Rev.* **176** (1968) 718.

[Sta93]   STAUDACHER M. 'Combinatorial solution of the two-matrix model'. *Phys. Lett.* **B305** (1993) 332.

[SV93]   SHURYAK E.V. AND VERBAARSCHOT J.J.M. 'Random matrix theory and spectral sum rules for the Dirac operator in QCD'. *Nucl. Phys.* **A560** (1993) 306.

[Tav93]   TAVARES J.N. 'Chen integrals, generalized loops and loop calculus'. *Int. J. Mod. Phys.* **A9** (1994) 4511.

[Tut62]   TUTTLE W.T. 'A census of planar triangulations'. *Can. J. Math.* **14** (1962) 21.

[VZ93]   VERBAARSCHOT J.J.M. AND ZAHED I. 'Spectral density of the QCD Dirac operator near zero virtuality'. *Phys. Rev. Lett.* **70** (1993) 3852.

[VDN95]   VOICULESCU D.V., DYKEMA K.J., AND NICA A. *Free random variables* (AMS, Providence, 1992).

[Ven76]   VENEZIANO G. 'Some aspects of a unified approach to gauge, dual and Gribov theories'. *Nucl. Phys.* **B117** (1976) 519.

[Wad81]   WADIA S.R. 'Dyson–Schwinger equations approach to the large-N limit: model systems and string representation of Yang–Mills theory'. *Phys. Rev.* **D24** (1981) 970.

[Wei79]   WEINGARTEN D. 'String equations for lattice gauge theories with quarks'. *Phys. Lett.* **B87** (1979) 97.

[Wig51]   WIGNER E.P. 'On a class of analytic functions from the quantum theory of collisions'. *Ann. Math.* **53** (1951) 36.

[Wil73]   WILSON K.G. 'Quantum field-theory models in less than four dimensions'. *Phys. Rev.* **D7** (1973) 2911.

[Wil74]   WILSON K.G. 'Confinement of quarks'. *Phys. Rev.* **D10** (1974) 2445.

[Wit79]   WITTEN E. 'The $1/N$ expansion in atomic and particle physics', in *Recent developments in gauge theories*, eds. G. 't Hooft *et al.* (Plenum, New York, 1980) p. 403.

[Wit90]   WITTEN E. 'On the structure of the topological phase of two-dimensional gravity'. *Nucl. Phys.* **B340** (1990) 281.

Part 4
Reduced Models

The large-N reduction was first discovered in 1982 by Eguchi and Kawai [EK82], who showed that the $SU(N)$ Yang–Mills theory on a d-dimensional space-time is equivalent at $N = \infty$ to the one at a point. This construction is based on an extra symmetry of the reduced model which should not be broken spontaneously.

Soon after that it was recognized that this symmetry is, in fact, broken for $d > 2$. Two ways were proposed to cure the construction: the quenching prescription [BHN82] and the twisting prescription [GO83a]. Each of these two prescriptions results in a reduced model which recovers multicolor QCD both on the lattice and in the continuum.

While the reduced models look like a great simplification, since the space-time is reduced to a point, they still involve an integration over d infinite matrices which is, in fact, a continual path integral. For some years it was not clear whether or not this is a real simplification of the original theory which can make it solvable, so the point of view on the reduced models was that they are just an elegant representation at large N.

The recent interest in reduced models has arisen from the matrix-model formulation [BFS97, IKK97] of M-theory combining all types of superstring theories. The novel point of view on the reduced models is that they are equivalent [CDS98] to gauge theories on noncommutative space. The gauge field is no longer matrix-valued but rather noncommutativity of matrices in the reduced models is transformed into noncommutativity of coordinates in the noncommutative gauge theory, which in the limit of large noncommutativity reproduces ordinary Yang–Mills theory at large N.

We shall start this part by describing the original Eguchi–Kawai model and its quenched version. Then we discuss twisted reduced models and their equivalence to noncommutative gauge theories. Finally, we concentrate on the properties of noncommutative gauge theories as such.

14

Eguchi–Kawai model

The large-N reduction was discovered by Eguchi and Kawai [EK82] who showed that the Wilson lattice gauge theory on a d-dimensional hypercubic lattice is equivalent at $N = \infty$ to that on a hypercube with periodic boundary conditions. This construction is based on an extra $U(1)^d$ symmetry which is present in the reduced model to each order of the strong-coupling expansion.

Soon after that it was recognized that a phase transition occurs in the reduced model with decreasing coupling constant, so this symmetry is broken in the weak-coupling regime. To cure the construction at weak coupling, a quenching prescription was proposed by Bhanot, Heller and Neuberger [BHN82] and elaborated by many authors. The quenching prescription results in a reduced model which recovers multicolor QCD both on the lattice and in the continuum.

We start this chapter with the simplest example of a matrix-valued scalar theory. The quenched reduced model for this case was advocated by Parisi [Par82] on the lattice and elaborated by Gross and Kitazawa [GK82] in the continuum. Then we consider the Eguchi–Kawai reduction of multicolor QCD both on the lattice and in the continuum.

14.1 Reduction of the scalar field (lattice)

Let us begin with the simplest example of a matrix-valued scalar theory on a lattice, the partition function of which is defined by the path integral

$$Z = \int \prod_x \prod_{i \geq j} \mathrm{d}\varphi_x^{ij} \, \mathrm{e}^{-S[\varphi]} \tag{14.1}$$

with the action

$$S[\varphi] \;=\; \sum_x N \operatorname{tr}\left[-\sum_\mu \varphi_x \varphi_{x+a\hat\mu} + V(\varphi_x) \right]. \tag{14.2}$$

Here $\varphi(x)$ is an $N \times N$ Hermitian matrix field and $V(\varphi)$ is a certain interaction potential, say

$$V(\varphi) \;=\; \frac{M}{2}\varphi^2 + \frac{\lambda_3}{3}\varphi^3 + \frac{\lambda_4}{4}\varphi^4. \tag{14.3}$$

The prescription of the large-N reduction is formulated as follows. We substitute

$$\varphi(x) \;\overset{\text{red.}}{\to}\; D^\dagger(x)\tilde\varphi D(x), \tag{14.4}$$

where

$$D(x) \;=\; e^{-iP_\mu x_\mu} \tag{14.5}$$

with

$$P^\mu \;=\; \operatorname{diag}\left(p_1^\mu, \ldots, p_N^\mu\right) \tag{14.6}$$

being a diagonal Hermitian matrix. Explicitly we have

$$\varphi^{kj}(x) \;\overset{\text{red.}}{\to}\; e^{i(p_k - p_j)^\mu x_\mu}\tilde\varphi^{kj}. \tag{14.7}$$

The matrix $D(x)$ in Eq. (14.4) subsumes the coordinate dependence, so that $\tilde\varphi$ does *not* depend on x.

The averaging of a functional $F[\varphi_x]$ which is defined with the same weight as in Eq. (14.1),

$$\left\langle F[\varphi_x] \right\rangle \;=\; Z^{-1} \int \prod_x \prod_{i \geq j} d\varphi_x^{ij}\, e^{-S[\varphi]}\, F[\varphi_x], \tag{14.8}$$

can be calculated at $N = \infty$ using

$$\left\langle F[\varphi_x] \right\rangle \;\overset{\text{red.}}{=}\; a^{Nd} \int_{-\pi/a}^{\pi/a} \prod_{\mu=1}^{d}\prod_{i=1}^{N} \frac{dp_i^\mu}{2\pi} \left\langle F[D^\dagger(x)\tilde\varphi D(x)] \right\rangle_{\mathrm{RM}}. \tag{14.9}$$

Here the RHS is calculated [Par82, GK82, DW82] for the *quenched reduced model*, for which the averages are defined by

$$\left\langle F[\tilde\varphi] \right\rangle_{\mathrm{RM}} \;\equiv\; Z_{\mathrm{RM}}^{-1} \int d\tilde\varphi\, e^{-S_{\mathrm{R}}[\tilde\varphi]}\, F[\tilde\varphi] \tag{14.10}$$

with the reduced action

$$S_R[\tilde{\varphi}] \;\; \stackrel{\text{red.}}{=} \;\; -N \sum_{ij} |\tilde{\varphi}_{ij}|^2 \sum_\mu \cos\left[(p_i^\mu - p_j^\mu)a\right] + N \operatorname{tr} V(\tilde{\varphi}). \quad (14.11)$$

We have put the symbol "red." on the top of the equality sign in Eq. (14.9) to emphasize that it holds as a result of the large-N reduction. The partition function of the reduced model is given by

$$Z_{\text{RM}} \;\; = \;\; \int d\tilde{\varphi}\, e^{-S_R[\tilde{\varphi}]} \quad (14.12)$$

which can be deduced, modulo the volume factor in the action (14.11), from the partition function (14.1) by the substitution (14.4). The measure $d\tilde{\varphi}$ in Eqs. (14.10) and (14.12) (as well as that in Eqs. (14.1) and (14.8)) is the one for integrating over $N \times N$ Hermitian matrices given by Eq. (13.2).

Similarly to Eq. (14.9), the free energy of the reduced model determines at large N the free energy per unit volume of the d-dimensional theory:

$$\frac{1}{N^2}\frac{\ln Z}{V} \;\; \stackrel{\text{red.}}{=} \;\; a^{d(N-1)} \int\limits_{-\pi/a}^{\pi/a} \prod_{\mu=1}^{d} \prod_{i=1}^{N} \frac{dp_i^\mu}{2\pi} \frac{1}{N^2} \ln Z_{\text{RM}}. \quad (14.13)$$

Note that the integration over the momenta p_i^μ on the RHS of Eq. (14.9) is taken *after* the calculation of averages in the reduced model. Analogously, the logarithm of Z_{RM} is integrated over p_i^μ on the RHS of Eq. (14.13), rather than Z_{RM} itself. Such variables are usually called *quenched* in statistical mechanics which clarifies the terminology.

Since $N \to \infty$, it is not necessary to integrate over the quenched momenta in Eq. (14.9) or Eq. (14.13). The integral should be recovered if p_i were uniformly distributed over a d-dimensional hypercube. This is analogous to the self-averaging phenomenon in condensed-matter physics.

In order to show how Eq. (14.9) works, let us demonstrate how the planar diagrams of perturbation theory for the matrix-valued scalar theory (14.1) are recovered in the quenched reduced model.

The quenched reduced model (14.12) is of the general type discussed in Chapter 11. The propagator is given by

$$\left\langle \tilde{\varphi}_{ij} \tilde{\varphi}_{kl} \right\rangle_{\text{Gauss}} \;\; = \;\; \frac{1}{N} G(p_i - p_j)\, \delta_{il}\delta_{kj} \quad (14.14)$$

with

$$G(p_i - p_j) \;\; = \;\; \frac{1}{M - 2\sum_\mu \cos\left[(p_i^\mu - p_j^\mu)a\right]}. \quad (14.15)$$

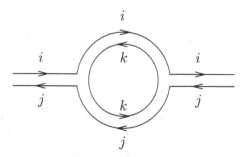

Fig. 14.1.   The simplest planar diagram of second order in λ_3 for the propagator in the quenched reduced model (14.12). The momentum p_i flows along the index line i. The momentum $p_i - p_j$ is associated with the double line ij.

It is convenient to associate the momenta p_i and p_j in Eq. (14.15) with each of the two index lines representing the propagator and carrying, respectively, indices i and j. Remember, that these lines are oriented for a Hermitian matrix $\tilde{\varphi}$ and their orientation can be associated naturally with the direction of the flow of the momentum. The total momentum carried by the double line is $p_i - p_j$.

The simplest diagram which represents the second order in λ_3 correction to the propagator is depicted Fig. 14.1. The momenta p_i and p_j flow along the index lines i and j, while the momentum p_k circulates along the index line k. The contribution of the diagram in Fig. 14.1 is given by

$$\text{Fig. 14.1} \; = \; \frac{\lambda_3^2}{N^2}\, G^2(p_i - p_j) \sum_k G(p_i - p_k)\, G(p_k - p_j)\,, \qquad (14.16)$$

where the summation over the index k is just a standard one over indices forming a closed loop.

In order to show that the quenched-model result (14.16) reproduces the second order in λ_3 correction to the propagator in the d-dimensional theory on an infinite lattice, we pass to the variables of the total momenta flowing along the double lines:

$$\left.\begin{aligned}
p_i - p_j &= p\,, \\
p_j - p_k &= q\,, \\
p_i - p_k &= p + q\,.
\end{aligned}\right\} \qquad (14.17)$$

This is obviously consistent with the momentum conservation at each of the two vertices of the diagram in Fig. 14.1.

Since p_k are uniformly distributed over the hypercube, the summation over k can be substituted as $N \to \infty$ by the integral

$$\frac{1}{N} \sum_k f(p_k) \quad \Rightarrow \quad a^d \int_{-\pi/a}^{\pi/a} \frac{d^d q}{(2\pi)^d} f(q) \,. \tag{14.18}$$

The prescription (14.9) then gives the correct expression

$$G^{(2)}(p) \quad = \quad a^d \frac{\lambda_3^2}{N} G^2(p) \int_{-\pi/a}^{\pi/a} \frac{d^d q}{(2\pi)^d} G(q) G(p+q) \tag{14.19}$$

for the second-order contribution of the perturbation theory for the propagator on the lattice.

It is now clear how a generic planar diagram is recovered by the reduced model. We first represent the diagram by the double lines and associate the momentum p_i to an index line carrying the index i. Then we write down an expression for the diagram in the reduced model with the propagator (14.15). Passing to momenta flowing along the double lines, similarly to Eq. (14.17), we obtain an expression which coincides with the *integrand* of the Feynman diagram for the theory on the d-dimensional lattice. It is crucial that such a change of variables can always be made for a planar diagram consistently with momentum conservation at each vertex. The last step is that a summation over indices of closed index lines reproduces an integration over momenta associated with each of the loops according to Eq. (14.18). It is assumed that the number of loops is much less than N which is always true for a given diagram since N is infinite.

Thus, we have shown how planar diagrams of the lattice theory defined by the partition function (14.1) are recovered in the reduced model (14.12). The lattice was needed only as a regularization to make all integrals well-defined and was not crucial in the consideration. In the next section we shall see how this construction can be formulated directly for the continuum theory.

Remark on large but finite N

If N is large but finite, the summation on the LHS of Eq. (14.18) runs over N different momenta. Similarly, if a theory is defined on a periodic lattice of size La, the momentum takes on L^d different values. One might think, therefore, that the quenched reduced model at very large but finite N can be associated with a quantum field theory on a periodic lattice with $L = N^{1/d}$.

14.2 Reduction of the scalar field (continuum)

The quenched reduced model can be formulated directly for the continuum theory. The proper formulas can be easily obtained from those of the previous section by setting $a \to 0$.

Equations (14.4)–(14.7) remain the same, while the derivative ∂_μ in the kinetic part of the continuum action

$$S[\varphi] \;=\; \int d^d x \, N \operatorname{tr} \left\{ \frac{1}{2}(\partial_\mu \varphi)^2 + V(\varphi) \right\} \tag{14.20}$$

is substituted by iP_μ acting in the adjoint representation.

The continuum reduced action

$$S_{\mathrm{R}} \;=\; v\, N \operatorname{tr} \left\{ -\frac{1}{2}[P_\mu, \tilde\varphi]^2 + \tilde V(\tilde\varphi) \right\} \tag{14.21}$$

determines the propagator of the matrix $\tilde\varphi_{ij}$ in the continuum quenched reduced model, which is given by Eq. (14.14) with

$$G(p_i - p_j) \;=\; \frac{v^{-1}}{(p_i - p_j)^2 + m^2}. \tag{14.22}$$

The dimension of the reduced field $\tilde\varphi$ is $[\text{mass}]^{(d-2)/2}$, i.e. the same as that of the field $\varphi(x)$ in the d-dimensional theory.

The normalizing factor of v on the RHS of Eq. (14.21) (and therefore Eq. (14.22)) plays the role of the volume element for a given regularization and depends on the region for integration over the momenta p_i. If p_i are restricted to a hypercube of size 2Λ, the proper formulas look like their lattice counterparts (cf. Eq. (14.18)) with

$$v \;=\; a^d \qquad \boxed{\text{lattice regularization}} \tag{14.23}$$

and

$$a \;=\; \frac{\pi}{\Lambda}. \tag{14.24}$$

Lorentz invariance is restored as $\Lambda \to \infty$ at least for renormalizable theories.

A Lorentz-invariant regularization can be achieved by choosing p_i inside a hypersphere. Alternatively, one can include the regularizing factor of $\exp\left(-p_i^2/\Lambda^2\right)$ in the integral over p_i [GK82] by defining

$$\int^\Lambda d^d p \cdots \;=\; \int \frac{d^d p}{(\Lambda\sqrt{\pi})^d}\, e^{-p^2/\Lambda^2} \cdots . \tag{14.25}$$

Then,

$$v = \left(\frac{2\pi}{\Lambda^2}\right)^{d/2} \qquad \boxed{\text{regularization (14.25)}} \qquad (14.26)$$

and

$$\left\langle F[\varphi(x)] \right\rangle \overset{\text{red.}}{=} \int^\Lambda \prod_{i=1}^N d^d p_i \left\langle F[D^\dagger(x)\tilde{\varphi}D(x)] \right\rangle_{\text{RM}} \qquad (14.27)$$

which is similar to Eq. (14.9) on the lattice.

Analogously to the lattice case, we expect the self-averaging phenomenon as $N \to \infty$ so that, instead of integration over p_i according to Eq. (14.25), we can simply choose them at $N = \infty$ to be distributed spherically symmetrically with Gaussian weight

$$\rho(p) = \left(\sqrt{\pi}\Lambda\right)^{-d} e^{-p^2/\Lambda^2}. \qquad (14.28)$$

A comment is needed concerning the normalization factors. A consideration similar to the topological analysis of Sect. 11.4 leads us to the conclusion that a planar diagram with n_2 loops possesses a factor of v^{-n_2} in the reduced model. It will normalize correctly the integral over momenta circulating along the n_2 loops, which remains to be done after the $N - n_2$ momenta p_i (which do not appear in the diagram) are integrated out. The analogous factor for the free energy is v^{-n_2+1} owing to the extra v in the definition (14.21).

Problem 14.1 Substituting (14.22) into Eq. (14.27), obtain a regularized propagator in the d-dimensional theory.

Solution Inserting (14.22) into Eq. (14.27), we find explicitly

$$
\begin{aligned}
G(x) &\equiv \left\langle \varphi_{ij}(x)\varphi_{ji}(0) \right\rangle \overset{\text{red.}}{=} \int^\Lambda \prod_{k=1}^N d^d p_k\, e^{i(p_i - p_j)x} \left\langle \tilde{\varphi}_{ij}\tilde{\varphi}_{ji} \right\rangle_{\text{RM}} \\
&= \left(\frac{\Lambda^2}{2\pi}\right)^{d/2} \int \frac{d^d p_i}{(\Lambda\sqrt{\pi})^d} \frac{d^d p_j}{(\Lambda\sqrt{\pi})^d}\, e^{-(p_i^2 + p_j^2)/\Lambda^2} \frac{e^{i(p_i - p_j)x}}{(p_i - p_j)^2 + m^2}.
\end{aligned}
$$
$$(14.29)$$

Introducing $p_\pm = p_i \pm p_j$ and accounting for a Jacobian, we have

$$
\begin{aligned}
G(x) &= \int \frac{d^d p_+}{(2\pi\Lambda^2)^{d/2}} e^{-p_+^2/2\Lambda^2} \int \frac{d^d p_-}{(2\pi)^d} e^{-p_-^2/2\Lambda^2} \frac{e^{i p_- x}}{p_-^2 + m^2} \\
&= \int \frac{d^d p}{(2\pi)^d} e^{-p^2/2\Lambda^2} \frac{e^{i p x}}{p^2 + m^2}
\end{aligned}
$$
$$(14.30)$$

which gives a regularized propagator with the correct normalization.

Remark on higher genera

The quenched reduced model reproduces only planar graphs of the original theory and fails to reproduce nonplanar graphs. This can be easily seen for the simplest nonplanar graph depicted in Fig. 11.3 where the same momentum circulates along the closed index line, so that the total momentum flowing along each of the two crossing double lines is zero in the reduced model. Of course, this is not the case for the original d-dimensional theory.

Similarly, the quenched reduced model reproduces only the factorized part of the correlators of "colorless" operators, for example $\mathrm{tr}\, \varphi^2(x_i)/N$, and cannot reproduce the connected correlators.

14.3 Reduction of the Yang–Mills field

The large-N reduction of the Yang–Mills fields has its specific features owing to gauge invariance. In order to make the results rigorous, we begin in this section with the lattice formulation of Yang–Mills theory introduced in Chapter 6 and then describe the continuum case in the next section.

The general prescription (14.4) and (14.5) of the large-N reduction is applicable for gauge fields. For the lattice gauge field $U_\mu(x)$, it gives

$$U_\mu(x) \stackrel{\mathrm{red.}}{\rightarrow} D^\dagger(x)\widetilde{U}_\mu D(x), \tag{14.31}$$

where d unitary $N \times N$ matrices $\widetilde{U}_\mu^{ij}$ ($\mu = 1, \ldots, d$) are x-independent.

It is easy to deduce what transformation of the reduced gauge field $\widetilde{U}_\mu$ is compatible with the lattice gauge transformation (6.13), where $\Omega(x)$ is to be reduced by

$$\Omega(x) \stackrel{\mathrm{red.}}{\rightarrow} D^\dagger(x)\widetilde{\Omega}D(x). \tag{14.32}$$

Here $\widetilde{\Omega}$ is again an x-independent unitary matrix.

If we first perform the gauge transformation (6.13) and then the reduction of the gauge-transformed field $U_\mu(x)$, we obtain

$$\Omega(x + a\hat{\mu})\, U_\mu(x)\, \Omega^\dagger(x)$$
$$\stackrel{\mathrm{red.}}{\rightarrow} D^\dagger(x + a\hat{\mu})\, \widetilde{\Omega}\, D(x + a\hat{\mu})\, D^\dagger(x)\widetilde{U}_\mu D(x)D^\dagger(x)\, \widetilde{\Omega}^\dagger D(x)$$
$$= D^\dagger(x)D_\mu^\dagger\widetilde{\Omega}D_\mu\widetilde{U}_\mu\widetilde{\Omega}^\dagger D(x), \tag{14.33}$$

where

$$D_\mu \stackrel{\mathrm{def}}{=} D(x + a\hat{\mu})\, D^\dagger(x) = \mathrm{e}^{-iP_\mu a} \tag{14.34}$$

for $D(x)$ given by Eqs. (14.5) and (14.6).

This determines the proper transformation of the reduced field $\widetilde{U}_\mu$ to be

$$\widetilde{U}_\mu \xrightarrow{\text{g.t.}} D_\mu^\dagger \widetilde{\Omega} D_\mu \widetilde{U}_\mu \widetilde{\Omega}^\dagger . \tag{14.35}$$

This transformation is referred to as the gauge transformation of the reduced gauge field.

The substitution of (14.31) into the Wilson action (6.18) results in the reduced action

$$
\begin{aligned}
S_{\mathrm{R}} &= \frac{1}{2} \sum_{\mu \neq \nu} \left\{ 1 - \frac{1}{N} \operatorname{tr} \left[\widetilde{U}_\nu^\dagger D_\nu^\dagger \widetilde{U}_\mu^\dagger D_\nu D_\mu^\dagger \widetilde{U}_\nu D_\mu \widetilde{U}_\mu \right] \right\} \\
&= \frac{1}{2} \sum_{\mu \neq \nu} \left\{ 1 - \frac{1}{N} \operatorname{tr} \left[\left(\widetilde{U}_\nu^\dagger D_\nu^\dagger \right) \left(\widetilde{U}_\mu^\dagger D_\mu^\dagger \right) \left(D_\nu \widetilde{U}_\nu \right) \left(D_\mu \widetilde{U}_\mu \right) \right] \right\},
\end{aligned}
\tag{14.36}
$$

where the equality between the first and second lines is because D_μ and $D_\nu^\dagger$ commute.

The structure of the RHS of Eq. (14.36) prompts us to introduce a new variable

$$U_\mu = D_\mu \widetilde{U}_\mu . \tag{14.37}$$

Then we obtain for the reduced action

$$S_{\mathrm{R}}[U] = \frac{1}{2} \sum_{\mu \neq \nu} \left(1 - \frac{1}{N} \operatorname{tr} U_\nu^\dagger U_\mu^\dagger U_\nu U_\mu \right) \tag{14.38}$$

and the gauge transformation (14.35) also simplifies to

$$U_\mu \xrightarrow{\text{g.t.}} \widetilde{\Omega} U_\mu \widetilde{\Omega}^\dagger . \tag{14.39}$$

If the measure $\mathrm{d}\widetilde{U}_\mu$ for averaging over $\widetilde{U}_\mu$ is the Haar measure, it is not changed under (left) multiplication by a unitary matrix D_μ: $\mathrm{d}\widetilde{U}_\mu = \mathrm{d}U_\mu$. Finally, we arrive at the reduced model discovered originally by Eguchi and Kawai [EK82].

Its partition function

$$Z_{\mathrm{EK}} = \int \prod_\mu \mathrm{d}U_\mu \, \mathrm{e}^{-N S_{\mathrm{R}}[U]/g^2} \tag{14.40}$$

is of the same type as Wilson's lattice gauge theory on a unit hypercube with periodic boundary conditions. There is no dependence on the quenched momenta p_i in the action of the Eguchi–Kawai model since the Ds have mutually canceled owing to the local gauge invariance of the lattice action (6.16).

In addition to the gauge symmetry (14.39), the Eguchi–Kawai model possesses symmetry under multiplication of U_μ by an element of $U(1)$, the center of $U(N)$, which depends on the direction μ:[*]

$$U_\mu \;\to\; Z_\mu U_\mu \qquad (Z_\mu \in U(1))\,. \tag{14.41}$$

Such a global symmetry is also present, of course, for the Wilson action (6.16) but plays no special role there because of local gauge invariance. It will be crucial in providing the equivalence of the d-dimensional theory and the Eguchi–Kawai model at large N.

The equivalence of the d-dimensional lattice gauge theory and the Eguchi–Kawai model at $N = \infty$ states

$$\Big\langle F[U_\mu(x)]\Big\rangle \;\overset{\text{red.}}{=}\; a^{Nd} \int_{-\pi/a}^{\pi/a} \prod_{\mu=1}^{d}\prod_{i=1}^{N} \frac{dp_i^\mu}{2\pi}\Big\langle F\Big[D^\dagger(x+a\hat\mu)\,U_\mu D(x)\Big]\Big\rangle_{\text{EK}}, \tag{14.42}$$

which is similar to Eq. (14.9) for scalars. Here the LHS is given by Eq. (6.39) and the RHS is calculated in the Eguchi–Kawai model:

$$\Big\langle F[U_\mu]\Big\rangle_{\text{EK}} \;=\; Z_{\text{EK}}^{-1} \int \prod_\mu dU_\mu\, e^{-N S_{\text{R}}[U]/g^2}\, F[U_\mu]\,. \tag{14.43}$$

The commutativity of D_μ was used in representing the argument of F on the RHS of Eq. (14.42) as $D^\dagger(x+a\hat\mu)\,U_\mu D(x)$. Note that it looks like a gauge transformation of a constant field in the d-dimensional theory.

For the latter reason Eq. (14.42) simplifies for the averages of gauge-invariant quantities when it takes the form

$$\Big\langle F[U_\mu(x)]\Big\rangle \;\overset{\text{red.}}{=}\; \Big\langle F[U_\mu]\Big\rangle_{\text{EK}} \qquad \boxed{\text{gauge invariant } F} \tag{14.44}$$

as $N \to \infty$. In this formula there is no dependence on $D(x)$ and correspondingly the quenched momenta p_i because $F[U_\mu(x)]$ is gauge invariant.

As has already been explained in Sect. 12.1, gauge-invariant observables in Yang–Mills theory can be expressed via the Wilson loops. Applying Eq. (14.44) for the Wilson loop averages (6.42), we obtain

$$\Big\langle \frac{1}{N}\operatorname{tr} U(C)\Big\rangle \;\overset{\text{red.}}{=}\; \Big\langle \frac{1}{N}\operatorname{tr} P \prod_i U_{\mu_i}\Big\rangle_{\text{EK}} \tag{14.45}$$

---

[*] For the gauge group $SU(N)$, it is an element of $Z(N)$ rather than $U(1)$ for the Haar measure dU_μ to be invariant.

as $N \rightarrow \infty$. In other words, the Wilson loop in the Eguchi–Kawai model is constructed as an (ordered) product of the constant matrices U_{μ_i} along the links forming the contour C. It is nontrivial since U_μ do not commute.

The equality (14.45) is possible since the Wilson loop averages in the original theory do not depend on the position of the beginning of the contour C owing to translational invariance.

The simplest example of the Wilson loop average is that for a rectangular contour depicted in Fig. 6.6 on p. 111. It is represented in the Eguchi–Kawai model by

$$W(R \times T) = \left\langle \frac{1}{N} \mathrm{tr}\, U_d^{\dagger T} U_1^{\dagger R} U_d^T U_1^R \right\rangle_{\mathrm{EK}}. \qquad (14.46)$$

There is an important difference between the averages of open Wilson loops in the d-dimensional theory and the Eguchi–Kawai model. In the former case, the averages of open Wilson loops always vanish because of the local gauge invariance which cannot be broken spontaneously owing to Elitzur's theorem, which was already mentioned in Sect. 7.3. In the latter case, open Wilson loops are invariant under the transformation (14.39) since $\widetilde{\Omega}$ is the same at the beginning and the end of the contour:

$$\frac{1}{N} \mathrm{tr}\, \boldsymbol{P} \prod_i U_{\mu_i} \xrightarrow{(14.39)} \frac{1}{N} \mathrm{tr}\, \left(\widetilde{\Omega}\, \boldsymbol{P} \prod_i U_{\mu_i} \widetilde{\Omega}^{\dagger} \right) = \frac{1}{N} \mathrm{tr}\, \boldsymbol{P} \prod_i U_{\mu_i}. \qquad (14.47)$$

They are *not* invariant however under the $U(1)^d$ transformation (14.41):

$$\frac{1}{N} \mathrm{tr}\, \boldsymbol{P} \prod_i U_{\mu_i} \xrightarrow{(14.41)} \prod_i Z_{\mu_i} \frac{1}{N} \mathrm{tr}\, \boldsymbol{P} \prod_i U_{\mu_i}. \qquad (14.48)$$

Only closed Wilson loops, where each link occurs with an equal number of positive and negative orientations, are invariant. This symmetry is global and can be broken spontaneously as $N \rightarrow \infty$.

It is easy to see that no such breaking occurs within the strong-coupling expansion of the Eguchi–Kawai model, which is pretty much similar to that described in Sect. 6.5. For this reason the equivalence (14.44) holds at least for large enough values of $g^2 N$. It was shown [BHN82] that the $U(1)^d$ symmetry *is* spontaneously broken for small values of $g^2 N$ and therefore in the continuum. We shall return to this point in Sect. 14.5.

Two modifications of the Eguchi–Kawai model were proposed: the quenched Eguchi–Kawai model (described later in Sect. 14.6) and the twisted Eguchi–Kawai model (described later in Sect. 15.3). These two models are equivalent, in the large-N limit, to the d-dimensional theory both on the lattice and in the continuum.

Problem 14.2 Derive the loop equation for the Eguchi–Kawai model.

Solution The derivation is similar to Problem 12.6 on p. 265. We perform the shift of U_μ:

$$U_\mu \to U_\mu \left(1 - i\epsilon_\mu\right), \qquad U_\mu^\dagger \to \left(1 + i\epsilon_\mu\right) U_\mu^\dagger \tag{14.49}$$

which is same as in Eq. (6.22) for x-independent ϵ. The resulting loop equation is given as [EK82]

$$\sum_p \left[W_{\mathrm{EK}}(C\,\partial p) - W_{\mathrm{EK}}(C\,\partial p^{-1}) \right] = g^2 N \sum_{l \in C} \tau_\nu(l)\, W_{\mathrm{EK}}(C_{yx})\, W_{\mathrm{EK}}(C_{xy}). \tag{14.50}$$

It is similar to Eq. (12.65) except the Kronecker symbol δ_{xy} is missing on the RHS of Eq. (14.50). It is restored if the averages of the open Wilson loops vanish, as prescribed by the unbroken $U(1)^d$ symmetry, since then we can substitute

$$W_{\mathrm{EK}}(C_{xy}) = \delta_{xy} W_{\mathrm{EK}}(C_{xx}). \tag{14.51}$$

The coincidence of the loop equations proves the equivalence of the two theories at $N = \infty$.

Problem 14.3 Verify Eq. (14.51) by an explicit calculation to zeroth order in g^2.

Solution Extrema of the Eguchi–Kawai action (14.38) are given modulo a gauge transformation by diagonal matrices

$$U_\mu^{\mathrm{cl}} = e^{-iP_\mu a}. \tag{14.52}$$

This determines the Wilson loop average to zeroth order in g^2 to be

$$W_{\mathrm{EK}}(C_{yx}) = a^{Nd} \int\limits_{-\pi/a}^{\pi/a} \prod_{i=1}^{N} \frac{d^d p_i}{(2\pi)^d} \frac{1}{N} \sum_{k=1}^{N} e^{ip_k(x-y)}$$

$$= a^d \int\limits_{-\pi/a}^{\pi/a} \frac{d^d p}{(2\pi)^d} e^{ip(x-y)} = \delta_{xy}, \tag{14.53}$$

where the integration over P_μ accounts for equivalent classical extrema. The Kronecker symbol in Eq. (14.53) appears because of the translational symmetry in momentum space.

14.4 The continuum Eguchi–Kawai model

The Eguchi–Kawai reduced model can be formulated directly for the continuum theory. The proper formulas can be derived from their lattice counterparts of the previous section by substituting

$$U_\mu = e^{iaA_\mu} \tag{14.54}$$

with $a \to 0$.

The continuum Eguchi–Kawai model describes a reduction of the d-dimensional Yang–Mills theory at $N = \infty$ to a point. The action of the continuum Eguchi–Kawai model is given by

$$S_{\mathrm{EK}}[A] = -\left(\frac{2\pi}{\Lambda^2}\right)^{d/2} \frac{1}{4g^2} \,\mathrm{tr}\, [A_\mu, A_\nu]^2 , \qquad (14.55)$$

where A_μ are d space-independent matrices.

The parameter Λ has the dimension of mass, same as has A_μ in $d = 4$. As we shall see in a moment, Λ is to be associated with a momentum-space ultraviolet cutoff in the spirit of Sect. 14.2. In this chapter we assume the Lorentz-invariant regularization (14.25) when the normalization factor in Eq. (14.55) is given by Eq. (14.26). For the lattice regularization, Λ is related to the lattice spacing a by Eq. (14.24) and the normalization factor in Eq. (14.55) is to be changed according to Eq. (14.23).

Therefore, the very formulation of the continuum Eguchi–Kawai model implies a regularization.

The action (14.55) is obviously invariant under the gauge transformation

$$A_\mu \overset{\text{g.t.}}{\longrightarrow} \widetilde{\Omega}\, A_\mu \,\widetilde{\Omega}^\dagger . \qquad (14.56)$$

It is worth noting that, owing to Eqs. (14.37), (6.10), and (14.34), A_μ is associated with the reduction of the covariant derivative $i\partial_\mu + \mathcal{A}_\mu(x)$ rather than the field $\mathcal{A}_\mu(x)$ itself:

$$i\partial_\mu + \mathcal{A}_\mu(x) \overset{\text{red.}}{\to} D^\dagger(x)\, A_\mu\, D(x) . \qquad (14.57)$$

This explains why Eq. (14.56) is consistent with the gauge transformation of the covariant derivative

$$i\partial_\mu + \mathcal{A}_\mu(x) \overset{\text{g.t.}}{\longrightarrow} \Omega(x)\,[i\partial_\mu + \mathcal{A}_\mu(x)]\,\Omega^\dagger(x) \qquad (14.58)$$

rather than $\mathcal{A}_\mu(x)$ itself.

Similarly to Eq. (14.44),

$$\Big\langle F[i\partial_\mu + \mathcal{A}_\mu(x)] \Big\rangle \overset{\text{red.}}{=} \Big\langle F[A_\mu] \Big\rangle_{\mathrm{EK}} \qquad \boxed{\text{gauge invariant } F} \qquad (14.59)$$

as $N \to \infty$ for gauge-invariant functionals F, where the LHS is calculated using the action (11.72) and the RHS is calculated using the Eguchi–Kawai action (14.55). For instance, the averages of closed Wilson loops coincide in both cases

$$\left\langle \frac{1}{N}\,\mathrm{tr}\,\boldsymbol{P}\, \mathrm{e}^{\mathrm{i}\oint \mathrm{d}\xi^\mu \mathcal{A}_\mu(\xi)} \right\rangle \overset{\text{red.}}{=} \left\langle \frac{1}{N}\,\mathrm{tr}\,\boldsymbol{P}\, \mathrm{e}^{\mathrm{i}\oint \mathrm{d}\xi^\mu A_\mu} \right\rangle_{\mathrm{EK}} . \qquad (14.60)$$

This is a continuum version of Eq. (14.45).

The continuum analog of the $U(1)^d$ symmetry (14.41) is the invariance of the Eguchi–Kawai action (14.55) under the shift of A_μ by a unit matrix:

$$A_\mu^{ij} \;\rightarrow\; A_\mu^{ij} + r_\mu \delta^{ij}, \tag{14.61}$$

where r_μ is a parameter of the transformation. It is often called the R^d symmetry.

Under the transformation (14.61), an open Wilson loop is transformed as

$$\frac{1}{N}\mathrm{tr}\left(\boldsymbol{P}\, \mathrm{e}^{\mathrm{i}\int_{C_{yx}}\mathrm{d}\xi^\mu A_\mu}\right) \;\rightarrow\; \mathrm{e}^{\mathrm{i}(y^\mu-x^\mu)r_\mu}\frac{1}{N}\mathrm{tr}\left(\boldsymbol{P}\, \mathrm{e}^{\mathrm{i}\int_{C_{yx}}\mathrm{d}\xi^\mu A_\mu}\right).$$

$$\tag{14.62}$$

This guarantees, if the symmetry is not broken, the vanishing of the averages of open Wilson loops

$$W_{\mathrm{EK}}(C_{yx}) \equiv \left\langle \frac{1}{N}\mathrm{tr}\,\boldsymbol{P}\, \mathrm{e}^{\mathrm{i}\int_{C_{yx}}\mathrm{d}\xi^\mu A_\mu}\right\rangle_{\mathrm{EK}} = 0 \quad \boxed{\text{for } y \neq x} \tag{14.63}$$

in the Eguchi–Kawai model.

Such vanishing in the d-dimensional theory is provided by the local gauge invariance under which

$$\left(\boldsymbol{P}\, \mathrm{e}^{\mathrm{i}\int_{C_{yx}}\mathrm{d}\xi^\mu A_\mu(\xi)}\right)_{ij} \;\rightarrow\; \left(\Omega(y)\,\boldsymbol{P}\, \mathrm{e}^{\mathrm{i}\int_{C_{yx}}\mathrm{d}\xi^\mu A_\mu(\xi)}\Omega^\dagger(x)\right)_{ij}. \tag{14.64}$$

In contrast, the global symmetry (14.56) does *not* guarantee the vanishing of the averages of open Wilson loops in the Eguchi–Kawai model.

When and only when the R^d symmetry (14.61) is not broken spontaneously, is the Eguchi–Kawai model equivalent to the d-dimensional Yang–Mills theory at large N.

The equivalence of the two theories can then be shown using the loop equation which is given for the Eguchi–Kawai model by

$$\begin{aligned}
\partial_\mu^x \frac{\delta}{\delta\sigma_{\mu\nu}(x)}W_{\mathrm{EK}}(C) &= \mathrm{i}\left\langle \frac{1}{N}\mathrm{tr}\,\boldsymbol{P}\,[A_\mu,[A_\mu,A_\nu]]\,\mathrm{e}^{\mathrm{i}\oint_{C_{xx}}\mathrm{d}\xi^\mu A_\mu}\right\rangle_{\mathrm{EK}} \\
&= -\mathrm{i}\lambda\left(\frac{\Lambda^2}{2\pi}\right)^{d/2}\left\langle \frac{1}{N}\mathrm{tr}\,\boldsymbol{P}\,\frac{\partial}{\partial A_\nu}\mathrm{e}^{\mathrm{i}\oint_{C_{xx}}\mathrm{d}\xi^\mu A_\mu}\right\rangle_{\mathrm{EK}} \\
&= \lambda\left(\frac{\Lambda^2}{2\pi}\right)^{d/2}\oint_C \mathrm{d}y_\nu\, W_{\mathrm{EK}}(C_{yx})\,W_{\mathrm{EK}}(C_{xy}),
\end{aligned}$$

$$\tag{14.65}$$

where $\lambda = g^2 N$. The RHS is pretty much similar to that in Eq. (12.59), while $(\Lambda^2/2\pi)^{d/2}$ is present instead of $\delta^{(d)}(x-y)$.

In order to show how the two RHSs are essentially equal to each other providing the R^d symmetry is not broken, we need to remember that the continuum Eguchi–Kawai model is, in fact, somehow regularized.

While the action (14.55) is formally invariant under the transformation (14.61) for arbitrary r_μ, admitable values of r_μ should be much smaller than the cutoff Λ. This is clear, in particular, from the lattice formula (14.41) where

$$Z_\mu = e^{iar_\mu} \qquad (14.66)$$

and to obtain Eq. (14.61) we expand in a which destroys the compactness.

For this reason we expect the average of an open Wilson loop to vanish in the continuum Eguchi–Kawai model only when the distance $|y - x|$ between the end points x and y is much larger than the ultraviolet cutoff $1/\Lambda$. Otherwise, we may regard the loop to be essentially closed since distances smaller than the cutoff make no sense in the theory.

Introducing a smeared delta-function $\delta_\Lambda^{(d)}(x - y)$, for example, by

$$\delta_\Lambda^{(d)}(x) = \left(\frac{\Lambda^2}{2\pi}\right)^{d/2} e^{-x^2\Lambda^2/2}, \qquad (14.67)$$

we therefore expect something like*

$$W_{\text{EK}}(C_{yx}) \approx \frac{\delta_{\Lambda/\sqrt{2}}^{(d)}(x - y)}{\delta_{\Lambda/\sqrt{2}}^{(d)}(0)} W_{\text{EK}}(C_{xx}) \qquad (14.68)$$

for the averages of open Wilson loops in the continuum Eguchi–Kawai model.

Finally, the delta function is recovered on the RHS of Eq. (14.65) as

$$\left(\frac{\Lambda^2}{2\pi}\right)^{d/2} \left[\frac{\delta_{\Lambda/\sqrt{2}}^{(d)}(x)}{\delta_{\Lambda/\sqrt{2}}^{(d)}(0)}\right]^2 = \left(\frac{\Lambda^2}{2\pi}\right)^{d/2} e^{-x^2\Lambda^2/2} = \delta_\Lambda^{(d)}(x) \rightarrow \delta^{(d)}(x),$$
$$(14.69)$$

reproducing the delta function on the RHS of Eq. (12.59).

This demonstrates the equivalence of the continuum Eguchi–Kawai model and the d-dimensional Yang–Mills theory at large N under the assumption that the R^d symmetry is not broken. The consideration simply repeats the proof of the equivalence given in Problem 14.2 on p. 336 by using the lattice regularization.

---

* Why it should be $\delta_{\Lambda/\sqrt{2}}^{(d)}$ rather than $\delta_\Lambda^{(d)}$ is clear from Eq. (14.69) and Problem 14.4.

Problem 14.4 Verify Eq. (14.68) by explicit calculation to zeroth order in g^2, regularizing the integral over the zero modes of A_μ by Eq. (14.25).

Solution The calculation is similar to that in Problem 14.3 for the lattice case. Extrema of the continuum Eguchi–Kawai action (14.55) are given modulo a gauge transformation by diagonal matrices $A_\mu^{\rm cl} = -P_\mu$. This determines the Wilson loop average to zeroth order in g^2 to be

$$
W_{\rm EK}(C_{yx}) = \int^\Lambda \prod_{i=1}^N d^d p_i \frac{1}{N} \sum_{k=1}^N e^{ip_k(x-y)}
$$

$$
= \int \frac{d^d p}{(\Lambda\sqrt{\pi})^d} e^{-p^2/\Lambda^2} e^{ip(x-y)} = \frac{\delta_{\Lambda/\sqrt{2}}^{(d)}(x-y)}{\delta_{\Lambda/\sqrt{2}}^{(d)}(0)}, \tag{14.70}
$$

where the integration over P_μ accounts for the zero modes of A_μ.

The R^d symmetry *is*, in fact, broken spontaneously in the continuum Eguchi–Kawai model for $d > 2$ as is discussed in the next section. For this reason the equivalence between the d-dimensional theory and the naive continuum Eguchi–Kawai model described in this section is valid, strictly speaking, only in $d = 2$. The reduced model should be slightly modified to be equivalent to the d-dimensional theory for $d > 2$. Such a modification, which is based on the quenched momentum prescription, is described later in Sect. 14.6.

14.5 R^d symmetry in perturbation theory

Since N is infinite, the R^d symmetry can be broken spontaneously. The point is that the large-N limit plays the role of a statistical average, as has already been mentioned in Sect. 11.8, and phase transitions are possible for an infinite number of degrees of freedom. This phenomenon occurs [BHN82] in perturbation theory for the naive Eguchi–Kawai model with $d > 2$.

A perturbation theory can be constructed by expanding the fields around solutions of the classical equation

$$
[A_\mu, [A_\mu, A_\nu]] = 0. \tag{14.71}
$$

An arbitrary diagonal matrix

$$
A_\mu^{\rm cl} = -P_\mu \tag{14.72}
$$

is a solution to Eq. (14.71) associated with the minimal value $S_{\rm EK} = 0$ of the action (14.55).

The perturbation theory of the reduced model can be constructed by expanding around the classical solution (14.72):

$$
A_\mu = A_\mu^{\rm cl} + g b_\mu, \tag{14.73}
$$

where b_μ is off-diagonal.

Substituting (14.73) into the action (14.55), we obtain

$$S_{\text{EK}} \;=\; -\left(\frac{2\pi}{\Lambda^2}\right)^{d/2} \operatorname{tr}\left\{\frac{1}{2}[P_\mu, b_\nu]^2 - \frac{1}{2}[P_\mu, b_\mu]^2\right\} + \text{higher orders} .$$

$$(14.74)$$

To fix the gauge symmetry (14.56), it is convenient to add

$$S_{\text{gf}} \;=\; -\left(\frac{2\pi}{\Lambda^2}\right)^{d/2} \operatorname{tr}\left\{\frac{1}{2}[P_\mu, b_\mu]^2 + [P_\mu, \bar c][P_\mu, c]\right\}, \qquad (14.75)$$

where c and $\bar c$ are ghosts.

The sum of (14.74) and (14.75) gives

$$S_2 \;=\; -\left(\frac{2\pi}{\Lambda^2}\right)^{d/2} \operatorname{tr}\left\{\frac{1}{2}[P_\mu, b_\nu]^2 + [P_\mu, \bar c][P_\mu, c]\right\} \qquad (14.76)$$

to quadratic order in b_μ.

Performing the Gaussian integral over b_ν, we find at the one-loop level:

$$\int dP_\mu \, db_\mu \; e^{-S_2} \cdots \;=\; \int \prod_{k=1}^{N} d^d p_k \prod_{i<j}\left[(p_i - p_j)^2\right]^{2-d} \cdots, \qquad (14.77)$$

where the integration over P_μ accounts for the moduli space of classical solutions.

For $d = 1$ the product on the RHS of Eq. (14.77) reproduces the square of the Vandermonde determinant (13.14). For $d = 2$ the exponent $2 - d$ vanishes so that the product equals unity and does not affect the dynamics. For $d \geq 3$ the measure is singular and the eigenvalues collapse. This leads us to a spontaneous breakdown of the R^d symmetry in perturbation theory.

Remark on supersymmetric case

In a supersymmetric gauge theory, there is an extra contribution from fermions to the exponent on the RHS of Eq. (14.77). Since the integration over fermions results in the extra factor of $[(p_i - p_j)^2]^{\operatorname{tr}\mathbb{I}}$, this finally yields the exponent $2 - d + \operatorname{tr}\mathbb{I}$. It vanishes in $d = 4$ for either Majorana or Weyl fermions and in $d = 10$ for the Majorana–Weyl fermions. This explicit calculation [IKK97] confirms, at first sight, the claim [MK83] that R^d symmetry is not broken perturbatively in supersymmetric Yang–Mills theory and no quenching is needed in the supersymmetric case. This statement seems, in fact, to be not quite correct because of fermionic zero modes [AIK00].

14.6 Quenched Eguchi–Kawai model

Soon after the breakdown of the R^d symmetry in perturbation theory was discovered for the Eguchi–Kawai model, a cure for the problem was proposed [BHN82]. The idea was to treat the eigenvalues of the Hermitian matrix A_μ as being quenched rather than dynamical variables.

In order to separate the degrees of freedom associated with the eigenvalues, we represent A_μ in a canonical form

$$A_\mu = -V_\mu P_\mu V_\mu^\dagger, \tag{14.78}$$

where P_μ is diagonal and V_μ is unitary. The measure for integration over A_μ is then represented in a standard Weyl form

$$dA_\mu = dP_\mu\, dV_\mu\, \Delta^2(P_\mu), \tag{14.79}$$

where dV_μ denotes the Haar measure* on $U(N)$ and $\Delta(P_\mu)$ is the Vandermonde determinant defined by Eq. (13.14). Equation (14.79) is the same as Eq. (13.13) for the one-matrix case.

Note that the substitution (14.78) is consistent with the gauge symmetry (14.56), which is equivalent to the left multiplication

$$V_\mu \;\to\; \tilde{\Omega} V_\mu. \tag{14.80}$$

The Haar measure dV_μ is invariant under such a multiplication.

In the quenched Eguchi–Kawai model, A_μ is substituted by Eq. (14.78) both in the reduced action (14.55) and in the averaging functionals. But the averaging is taken only with respect to the V_μ variables considering P_μ as quenched variables. The averages are then integrated over P_μ which is quite analogous to Eq. (14.27):

$$\Big\langle F[i\partial_\mu + \mathcal{A}_\mu(x)] \Big\rangle \overset{\text{red.}}{=} \int^\Lambda \prod_{i=1}^N d^d p_i \, \Big\langle F\Big[-D^\dagger(x)\, V_\mu P_\mu V_\mu^\dagger D(x)\Big] \Big\rangle_{\text{QEK}}. \tag{14.81}$$

The average on the RHS of Eq. (14.81) is defined for the quenched Eguchi–Kawai model by

$$
\Big\langle F\Big[V_\mu P_\mu V_\mu^\dagger\Big] \Big\rangle_{\text{QEK}}
$$

$$
= Z_{\text{QEK}}^{-1} \int \prod_\nu dV_\nu\, \Delta^2(P_\nu)\, e^{-S_{\text{EK}}[V_\mu P_\mu V_\mu^\dagger]}\, F\Big[V_\mu P_\mu V_\mu^\dagger\Big] \tag{14.82}
$$

---

* Strictly speaking, V_μ in Eq. (14.78) should be off-diagonal to match the number of degrees of freedom, so the measure dV_μ should be the Haar measure on the coset $U(N)/U(1)^N$. But nothing depends on these diagonal degrees of freedom of V_μ since P_μ is diagonal. We simply normalize the proper (compact) integrals over these diagonal degrees of freedom of a unitary matrix to unity.

and

$$Z_{\text{QEK}} = \int \prod_\nu dV_\nu \, \Delta^2(P_\nu) \, e^{-S_{\text{EK}}[V_\mu P_\mu V_\mu^\dagger]} \tag{14.83}$$

is the partition function of the quenched Eguchi–Kawai model.

Similarly to Eq. (14.13), the free energy per unit volume is given as $N \to \infty$ by

$$\frac{1}{N^2} \frac{\ln Z}{V} = \left(\frac{\Lambda^2}{2\pi}\right)^{d/2} \int^\Lambda \prod_{i=1}^N d^d p_i \, \frac{1}{N^2} \ln Z_{\text{QEK}}. \tag{14.84}$$

This prescription for constructing the quenched Eguchi–Kawai model is very similar to what is described in Sect. 14.2 for scalars. The measure dA_μ is split according to Eq. (14.79) but the integration in Eq. (14.82) or Eq. (14.83) is solely over V_μ, keeping P_μ quenched. Only these averages of the quenched Eguchi–Kawai model (or the logarithm of the partition function in Eq. (14.84)) are integrated over the quenched momenta p_i according to Eq. (14.81).

This is crucial to cure the breakdown of the R^d symmetry in perturbation theory. The perturbative calculation in the quenched Eguchi–Kawai model looks like that of the previous section since now the classical vacuum is associated with $V_\mu = 1$ (modulo a gauge transformation). Instead of integrating over the distinct classical vacua as in the naive Eguchi–Kawai model, we have in the quenched Eguchi–Kawai model integration over the quenched variables p_i which enters differently. The factor of $\prod_{i<j}[(p_i - p_j)^2]^{2-d}$, which resulted in the breaking of the R^d symmetry in Eq. (14.77), appears now both in the numerator and denominator of the averages and thus cancels. Similarly, its logarithm is integrated over p_i in Eq. (14.84) which does not result in a collapse of eigenvalues in the quenched Eguchi–Kawai model. The R^d symmetry is *not* broken perturbatively in the quenched Eguchi–Kawai model and it is equivalent to the d-dimensional Yang–Mills theory in the $N = \infty$ limit.

Just as in the scalar case, we can substitute the integration over p_i^μ in Eq. (14.81) at $N = \infty$ by distributing them with a proper weight. It is again convenient to choose the weight (14.28) as is prescribed by Eq. (14.25). In contrast to the momentum regularization in the d-dimensional gauge theory, this results in a gauge-invariant regularization of perturbation theory since the eigenvalues of A_μ are gauge invariant (cf. Eq. (14.57)).

In fact, the precise form of the measure for integrating over p_i on the RHS of Eq. (14.81) is not essential as $N \to \infty$. All that is needed from the measure is for the integral over p_i to converge, which would protect the eigenvalues from collapsing in perturbation theory. Any other measure performing the same job is as good as this one.

For the same reason, the precise form of the distribution of the quenched momenta, substituting the integration at $N = \infty$, is not essential if it is smooth. The distribution (14.28) simply provides a nice gauge-invariant regularization of perturbation theory which is of the same type as the proper-time regularization.

Given a distribution of the eigenvalues p_i^μ, Eq. (14.81) simplifies to

$$\Big\langle F[\mathrm{i}\partial_\mu + \mathcal{A}_\mu(x)] \Big\rangle \overset{\text{red.}}{=} \Big\langle F\Big[-D^\dagger(x)V_\mu P_\mu V_\mu^\dagger D(x)\Big] \Big\rangle_{\text{QEK}}. \qquad (14.85)$$

In particular, the averages of closed Wilson loops are given by

$$\left\langle \frac{1}{N}\operatorname{tr}\boldsymbol{P}\,\mathrm{e}^{\mathrm{i}\oint \mathrm{d}\xi^\mu \mathcal{A}_\mu(\xi)} \right\rangle \overset{\text{red.}}{=} \left\langle \frac{1}{N}\operatorname{tr}\boldsymbol{P}\,\mathrm{e}^{-\mathrm{i}\oint \mathrm{d}\xi^\mu V_\mu P_\mu V_\mu^\dagger} \right\rangle_{\text{QEK}}. \qquad (14.86)$$

The averages of open Wilson loops in the quenched Eguchi–Kawai model obey Eq. (14.68).

A formal proof of the equivalence of the d-dimensional Yang–Mills theory in the large-N limit and the quenched Eguchi–Kawai model can be given [GK82, Mig82] using the loop equation. To derive the equation for the Wilson loops in the quenched Eguchi–Kawai model, which are defined by the RHS of Eq. (14.86), we perform the right shift of the unitary matrix V_μ:

$$\delta V_\mu \;=\; \mathrm{i}V_\mu \epsilon_\mu\,, \qquad (14.87)$$

where ϵ_μ is Hermitian. Substituting into Eq. (14.78), we obtain

$$\delta A_\mu \;=\; \mathrm{i}V_\mu\,[P_\mu, \epsilon_\mu]\,V_\mu^\dagger \qquad (14.88)$$

under the shift (14.87).

Using the gauge symmetry (14.80), we can always choose the gauge where $V_\mu = 1$ for the given μ. Then

$$\delta A_\mu \;=\; \mathrm{i}\,[P_\mu, \epsilon_\mu] \qquad (14.89)$$

does not depend on V_μ.

The variation (14.89) is almost the same as that which resulted in the loop equation (14.65) of the Eguchi–Kawai model. The only difference resides in the fact that the variation (14.87) does not change the eigenvalues of A_μ. When we expand the induced variation of A_μ, given by Eq. (14.89), in the Lie algebra basis, no diagonal generators appear. But their number is $\sim N$ and hence $\mathcal{O}(N^{-1})$ of the total number of generators. For this reason, the Wilson loop averages in the quenched Eguchi–Kawai model obey at $N = \infty$ the same loop equation as in the naive Eguchi–Kawai model. Additional terms of order $1/N$ appear in the loop equation of the quenched Eguchi–Kawai model since diagonal generators, which are needed for the completeness condition (11.6), are missing. Hence, corrections to the $N = \infty$ loop equation of the quenched Eguchi–Kawai model are $\sim 1/N$ rather than $\sim 1/N^2$ as in the d-dimensional Yang–Mills theory.

This demonstrates once again that quenched reduced models can reproduce only planar diagrams of the d-dimensional theories but cannot reproduce diagrams of higher genera.

The representation (14.82) of the averages in the quenched Eguchi–Kawai model does not look like that in gauge theories where the averaging is over quantum fluctuations of A_μ. The quenched Eguchi–Kawai model can, however, be represented in such a form as is shown by Gross and Kitazawa [GK82].

Let us introduce

$$1 = \int dA_\mu\, \delta\left(A_\mu + V_\mu P_\mu V_\mu^\dagger\right) \tag{14.90}$$

into the numerator and denominator on the RHS of Eq. (14.82).

Changing the order of integration over dA_μ and dV_μ, we obtain

$$\left\langle F[A] \right\rangle_{\mathrm{QEK}} = \frac{\int \prod_\mu dA_\mu\, C(A,P)\, \mathrm{e}^{-S_{\mathrm{EK}}[A]}\, F[A]}{\int \prod_\mu dA_\mu\, C(A,P)\, \mathrm{e}^{-S_{\mathrm{EK}}[A]}}, \tag{14.91}$$

where

$$C(A;P) = \int \prod_\mu dV_\mu\, \delta\left(A_\mu + V_\mu P_\mu V_\mu^\dagger\right) \Delta^2(P_\mu). \tag{14.92}$$

And analogously,

$$Z_{\mathrm{QEK}} = \int \prod_\mu dA_\mu\, C(A,P)\, \mathrm{e}^{-S_{\mathrm{EK}}[A]} \tag{14.93}$$

for the partition function of the quenched Eguchi–Kawai model from Eq. (14.83).

Substituting

$$A_\mu = -P_\mu + g b_\mu, \tag{14.94}$$

we can calculate $C(A,P)$ at least perturbatively in g. Evaluating the integral on the RHS of Eq. (14.92), we find

$$C(A,P) = \prod_\mu \prod_{i=1}^N \delta\left(b_{ii}^\mu + g \sum_{j \neq i} \frac{|b_{ij}^\mu|^2}{p_i^\mu - p_j^\mu} + \mathcal{O}(b_\mu^3) \right) \tag{14.95}$$

to quadratic order in b_μ.

The meaning of this constraint is obvious: diagonal elements of b_μ are expressed via off-diagonal elements for the eigenvalues of A_μ to coincide with $-p_i^\mu$. In particular, the diagonal elements of b_μ vanish to the leading order. This vanishing of b_{ii}^μ is however not a gauge-invariant condition to higher orders in g. The role of the higher terms in the argument of the delta-function in Eq. (14.95) is to ensure gauge invariance to all orders in g as $C(A, P)$ is gauge invariant according to the definition (14.92).

The constraint (14.95) restricts only N out of N^2 degrees of freedom, which explains why it is inessential, say, in the large-N limit of the loop equations in the quenched Eguchi–Kawai model.

The presence of the delta-function affects, however, the dynamics of the degrees of freedom associated with the diagonal elements A_{ii}. In particular, the analog of the continuum propagator (14.22) is given by

$$\left\langle b_{ij}^\mu b_{ji}^\nu \right\rangle_{\mathrm{QEK}} = \begin{cases} \left(\dfrac{\Lambda^2}{2\pi}\right)^{d/2} \dfrac{\delta_{\mu\nu}}{(p_i - p_j)^2} & i \neq j \\[4mm] 0 & i = j \end{cases} \tag{14.96}$$

which cures the divergence of a massless propagator at $i = j$.

If the constraint (14.95) is solved for b_{ii}^μ versus off-diagonal components and the result is substituted into the action, this will generate new interactions. The diagrams of perturbation theory in the quenched Eguchi–Kawai model coincide with the integrands of the planar Feynman graphs in the d-dimensional Yang–Mills theory except for diagrams with the new vertices which are needed for gauge invariance of the quenched Eguchi–Kawai model. The sum of these additional diagrams vanishes [GK82] after averaging over the quenched momenta.

Problem 14.5 Derive Eq. (14.95) to quadratic order in b_μ.

Solution We need to solve the equation

$$P_\mu - g b_\mu = V_\mu P_\mu V_\mu^\dagger \tag{14.97}$$

for V_μ iteratively in g. Substituting

$$V_\mu = e^{igh_\mu}, \tag{14.98}$$

we find that Eq. (14.97) is reduced to the linear order in g to

$$b_{ij}^\mu = i\left(p_i^\mu - p_j^\mu\right) h_{ij}^\mu. \tag{14.99}$$

This equation requires $b_{ii}^\mu = 0$ and fixes off-diagonal components of h_{ij}^μ to be

$$h_{ij}^\mu = -i\frac{b_{ij}^\mu}{p_i^\mu - p_j^\mu}. \tag{14.100}$$

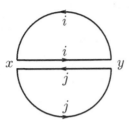

Fig. 14.2. Index-space diagram for the average of closed Wilson loop to order g^2. The momentum p_i or p_j flows along the index line i or j. The momentum $p_i - p_j$ is associated with the double line ij. The diagram is associated with an analytic formula given in Eqs. (14.102) and (14.103).

To the quadratic order in g, only h_{ij}^μ to the linear order contributes to the diagonal components of Eq. (14.97) since a commutator with the diagonal matrix P_μ has no diagonal components. Then Eq. (14.97) yields

$$b_{ii}^\mu = g \sum_{j \neq i} \left(p_i^\mu - p_j^\mu \right) h_{ij}^\mu h_{ji}^\mu = -g \sum_{j \neq i} \frac{|b_{ij}^\mu|^2}{p_i^\mu - p_j^\mu} \tag{14.101}$$

which reproduces the argument of the delta-function in Eq. (14.95).

Problem 14.6 Calculate the average of a closed Wilson loop in the quenched Eguchi–Kawai model to order g^2.

Solution The calculation is similar to that in Problem 14.1 on p. 331. Substituting Eq. (14.94) into the RHS of Eq. (14.86) and expanding to order g^2, we obtain

$$W_{\text{QEK}}^{(2)}(C) = -\frac{\lambda}{2} \oint_C dx_\mu \oint_C dy_\nu \frac{1}{N^2} \sum_{i,j=1}^{N} e^{i(p_i - p_j)(y-x)} \left\langle b_{ij}^\mu b_{ji}^\nu \right\rangle_{\text{QEK}} \tag{14.102}$$

since P_μ and b_ν do not commute. The associated index-space diagram is depicted in Fig. 14.2. For the distribution of eigenvalues given as $N \to \infty$ by the Gaussian weight (14.28), we have using Eq. (14.96)

$$W_{\text{QEK}}^{(2)}(C) = -\frac{\lambda}{2} \oint_C dx_\mu \oint_C dy_\mu$$

$$\times \left(\frac{\Lambda^2}{2\pi} \right)^{d/2} \int \frac{d^d p_i}{(\Lambda\sqrt{\pi})^d} \frac{d^d p_j}{(\Lambda\sqrt{\pi})^d} e^{-p_i^2/\Lambda^2 - p_j^2/\Lambda^2} \frac{e^{i(p_i - p_j)(y-x)}}{(p_i - p_j)^2}. \tag{14.103}$$

Introducing the variables $p_\pm = p_i \pm p_j$ and accounting for a Jacobian, we have quite similarly to Problem 14.1:

$$W_{\text{QEK}}^{(2)}(C) \;=\; -\frac{\lambda}{2} \oint_C \mathrm{d}x_\mu \oint_C \mathrm{d}y_\mu \int \frac{\mathrm{d}^d p}{(2\pi)^d}\, \mathrm{e}^{-p^2/2\Lambda^2} \frac{\mathrm{e}^{\mathrm{i}p(y-x)}}{p^2} \tag{14.104}$$

which reproduces (a regularized version of) Eq. (12.17) with the correct normalization.

Remark on the quenched Eguchi–Kawai model on the lattice

The quenched Eguchi–Kawai model was originally formulated on a lattice [BHN82]. All the formulas are analogous to those given above in this section, while taking into account the fact that U_μ is compact on the lattice.

The analogs of Eqs. (14.78) and (14.79) are given by

$$U_\mu \;=\; V_\mu\, \mathrm{e}^{-\mathrm{i}P_\mu a}\, V_\mu^\dagger \tag{14.105}$$

and

$$\mathrm{d}U_\mu \;=\; \mathrm{d}P_\mu\, \mathrm{d}V_\mu\, \Delta^2\!\left(\mathrm{e}^{-\mathrm{i}P_\mu a}\right), \tag{14.106}$$

where explicitly

$$\Delta\!\left(\mathrm{e}^{-\mathrm{i}P_\mu a}\right) \;=\; \prod_{i<j} 2\sin\!\left(\frac{p_i^\mu - p_j^\mu}{2} a\right). \tag{14.107}$$

The quenched variables $p_i^\mu \in (-\pi/a, +\pi/a]$ play the role of the lattice momenta restricted to the Brillouin zone.

The measure $\mathrm{d}U_\mu$ of the naive Eguchi–Kawai model (14.43) is multiplied by

$$C(U, P) \;=\; \int \prod_\mu \mathrm{d}V_\mu\, \delta\!\left(U_\mu - V_\mu\, \mathrm{e}^{-\mathrm{i}P_\mu a} V_\mu^\dagger\right) \Delta^2\!\left(\mathrm{e}^{-\mathrm{i}P_\mu a}\right). \tag{14.108}$$

Correspondingly, the partition function of the lattice quenched Eguchi–Kawai model is given by

$$Z_{\text{QEK}} \;=\; \int \prod_\mu \mathrm{d}U_\mu\, C(U, P)\, \mathrm{e}^{-N S_{\text{R}}[U]/g^2}, \tag{14.109}$$

where the eigenvalues p_i^μ are distributed uniformly over the hypercube.

The $U(1)^d$ symmetry is not broken in the lattice version of the quenched Eguchi–Kawai model for all values of the coupling $g^2 N$. This is illustrated by the one-loop calculation in Problem 14.7. The lattice quenched Eguchi–Kawai model is equivalent to an $N = \infty$ Wilson lattice gauge theory on a d-dimensional lattice for all values of $g^2 N$. This is verified, in particular, by numerical simulations.

Problem 14.7 Calculate the partition function (14.109) to the leading order in g^2, fixing the gauge by $V_d = 1$.

Solution Since the gauge is fixed by $V_d = 1$, the vacuum state is simply

$$U_\mu^{\text{cl}} = e^{-iP_\mu a} \tag{14.110}$$

or $V_\mu = 1$. This can be seen representing the action (14.38) in an equivalent form by rewriting

$$S_{\text{R}} = \frac{1}{4N} \sum_{\mu \neq \nu} \text{tr} \left| [U_\mu, U_\nu] \right|^2 . \tag{14.111}$$

We expand

$$V_{ij}^\mu = \delta_{ij} - iga \frac{b_{ij}^\mu}{S_{ij}^\mu} \qquad \mu = 1, \ldots, d-1 , \tag{14.112}$$

where

$$S_{ij}^\mu = 2 \sin \left(\frac{p_i^\mu - p_j^\mu}{2} a \right) . \tag{14.113}$$

Here b^μ is the off-diagonal Hermitian matrix as has already been explained. Equation (14.112) reproduces the continuum equations (14.98) and (14.100) as $a \to 0$.

Keeping the terms which are quadratic in b_μ in the action, we have

$$S_2 = \frac{1}{2} \sum_{\mu,\nu=1}^{d-1} \sum_{i,j} \left| S_{ij}^\mu b_{ij}^\nu - S_{ij}^\nu b_{ij}^\mu \right|^2 , \tag{14.114}$$

while the measure is

$$\prod_{\mu=1}^{d-1} dV_\mu = \prod_{\mu=1}^{d-1} \Delta^{-2} \left(e^{-iP_\mu a} \right) db_\mu \tag{14.115}$$

to this level of accuracy.

The calculation of the Gaussian integral over b_μ reduces for the given indices i and j to a calculation of the determinant of the $(d-1) \times (d-1)$ matrix

$$R_{\mu\nu} = \sum_{\rho=1}^{d} S_\rho^2 \delta_{\mu\nu} - S_\mu S_\nu , \tag{14.116}$$

which has one eigenvalue S_d^2 and $d-2$ eigenvalues $\sum_{\rho=1}^{d} S_\rho^2$. This can be easily seen using the rotational symmetry, which allows us to choose $S_\mu = 0$ for $\mu = 2, \ldots, d-1$. Therefore, we have

$$\det_{\mu\nu} R_{\mu\nu} = S_d^2 \left(\sum_{\rho=1}^{d} S_\rho^2 \right)^{d-2} . \tag{14.117}$$

Finally, we obtain

$$\int \prod_{\mu=1}^{d} \mathrm{d}V_\mu \Delta^2\left(\mathrm{e}^{-iP_\mu a}\right) \mathrm{e}^{-S_2} \;=\; \prod_{i<j}\left(\sum_\mu 4\sin^2 \frac{p_i^\mu - p_j^\mu}{2}a\right)^{2-d} , \qquad (14.118)$$

which reproduces the integrand in Eq. (14.77) as $a \to 0$. There is no collapse of eigenvalues of U_μ thanks to the quenching procedure.

In this Problem we have followed the calculation of [KM82].

15

Twisted reduced models

There is an elegant alternative to the quenched Eguchi–Kawai model, described in the previous chapter, which also preserves the $U(1)^d$ symmetry. It was proposed by González-Arroyo and Okawa [GO83a, GO83b] on the basis of a twisting reduction prescription. The corresponding lattice version of the reduced model lives on a hypercube with twisted boundary conditions. The twisted reduced model for a scalar field was constructed by Eguchi and Nakayama [EN83].

The twisted reduced models reveal interesting mathematical structures associated with representations of the Heisenberg commutation relation (in the continuum) or its finite-dimensional approximation by unitary matrices (on the lattice). In contrast to the quenched reduced models which describe only planar graphs, the twisted reduced models make sense order by order in $1/N$ and even at finite N. In this case they are associated with gauge theories on noncommutative space, whose limit of large noncommutativity is given by planar graphs thereby reproducing a d-dimensional Yang–Mills theory at large N.

We begin this chapter with a description of the twisted reduced models first on the lattice and then in the continuum and show how they describe planar graphs of a d-dimensional theory.

15.1 Twisting prescription

We start by working on a lattice to make the results rigorous and then repeat them for the continuum.

The twisting reduction prescription is a version of the general reduction prescription described in Sect. 14.1. We again perform the unitary transformation (14.4), where the matrices $D(x)$ are now expressed via a

351

set of d (unitary) $N \times N$ matrices Γ_μ by

$$D(x) = \Gamma_1^{x_1/a} \Gamma_2^{x_2/a} \cdots \Gamma_d^{x_d/a}, \qquad (15.1)$$

and the coordinates of the (lattice) vector x_μ are measured in the lattice units so that all the exponents are integral.

The matrices Γ_μ obey the Weyl–'t Hooft commutation relation

$$\Gamma_\mu \Gamma_\nu = Z_{\mu\nu} \Gamma_\nu \Gamma_\mu \qquad (15.2)$$

with $Z_{\mu\nu} = Z_{\nu\mu}^\dagger$ being elements of $Z(N)$ and d is assumed to be even. These matrices Γ_μ, which are called *twist eaters*, will be constructed explicitly in a moment.

Substituting (14.4) with $D(x)$ given by Eq. (15.1) into Eq. (14.1), we obtain the following partition function of the *twisted reduced model* [EN83]

$$Z_{\text{TRM}} = \int d\tilde{\varphi}\, e^{-S_{\text{TRM}}[\tilde{\varphi}]} \qquad (15.3)$$

with the action

$$S_{\text{TRM}}[\tilde{\varphi}] = -N \sum_\mu \text{tr}\, \Gamma_\mu \tilde{\varphi} \Gamma_\mu^\dagger \tilde{\varphi} + N\, \text{tr}\, \tilde{V}(\tilde{\varphi}). \qquad (15.4)$$

The "derivation" is again modulo the volume factor in the action.

Correspondingly, an analog of Eq. (14.9) for the twisting reduction prescription is given by

$$\left\langle F[\varphi_x] \right\rangle \stackrel{\text{red.}}{=} \left\langle F\left[D^\dagger(x) \tilde{\varphi} D(x) \right] \right\rangle_{\text{TRM}}, \qquad (15.5)$$

where the average on the RHS is calculated for the twisted reduced model:

$$\left\langle F[\tilde{\varphi}] \right\rangle_{\text{TRM}} = Z_{\text{TRM}}^{-1} \int d\tilde{\varphi}\, e^{-S_{\text{TRM}}[\tilde{\varphi}]} F[\tilde{\varphi}]. \qquad (15.6)$$

The equality of the LHS of Eq. (15.5) (calculated for the d-dimensional theory (14.1)) and the RHS (calculated for the twisted reduced model) takes place in the planar limit, i.e. for $N = \infty$, owing to the explicit form of $D(x)$ given by Eq. (15.1).

Problem 15.1 Demonstrate that the order of Γ_μ in Eq. (15.1) is inessential.

Solution Let us define a more general path-dependent factor

$$D(C_{x0}) = P \prod_i \Gamma_{\mu_i}, \qquad (15.7)$$

where the path-ordered product runs over all links forming a path C_{x0} from the origin to the point x. Owing to Eq. (15.2), a change of the path multiplies $D(x)$ by the Abelian factor

$$Z(C) \;=\; \prod_{p\in S:\partial S=C} Z^*_{\mu\nu}\,, \tag{15.8}$$

where (μ, ν) is the orientation of the plaquette p. The product runs over any surface spanned by the closed loop C, which is obtained by passing the original path forward and the new path backward. Owing to the Bianchi identity

$$\prod_{p\in\text{cube}} Z_{\mu\nu} \;=\; 1\,, \tag{15.9}$$

where the product goes over six plaquettes forming a three-dimensional cube on the lattice, the product on the RHS of Eq. (15.8) does not depend on the form of the surface S and is a functional solely of the loop C.

It is now easy to see that under this change of the path we obtain

$$D^\dagger_{ij}(x)D_{kl}(x) \;\longrightarrow\; |Z(C)|^2\, D^\dagger_{ij}(x)D_{kl}(x) \tag{15.10}$$

and the path-dependence is canceled because $|Z(C)|^2 = 1$. This is a general property which holds for the twisting reduction prescription of any even representation of $SU(N)$ (i.e. invariant under transformations from the center $Z(N)$).

Let us now explicitly construct the matrices Γ_μ which obey Eq. (15.2). We begin with the case of $d = 2$, where Γ_1 and Γ_2 can be chosen to coincide with the $L \times L$ Weyl "clock" and "shift" matrices [Wey31]:

$$\mathcal{Q} \;=\; \text{diag}\left(1, \omega, \omega^2, \ldots, \omega^{L-1}\right) \tag{15.11}$$

and

$$\mathcal{P} \;=\; \begin{pmatrix} 0 & 1 & 0 & 0 & \cdots & 0 \\ 0 & 0 & 1 & 0 & \cdots & 0 \\ 0 & 0 & 0 & 1 & \cdots & 0 \\ \vdots & \vdots & \vdots & \vdots & \ddots & \vdots \\ 0 & 0 & 0 & 0 & \cdots & 1 \\ 1 & 0 & 0 & 0 & \cdots & 0 \end{pmatrix}, \tag{15.12}$$

which are unitary and obey

$$\mathcal{Q}^L \;=\; 1 \;=\; \mathcal{P}^L\,, \tag{15.13}$$
$$\mathcal{P}\mathcal{Q} \;=\; \omega\,\mathcal{Q}\mathcal{P} \tag{15.14}$$

with $\omega \in Z(L)$. A solution to Eq. (15.2) in $d = 2$ is obviously given by $\Gamma_1 = \mathcal{P}$, $\Gamma_2 = \mathcal{Q}$ providing $L = N$ and $\omega = Z_{12} = e^{2\pi i/L}$.

For even $d > 2$, the factor of $Z_{\mu\nu}$ can always be written as

$$Z_{\mu\nu} \;=\; e^{2\pi i n_{\mu\nu}/N} \in Z(N) \qquad (n_{\mu\nu} = -n_{\nu\mu} \in \mathbb{Z}_N)\,, \tag{15.15}$$

where $n_{\mu\nu}$ can be represented in a canonical skew-diagonal form

$$
n_{\mu\nu} = \begin{pmatrix} 0 & +n_1 & & & \\ -n_1 & 0 & & & \\ & & \ddots & & \\ & & & 0 & +n_{d/2} \\ & & & -n_{d/2} & 0 \end{pmatrix}. \tag{15.16}
$$

Although a solution to Eq. (15.2) is known for an arbitrary set of $\{n_1, \ldots, n_{d/2}\}$ (it is described in Problem 15.3), it is enough for our purposes to consider the simplest case of $n_1 = n_2 = n$ in $d = 4$ dimensions.

The idea is now to combine $\Gamma_1, \ldots, \Gamma_4$ into two pairs: Γ_1, Γ_2 and Γ_3, Γ_4, so the commutator of the two matrices from the same pair is similar to that in two dimensions, while the matrices from different pairs commute. These rules are prescribed by an explicit form of $n_{\mu\nu}$ given for this simplest twist by

$$
n_{\mu\nu} = \begin{pmatrix} 0 & +n & & \\ -n & 0 & & \\ & & 0 & +n \\ & & -n & 0 \end{pmatrix}. \tag{15.17}
$$

The solution to Eq. (15.2) can then be represented by a direct product of the $L \times L$ Weyl matrices (15.11) and (15.12):

$$
\left.\begin{array}{ll} \Gamma_1 = \mathcal{P} \otimes \mathbb{I}, & \Gamma_2 = \mathcal{Q} \otimes \mathbb{I}, \\ \Gamma_3 = \mathbb{I} \otimes \mathcal{P}, & \Gamma_4 = \mathbb{I} \otimes \mathcal{Q}. \end{array}\right\} \tag{15.18}
$$

In other words, the solution is given on a subgroup $SU(L) \otimes SU(L)$ of the group $SU(N)$, which is possible only if $N = L^2$ and $n = L$. Once again, this solution is not the most general one, but it is enough for our purposes. Note that $\Gamma_\mu^L = 1$ for this simplest solution.

Problem 15.2 Extend the solution (15.18) to d dimensions assuming that all $n_i = L^{d/2-1}$ in Eq. (15.16).

Solution For such $n_{\mu\nu}$ the solution may be given on a subgroup $\prod_1^{d/2} \otimes SU(L)$ of $SU(N)$ so that $N = L^{d/2}$ and Γ_i, Γ_{i+1} (odd $i = 1, 3, \ldots, d-1$) can be chosen as a direct product of the Weyl matrices for the ith of $SU(L)$s and the unit matrices for the others. Again $\Gamma_\mu^L = 1$ for this simplest twist.

Problem 15.3 Find a solution to Eq. (15.2) for a general matrix $n_{\mu\nu}$.

Solution We proceed similarly to the previous Problem. Given N and the $d/2$ numbers $n_i \in \mathbb{Z}_N$, we introduce the integers $p_i = N/\gcd(n_i, N)$. A solution to

Eq. (15.2) exists if the product $p_1 \cdots p_{d/2}$, which plays the role of the dimension of irreducible representation of the algebra, divides N. In that case we write

$$N = p_0 \prod_{i=1}^{d/2} p_i \qquad (p_0 \in \mathbb{Z}) \qquad (15.19)$$

and the solution may be given on the subgroup $SU(p_0) \otimes SU(p_1) \otimes \cdots \otimes SU(p_{d/2})$ of $SU(N)$ such that Γ_i, Γ_{i+1} are constructed as a direct product of the Weyl matrices on $SU(p_i)$ and unit matrices for the others. The subgroup of $GL(p, \mathbb{C})$ consisting of matrices which commute with the twist eaters Γ_μ is then $GL(p_0, \mathbb{C})$.

This solution was found in [BG86, LP86], where it was shown that Eq. (15.19) is a necessary and sufficient condition for the existence of solutions to Eq. (15.2). The simplest solution from the previous Problem is associated with $p_0 = 1$, $p_1 = \cdots = p_{d/2} = L$. Another simple example is $N = p_0 L^{d/2}$, $n_i = p_0 L^{d/2-1}$, $p_1 = \cdots = p_{d/2} = L$ when $\Gamma_\mu^L = 1$.

15.2 Twisted reduced model for scalars

We shall now demonstrate how the twisted reduced model, which is defined for a scalar field by Eqs. (15.3) and (15.6), reproduces [EN83] at large N the planar graphs of the d-dimensional theory.

In principle, we may try to simply repeat the perturbative analysis of Sect. 14.1, representing the propagator of $\tilde{\varphi}_{ij}$ via the Γ_μ and expecting that momentum integrals will be recovered after summing over indices circulating in closed loops owing to the explicit form of the twist eaters. This is indeed the case.

It is more instructive, however, to proceed in a slightly different way explicitly introducing the momentum variable via a sort of a Fourier transformation on $gl(N; \mathbb{C})$ (general complex $N \times N$ matrices).

A convenient Weyl basis on $gl(L; \mathbb{C})$ is given [Hoo78, Hoo81] by (symmetric) products of the "clock" and "shift" matrices (15.11) and (15.12).

Let us introduce L^2 matrices

$$J_{m_1,m_2} = \mathcal{P}^{m_1} \mathcal{Q}^{m_2} \omega^{-m_1 m_2/2}, \qquad (15.20)$$

where $m_1, m_2 \in \mathbb{Z}_L$. An arbitrary $L \times L$ matrix M can be expanded in this basis:

$$M^{ij} = \frac{1}{N^2} \sum_{m_1,m_2=1}^{L} J_{m_1,m_2}^{ij} M(m_1, m_2), \qquad (15.21)$$

where $M(m_1, m_2)$ are certain expansion coefficients.

We see that a pair of integers m_1 and m_2, forming a two-dimensional vector, $m = \{m_1, m_2\} \in \mathbb{Z}_L^2$, is naturally associated with this construction. As we shall see in a moment, these integers label momenta on an $L \times L$ periodic lattice.

An extension of this construction to arbitrary (even) d dimensions is obvious for the simplest twist given by the matrix (15.16) with $n_i = L^{d/2-1}$ for all $i = 1, \ldots, d/2$, which is considered in Problem 15.2 on p. 354. We introduce the basis on $gl(N; \mathbb{C})$:

$$ J_m = \Gamma_1^{m_1} \cdots \Gamma_d^{m_d}\, e^{-\pi i \sum_{\mu < \nu} m_\mu n_{\mu\nu} m_\nu / N}, \tag{15.22} $$

where $m = \{m_1, \ldots, m_d\} \in \mathbb{Z}_L^d$ is a d-dimensional vector (remember that $N = L^{d/2}$). The last factor* again makes the product symmetric.

There exist $L^d = N^2$ independent orthogonal generators J_m which obey

$$ J_m^\dagger = J_{-m} \quad (\text{mod } L), \tag{15.23} $$

$$ \frac{1}{N}\, \text{tr}\, J_m J_n^\dagger = \delta_{mn}, \tag{15.24} $$

$$ \sum_{m \in \mathbb{Z}_L^d} J_m^{ij} J_m^{\dagger kl} = N \delta^{il} \delta^{kj}, \tag{15.25} $$

$$ J_m J_n = J_{m+n}\, e^{\pi i \sum_{\mu,\nu} m_\mu n_{\mu\nu} n_\nu / N} \quad (\text{mod } L). \tag{15.26} $$

Equations (15.24) and (15.25) represent, respectively, orthogonality and completeness of the generators. The product of two generators is given explicitly by Eq. (15.26).

An arbitrary $N \times N$ complex matrix M^{ij} can be expanded as

$$ M^{ij} = \frac{1}{N^2} \sum_{m \in \mathbb{Z}_L^d} J_m^{ij}\, M(m) \tag{15.27} $$

and $M(-m) = M^*(m)$ if M is Hermitian as a consequence of Eq. (15.23). Using Eq. (15.24), the coefficient $M(m)$ is given by

$$ M(m) = N\, \text{tr}\left(M\, J_m^\dagger \right). \tag{15.28} $$

Equation (15.27) simply extends Eq. (15.21) to the multidimensional case.

A mapping of the twisted reduced model into a d-dimensional field theory can be established as follows.

We expand the matrix $\tilde{\varphi}$ in the basis of J_m:

$$ \tilde{\varphi} = \frac{1}{N^2} \sum_{m \in \mathbb{Z}_L^d} J_m\, \varphi(m). \tag{15.29} $$

---

* Strictly speaking, we assume that $n_{\mu\nu}$ is even and L is odd for the J_m to obey a periodicity property $J_{\ldots, m_i + L, \ldots} = J_{\ldots, m_i, \ldots}$. This is necessary only for finite N since this periodicity is lost as $N \to \infty$.

The measure (13.2) for the averaging over the matrices $\tilde{\varphi}$ is then represented by

$$d\tilde{\varphi} = \prod_{m \in \mathbb{Z}_L^d} d\varphi(m). \tag{15.30}$$

The substitution of the expansion (15.29) into the kinetic part of the action of the twisted reduced model yields

$$
\begin{aligned}
S_{\text{TRM}}^{(2)} &\equiv N \operatorname{tr} \left(\frac{M}{2} \tilde{\varphi}^2 - \sum_\mu \Gamma_\mu \tilde{\varphi} \Gamma_\mu^\dagger \tilde{\varphi} \right) \\
&= \frac{1}{N^2} \sum_{m \in \mathbb{Z}_L^d} \left(\frac{M}{2} - \sum_\mu \cos \frac{2\pi \sum_\nu n_{\mu\nu} m_\nu}{N} \right) \varphi(-m)\varphi(m)
\end{aligned}
\tag{15.31}
$$

which coincides with the kinetic part of the action of a single-component scalar field on a d-dimensional lattice of spatial extent $L^d = N^2$ with periodic boundary conditions.

In the latter case, we substitute the Fourier expansion

$$\tilde{\varphi}_x = \frac{1}{L^d} \sum_{m \in \mathbb{Z}_L^d} e^{2\pi i \sum_{\mu,\nu} x_\mu n_{\mu\nu} m_\nu / aN} \varphi(m), \tag{15.32}$$

which yields

$$
\begin{aligned}
S^{(2)} &\equiv \sum_x \left(\frac{M}{2} \tilde{\varphi}_x^2 - \sum_\mu \tilde{\varphi}_x \tilde{\varphi}_{x+a\hat{\mu}} \right) \\
&= \frac{1}{L^d} \sum_{m \in \mathbb{Z}_L^d} \left(\frac{M}{2} - \sum_\mu \cos \frac{2\pi \sum_\nu n_{\mu\nu} m_\nu}{N} \right) \varphi(-m)\varphi(m).
\end{aligned}
\tag{15.33}
$$

The number of degrees of freedom is the same in both cases: there are N^2 elements of the matrix $\tilde{\varphi}$ in the twisted reduced model which matches the $L^d = N^2$ sites of the lattice. The expansion coefficients in Eqs. (15.29) and (15.32) are just the same! Correspondingly, the measure for path integration over φ_x is

$$\prod_x d\varphi_x = \prod_{m \in \mathbb{Z}_L^d} d\varphi(m) \tag{15.34}$$

which coincide with the measure (15.30).

In other words, the degrees of freedom described by the matrix (15.29) or the field (15.32) are the same.

The coincidence of the actions of the two theories at finite N is only for the kinetic part of the actions which is quadratic in fields. This is no longer the case for interaction terms. For the simplest cubic interaction, we have, using Eq. (15.26),

$$
\begin{aligned}
N \operatorname{tr} \tilde{\varphi}^3 &= \frac{1}{N^6} \sum_{m_1, m_2, m_3} \varphi(m_1)\varphi(m_2)\varphi(m_3) N \operatorname{tr} J_{m_1} J_{m_2} J_{m_3} \\
&= \frac{1}{N^4} \sum_{m,n} \varphi(-m-n)\varphi(m)\varphi(n) \, e^{\pi i \sum_{\mu,\nu} m_\mu n_{\mu\nu} n_\nu / N}
\end{aligned}
$$

(15.35)

which has an extra phase in contrast to the cubic interaction of a single-component scalar field outlined in Sect. 2.3.*

The presence of this factor is crucial in showing that the twisted reduced model at $N = \infty$ correctly reproduces planar graphs of the d-dimensional theory. This happens because of a remarkable *theorem* proven in [EN83, GO83b] which states that

(1) the phases cancel out in planar graphs,
(2) the phases remain in nonplanar graphs suppressing their contribution as $N \to \infty$.

In order to sketch a proof of the theorem, we introduce the momentum variables $p_\mu \equiv 2\pi \sum_\nu n_{\mu\nu} m_\nu / aN$ and $q_\mu \equiv 2\pi \sum_\nu n_{\mu\nu} n_\nu / aN$, which become continuous momenta from the first Brillouin zone $(-\pi/a, \pi/a]$ as $N \to \infty$, and pass to the momenta p_i, p_j, and p_k associated with the single lines carrying the indices i, j, and k as in Eq. (14.17).

The phase factor in Eq. (15.35) can then be rewritten in the form

$$
e^{\pi i \sum_{\mu,\nu} m_\mu n_{\mu\nu} n_\nu / N} = e^{-i p \theta q / 2} = e^{-i(p_i \theta p_j + p_j \theta p_k + p_k \theta p_i)/2},
$$

(15.36)

where we have used the rules of matrix multiplication of the Lorentz indices so that

$$
p_i \theta p_j = \sum_{\mu,\nu} p_i^\mu \theta_{\mu\nu} p_j^\nu
$$

(15.37)

and

$$
\theta_{\mu\nu} = \frac{a^2 N}{2\pi} n_{\mu\nu}^{-1}.
$$

(15.38)

---

* We shall demonstrate in the next chapter that the twisted reduced model at finite N is precisely equivalent to a theory on a noncommutative lattice.

A proof of the theorem simplifies [IIK00] after rewriting the phase factor according to Eq. (15.36). Now each factor of $\exp\left(-ip_i\theta p_j/2\right)$ can be assigned to each of the three propagators which meet at a vertex. The overall phase of a graph can be computed by summing up the phases associated with both ends of each of the propagator lines. Since the two ends of an internal double line are oriented in an opposite way for a planar graph, the contributions of the two ends mutually cancel. Therefore, the overall phase of a planar graph involves only external momenta and is the same to all orders of perturbation theory. For example, there is no such phase for vacuum graphs, while the RHS of Eq. (15.36) is reproduced for each planar graph contributing to the three-point vertex. This phase depending on external momenta is simply related to the mapping (15.5) of observables.

The cancellation of the phases does not occur for crossing lines which are inevitably present for nonplanar graphs. For example, the nonplanar graph depicted in Fig. 11.3 has the extra factor of $\exp\left(ip\theta q\right)$ where p and q are momenta associated with the two lines which cross over each other. This is because if these two lines were to form a four-gluon vertex, instead of crossing, it would produce the additional phase factor $\exp\left(-ip\theta q\right)$ and the graph would then be planar.

A nonplanar graph possesses such an extra momentum-dependent phase factor in the integrand, whose rapid oscillations suppress the integral over internal momenta. Note that $\theta_{\mu\nu}$ given by Eq. (15.38) is $\sim L$ so that the integral for a nonplanar diagram of genus h is suppressed by

$$\left(p^{2d}\det_{\mu\nu}\theta_{\mu\nu}\right)^{-h} \sim L^{-hd} \sim N^{-2h} \tag{15.39}$$

in accord with the topological expansion in $1/N$. Here p^2 is typical external momentum associated with the diagram.

If $N \to \infty$ then only planar diagrams survive in the twisted reduced model, thereby reproducing the planar limit of the d-dimensional theory.

Remark on twisted versus quenched models at large but finite N

The size of the lattice associated with the twisted reduced model at large but finite N is $L = N^{2/d}$. This is to be compared with its counterpart for the quenched reduced model where $L = N^{1/d}$ as discussed in the Remark on p. 329. For a given N, the value of L for the twisted reduced model is much larger than for the quenched reduced model. The former therefore provides a more economic approach to the limit of infinite volume.

We shall see in the next chapter that yet another continuum limit, associated with noncommutative theories, can be obtained in the twisted reduced models at the distances $\sim \sqrt{|\theta|} \sim aN^{1/d}$. The corresponding momenta are $p^2 \sim 1/|\theta|$ so that the dimensionless parameter on the LHS of Eq. (15.39) is ~ 1 and each term of the genus expansion is of order one in this continuum limit.

Remark on mapping between matrices and fields

The transition from matrices to functions on a periodic L^d lattice can be formalized by introducing the matrix-valued function [Bar90]

$$\Delta^{ij}(x) = \frac{1}{N^2} \sum_{m \in \mathbb{Z}_L^d} j_m^*(x) J_m^{ij}, \tag{15.40}$$

where the functions

$$j_m(x) = e^{2\pi i \sum_{\mu,\nu} x_\mu n_{\mu\nu} m_\nu / aN} \tag{15.41}$$

form a Fourier basis.

Thus defined $\Delta^{ij}(x)$ possesses the properties

$$\Delta^\dagger(x) = \Delta(x), \tag{15.42}$$

$$N \operatorname{tr}\left[J_m \Delta(x) \right] = j_m(x), \tag{15.43}$$

$$\sum_x \Delta^{ij}(x) \Delta^{kl}(x) = \frac{1}{N} \delta^{il} \delta^{kj}, \tag{15.44}$$

$$\Gamma_\mu \Delta(x) \Gamma_\mu^\dagger = \Delta(x - a\hat{\mu}), \tag{15.45}$$

$$N \operatorname{tr}\left[\Delta(x)\Delta(y) \right] = \delta_{xy}. \tag{15.46}$$

Equation (15.46) represents completeness of the Fourier basis (15.41) in the space of functions on a discrete torus.

Given the definitions (15.29), (15.32), and (15.40), we can directly relate matrices with functions of x by

$$\tilde{\varphi} = \sum_x \Delta(x)\varphi(x) \tag{15.47}$$

and vice versa

$$\varphi(x) = N \operatorname{tr}\left[\tilde{\varphi} \Delta(x) \right]. \tag{15.48}$$

In particular, the equivalence of the actions (15.31) and (15.33) is a consequence of the general formula

$$N \operatorname{tr} \tilde{\mathcal{L}} = \sum_x \mathcal{L}(x), \tag{15.49}$$

where $\tilde{\mathcal{L}}$ is arbitrary and $\mathcal{L}(x)$ is related to it by Eqs. (15.47) and (15.48).

As $N \to \infty$, we approach the limit of an infinite lattice since $aL \to \infty$. Then the discrete variable m_μ is to be substituted by a continuum momentum variable from the first Brillouin zone:

$$k_\mu = \frac{2\pi \sum_\nu n_{\mu\nu} m_\nu}{aN} \in \left(-\frac{\pi}{a}, \frac{\pi}{a}\right]. \tag{15.50}$$

The summation over m_μ is to be substituted in all the formulas above by an integration over k_μ:

$$\frac{1}{N^2} \sum_{m \in \mathbb{Z}_L^d} \cdots \longrightarrow \int \prod_{\mu=1}^d \frac{\mathrm{d}k_\mu}{2\pi} \cdots . \tag{15.51}$$

For *smooth* configurations when only modes with $|k_\mu| \ll 1/a$ are essential, we can additionally substitute the summation over the lattice sites x by an integration over the continuum variable $x \in \mathbb{R}^d$ (d-dimensional Euclidean space):

$$a^d \sum_x \cdots \implies \int \mathrm{d}^d x \cdots . \tag{15.52}$$

Then Eq. (15.49) becomes

$$a^d N \operatorname{tr} \widetilde{\mathcal{L}} \implies \int \mathrm{d}^d x \, \mathcal{L}(x). \tag{15.53}$$

In particular, we have

$$a^d N \operatorname{tr} \mathbb{I} \implies \int \mathrm{d}^d x = V \tag{15.54}$$

which relates the (infinite) trace of a unit matrix with the (infinite) volume.

We shall return to the relation between infinite-dimensional matrices (= operators) and functions on $\mathbb{R}^d$ in Sect. 15.4 when discussing a continuum limit of the twisted reduced models.

Remark on $SU(\infty)$ and symplectic diffeomorphisms in $d = 2$

The group $SU(N)$ can be approximated at large N by the group $SL(2; \mathbb{R})$ of *area-preserving* or *symplectic diffeomorphisms* (SDiff) in two dimensions.

It follows from Eq. (15.26) that

$$[J_m, J_n] = 2\mathrm{i} \sin\left[\frac{\pi}{N}(m_\mu \varepsilon^{\mu\nu} n_\nu)\right] J_{m+n}, \tag{15.55}$$

where $m_\mu \varepsilon^{\mu\nu} n_\nu = m_1 n_2 - m_2 n_1$.

At large N and $m_\mu \varepsilon^{\mu\nu} n_\nu \ll N$ the sin can be expanded, which yields

$$[J_m, J_n] \approx i\frac{2\pi}{N}(m_\mu \varepsilon_{\mu\nu} n_\nu) J_{m+n}. \tag{15.56}$$

Equation (15.56) is to be compared with the Poisson bracket

$$\begin{aligned}
\{j_m, j_n\}_{\text{PB}} &\equiv \varepsilon^{\mu\nu} \partial_\mu j_m \partial_\nu j_n \\
&\propto (m_\mu \varepsilon^{\mu\nu} n_\nu) j_{m+n}
\end{aligned} \tag{15.57}$$

of the basis functions j_m given by Eq. (15.41). The group $SL(2;\mathbb{R})$ of symplectic diffeomorphisms arose since it is a symmetry of the Poisson structure.

For *smooth* matrices $\tilde{\varphi}^{ij}$ when the low modes dominate in Eq. (15.29), the commutator can be substituted by the Poisson bracket

$$[\cdot,\cdot] \implies i\{\cdot,\cdot\}_{\text{PB}}. \tag{15.58}$$

This looks like a semiclassical approximation of commutators in quantum mechanics by the Poisson brackets. It is now justified by the large value of N.

This clarifies the relation between the group $SL(2;\mathbb{R})$ of symplectic diffeomorphisms and the group $SU(\infty)$ for smooth fields.

There is a vast literature on this issue starting from unpublished works by J. Goldstone and J. Hoppe (see [Hop89]) in the early 1980s who approximated symplectic diffeomorphisms of a spherical membrane by $SU(N)$. This relation was applied [FIT89] to classical Yang–Mills theory and formulated [FFZ89, FZ89] in an elegant way for a torus. These two cases describe two *different* $N = \infty$ limits of $SU(N)$ [PS89]. It was conjectured [Bar90] that strings could appear from the reduced models owing to this mechanism, which also seems to explain early results [Cre84] on an $SL(2;\mathbb{R})$ invariance of the $SU(\infty)$ Yang–Mills theory in two dimensions.

More on the relation between symplectic diffeomorphisms and $SU(\infty)$ can be found in the review [Ran92].

15.3 Twisted Eguchi–Kawai model

In order to construct the twisted Eguchi–Kawai model (TEK), we proceed quite similarly to Sect. 14.3 by substituting $D(x)$ given by Eq. (15.1). Equation (14.31) remains the same but the difference is that now

$$D_\mu \equiv D(x + a\hat{\mu})\, D^\dagger(x) = \Gamma_\mu \tag{15.59}$$

and, hence, the D_μ do not commute.

Reordering the matrices in $D(x)$ produces an Abelian factor which depends on the ordering prescription. It is possible to use a symmetric ordering (15.22) instead of the normal ordering (15.1) when

$$D(x) = J_{x/a},$$
(15.60)

and

$$D_\mu = \Gamma_\mu \prod_{\nu=1}^{d} Z_{\mu\nu}^{x_\nu/2a}.$$
(15.61)

This extra Abelian factor cancels in most of the formulas.

The substitution of (14.31) with $D(x)$ given by Eq. (15.1) (or Eq. (15.60)) into the Wilson action (6.18) results in

$$
\begin{aligned}
S_{\text{TEK}} &= \frac{1}{2} \sum_{\mu \neq \nu} \left\{ 1 - \frac{1}{N} \operatorname{tr} \left[\tilde{U}_\nu^\dagger \Gamma_\nu^\dagger \tilde{U}_\mu^\dagger \Gamma_\nu \Gamma_\mu^\dagger \tilde{U}_\nu \Gamma_\mu \tilde{U}_\mu \right] \right\} \\
&= \frac{1}{2} \sum_{\mu \neq \nu} \left\{ 1 - Z_{\mu\nu} \frac{1}{N} \operatorname{tr} \left[\left(\tilde{U}_\nu^\dagger \Gamma_\nu^\dagger \right) \left(\tilde{U}_\mu^\dagger \Gamma_\mu^\dagger \right) \left(\Gamma_\nu \tilde{U}_\nu \right) \left(\Gamma_\mu \tilde{U}_\mu \right) \right] \right\},
\end{aligned}
$$
(15.62)

where the factor of $Z_{\mu\nu}$ emerged because of the commutation relation (15.2).

Introducing the new variable

$$U_\mu = \Gamma_\mu \tilde{U}_\mu$$
(15.63)

as in Eq. (14.37), we finally obtain

$$S_{\text{TEK}}[U] = \frac{1}{2} \sum_{\mu \neq \nu} \left(1 - Z_{\mu\nu} \frac{1}{N} \operatorname{tr} U_\nu^\dagger U_\mu^\dagger U_\nu U_\mu \right)$$
(15.64)

for the action of the twisted Eguchi–Kawai model.

Noting that the Haar measure $d\tilde{U}_\mu = dU_\mu$ is not changed* under the (left) multiplication (15.63) by a unitary matrix Γ_μ, we arrive at the partition function of the twisted Eguchi–Kawai model

$$Z_{\text{TEK}} = \int \prod_\mu dU_\mu \, e^{-N S_{\text{TEK}}[U]/g^2}.$$
(15.65)

The only difference from the original Eguchi–Kawai model (14.40) resides in the twisting factor of $Z_{\mu\nu}$ in the action (15.64).

---

* We shall see in Sect. 16.6 some examples when this is not the case.

The twisted Eguchi–Kawai model possesses the gauge symmetry (14.39) and the $U(1)^d$ symmetry (14.41). The second one is not broken for all values of the coupling g^2N owing to the presence of the twisting factor as will be demonstrated in a moment. For this reason, the twisted Eguchi–Kawai model is equivalent at large N to the planar limit of d-dimensional Yang–Mills theory for all values of the coupling g^2N.

The vacuum state of the twisted Eguchi–Kawai model is given modulo a gauge transformation by

$$U_\mu^{\text{cl}} = \Gamma_\mu, \tag{15.66}$$

where the twist eaters Γ_μ were constructed explicitly in Sect. 15.1. The value of the action (15.64) of the twisted Eguchi–Kawai model is 0 for this configuration which is smaller, say, than the value of $\sum_{\mu\nu}(1 - \text{Re}\, Z_{\mu\nu})$ of the action for a configuration given by diagonal matrices.

An analog of Eq. (14.42) for the twisted Eguchi–Kawai model is given by

$$\left\langle F[U_\mu(x)] \right\rangle \stackrel{\text{red.}}{=} \left\langle F\!\left[D^\dagger(x + a\hat\mu)\, U_\mu D(x)\right] \right\rangle_{\text{TEK}}. \tag{15.67}$$

But the fact that D_μ no longer commute results in subtleties in representing the averages, in particular the Wilson loops, in the language of the twisted Eguchi–Kawai model.

The Wilson loop averages in the twisted Eguchi–Kawai model are defined by

$$W_{\text{TEK}}(C) = \left\langle \frac{1}{N}\,\text{tr}\, D^\dagger(C)\, \frac{1}{N}\,\text{tr}\, \boldsymbol{P}\prod_i U_{\mu_i} \right\rangle_{\text{TEK}}, \tag{15.68}$$

where

$$D(C) = \boldsymbol{P}\prod_i \Gamma_{\mu_i} \tag{15.69}$$

and the product runs over links forming the contour C. This is an analog of Eq. (14.45).

Note that $W_{\text{TEK}}(C) = 1$ for the vacuum configuration (15.66) when C is closed.

For closed loops $D(C) \in Z(N)$ which can be proven using the commutation relation (15.2). For instance,

$$D(\partial p) = \Gamma_\nu^\dagger \Gamma_\mu^\dagger \Gamma_\nu \Gamma_\mu = Z_{\mu\nu}^*. \tag{15.70}$$

The value of $D(C)$ for a closed loop is the same as that prescribed by Eq. (15.67).

The first trace on the RHS of Eq. (15.68) vanishes for open loops, thereby providing the vanishing of the open Wilson loop averages themselves. Strictly speaking, this vanishing does not hold, say, for the loops in the form of a straight line consisting of L links for the Γ_μ given by Eq. (15.18) when $\Gamma_\mu^L = 1$. But this will be inessential for the purposes of this section since such loops are infinitely long as $N \to \infty$. We shall return to this point below when considering the twisted Eguchi–Kawai model at finite N and associating it with a theory on a finite lattice.

Because the averages of the open Wilson loops vanish in the twisted Eguchi–Kawai model as $N \to \infty$ by construction, all that was said in Sect. 14.3 concerning the equivalence with the d-dimensional lattice gauge theory is applicable for the twisted Eguchi–Kawai model as well.

Problem 15.4 Extend Eq. (15.70) to arbitrary closed contours.

Solution The calculation is similar to that in Problem 15.1 on p. 352. The result is

$$D(C) \;=\; \prod_{p \in S:\partial S = C} Z^*_{\mu\nu}\,, \tag{15.71}$$

where (μ, ν) is the orientation of the plaquette p. The product runs over any surface spanned by the closed loop C and is surface-independent owing to the Bianchi identity (15.9).

Problem 15.5 Derive the loop equation for the twisted Eguchi–Kawai model.

Solution The derivation is quite similar to Problem 14.2 on p. 336. Performing the shift (14.49) in the action (15.64), we obtain an extra factor of $Z^*_{\mu\nu}$ which is absorbed by $D(C)$ in the definition (15.68) of the Wilson loop averages in the twisted Eguchi–Kawai model owing to Eq. (15.70):

$$D^\dagger(C)\, Z_{\mu\nu} \;=\; D^\dagger(C\,\partial p) \tag{15.72}$$

and similarly

$$D^\dagger(C)\, Z^*_{\mu\nu} \;=\; D^\dagger(C\,\partial p^{-1}) \tag{15.73}$$

for the Hermitian conjugate term in the action.

The resulting loop equation reads [GO83b]

$$\sum_p \left[W_{\mathrm{TEK}}(C\,\partial p) - W_{\mathrm{TEK}}(C\,\partial p^{-1}) \right]$$
$$= \; g^2 N \sum_{l \in C} \tau_\nu(l)\, W_{\mathrm{TEK}}(C_{yx})\, W_{\mathrm{TEK}}(C_{xy})\,. \tag{15.74}$$

The Kronecker symbol δ_{xy} is again restored on the RHS of Eq. (14.50) since the averages of the open Wilson loops vanish:

$$W_{\mathrm{TEK}}(C_{xy}) \;=\; \delta_{xy} W_{\mathrm{TEK}}(C_{xx})\,. \tag{15.75}$$

This reproduces at $N = \infty$ the loop equation (12.65) of the d-dimensional lattice gauge theory which proves the equivalence.

Remark on twisted boundary conditions

When gauge theory is defined in a box, the boundary conditions are not necessarily periodic. The values of the gauge field at opposite sides of the box can rather coincide modulo an $SU(N)$ gauge transformation:

$$A_\mu(x + \ell_\nu) = \Omega_\nu(x) A_\mu(x) \Omega_\nu^\dagger(x) + i \Omega_\nu(x) \partial_\mu \Omega_\nu^\dagger(x). \qquad (15.76)$$

Here ℓ_ν denotes the spatial extent of the box in direction ν. This equation represents a *twisted boundary condition*.

A box with periodic boundary conditions looks geometrically like a torus $\mathbb{T}^d$ with the period matrix $\ell_{\mu\nu} = \ell_\nu \delta_{\mu\nu}$. Similarly, a box with the twisted boundary conditions (15.76) is often called a *twisted torus*.

The transition matrices Ω_ν in Eq. (15.76) obey the consistency condition [Hoo79]

$$\Omega_\mu(x + \ell_\nu) \Omega_\nu(x) = Z_{\mu\nu} \Omega_\nu(x + \ell_\mu) \Omega_\mu(x), \qquad (15.77)$$

where $Z_{\mu\nu} \in Z_N$.

This factor of $Z_{\mu\nu}$ cannot be removed in a pure Yang–Mills theory since the gauge group is actually $SU(N)/Z(N)$. Therefore, there are N distinct choices of boundary conditions per plane of a box, which are not related by a gauge transformation. The factor of $Z_{\mu\nu}$ is associated with an additive flux known as the *'t Hooft flux*, which is a feature of non-Abelian field configurations.

A lattice counterpart of Eq. (15.76) reads

$$U_\mu(x + \ell_\nu) = \Omega_\nu(x + a\hat{\mu}) U_\mu(x) \Omega_\nu^\dagger(x). \qquad (15.78)$$

Correspondingly, a periodic lattice of finite size L^d is called a *discrete torus* $\mathbb{T}_L^d$.

The twisted Eguchi–Kawai model is of the same type as Wilson's lattice gauge theory on a unit hypercube with the twisted boundary condition and $\Omega_\mu = \Gamma_\mu^\dagger$. This explains the terminology.

Problem 15.6 Show the equivalence between the twisted Eguchi–Kawai model and Wilson's lattice gauge theory on a unit hypercube with the twisted boundary condition.

Solution The twisted boundary condition (15.78) for a hypercube with the corner at $x = 0$ is given generically as

$$U_\mu(a\hat{\nu}) = \Omega_\nu(a\hat{\mu}) U_\mu(0) \Omega_\nu^\dagger(0), \qquad (15.79)$$

or

$$U_\mu(a\hat{\nu}) = \Gamma_\nu^\dagger U_\mu(0) \Gamma_\nu \qquad (15.80)$$

for $\Omega_\nu(0) = \Omega_\nu(a\hat{\mu}) = \Gamma_\nu^\dagger$.

The action of Wilson's lattice gauge theory on a unit hypercube with the twisted boundary condition can be transformed using Eq. (15.80) to the form

$$\frac{1}{2} \sum_{\mu \neq \nu} \left\{ 1 - \frac{1}{N} \operatorname{tr} \left[U_\nu^\dagger(0) U_\mu^\dagger(a\hat\nu) U_\nu(a\hat\mu) U_\mu(0) \right] \right\}$$

$$= \frac{1}{2} \sum_{\mu \neq \nu} \left\{ 1 - \frac{1}{N} \operatorname{tr} \left[U_\nu^\dagger(0) \left(\Gamma_\nu^\dagger U_\mu^\dagger(0) \Gamma_\nu \right) \left(\Gamma_\mu^\dagger U_\nu(0) \Gamma_\mu \right) U_\mu(0) \right] \right\}$$

$$= \frac{1}{2} \sum_{\mu \neq \nu} \left\{ 1 - Z_{\mu\nu} \frac{1}{N} \operatorname{tr} \left[\left(U_\nu^\dagger(0) \Gamma_\nu^\dagger \right) \left(U_\mu^\dagger(0) \Gamma_\mu^\dagger \right) \left(\Gamma_\nu U_\nu(0) \right) \left(\Gamma_\mu U_\mu(0) \right) \right] \right\},$$

$$(15.81)$$

where we have used Eq. (15.2). Introducing the variable $U_\mu = \Gamma_\mu U_\mu(0)$, we arrive at the action (15.64) and the partition function (15.65) of the twisted Eguchi–Kawai model.

The consideration of this Problem was the original motivation of [GO83b] for introducing the twisting factor of $Z_{\mu\nu}$ in the action of the naive Eguchi–Kawai model.

Remark on $U(N)$ gauge fields

The consistency condition for the $U(N)$ gauge group is simply

$$\Omega_\mu(x + \ell_\nu) \Omega_\nu(x) = \Omega_\nu(x + \ell_\mu) \Omega_\mu(x) \quad \boxed{U(N) \text{ matrices}} \quad (15.82)$$

in order for a field in the fundamental representation to be single-valued on a twisted torus.

But now field configurations are characterized by the first Chern class

$$q_{\mu\nu} = \frac{1}{2\pi} \int \mathrm{d}x_\mu \mathrm{d}x_\nu \frac{1}{N} \operatorname{tr} \mathcal{F}_{\mu\nu} \quad \boxed{\text{no sum over } \mu, \nu} \quad (15.83)$$

which is nothing but the (magnetic) $U(1)$ flux through the (μ, ν)-plane of the torus. It is quantized since the homotopy group $\pi_1(U(N)) = \mathbb{Z}$.

Given a $U(N)$ field configuration with a constant $U(1)$ flux and subtracting it, we arrive at an $SU(N)$ part of the gauge field:

$$\mathcal{A}_\mu^{SU(N)} = \mathcal{A}_\mu + \frac{\pi q_{\mu\nu} x_\nu}{\ell^2 N}, \quad (15.84)$$

which obeys Eq. (15.76) with

$$\Omega_\mu^{SU(N)} = e^{-\pi i q_{\mu\nu} x_\nu / \ell N} \Omega_\mu \quad (15.85)$$

satisfying Eq. (15.77) with $Z_{\mu\nu} = e^{-2\pi i q_{\mu\nu}/N}$. Therefore, the $U(1)$ (magnetic) flux induces [LPR89] the 't Hooft flux for the $SU(N)$ part of the $U(N)$ gauge group.

15.4 Twisting prescription in the continuum

The twisting reduction prescription can be formulated directly for the continuum theory [GK83] by substituting

$$\tilde{\varphi} \rightarrow a^{d/2-1}\tilde{\varphi}, \qquad \Gamma_\mu = e^{-iaP_\mu}, \tag{15.86}$$

with the lattice spacing $a \rightarrow 0$ and $N \rightarrow \infty$. The $N \times N$ Hermitian matrices $\tilde{\varphi}$ and P_μ become Hermitian operators $\tilde{\varphi}$ and $\boldsymbol{P}_\mu$ as $N \rightarrow \infty$.

While the Γ_μ in Eq. (15.86) look like Eq. (14.34), $\boldsymbol{P}_\mu$ are no longer diagonal and do not commute. As a consequence of the Weyl–'t Hooft relation (15.2), they obey the Heisenberg commutation relation

$$[\boldsymbol{P}_\mu, \boldsymbol{P}_\nu] = -iB_{\mu\nu}\mathbf{1}, \tag{15.87}$$

where

$$B_{\mu\nu} = \frac{2\pi n_{\mu\nu}}{Na^2} \tag{15.88}$$

from the matrix approximation. However, we shall not refer to the matrix approximation during most of this section and consider $B_{\mu\nu}$ as an independent variable.

The commutator (15.87) is similar to that for the coordinate and momentum operators in quantum mechanics. For this reason, the formulation of the continuum twisted reduced model uses operator calculus of quantum mechanics.

Let us mention once again that a solution to Eq. (15.87) for $\boldsymbol{P}_\mu$ exists only for infinite-dimensional Hermitian matrices (representing operators). This is a well-known property of the Heisenberg commutation relation. It can be seen by taking the trace of both sides of Eq. (15.87). If $\boldsymbol{P}_\mu$ were finite-dimensional matrices, the trace of the LHS would vanish owing to the cyclic property of the trace, while that of the RHS would not. In contrast, Eq. (15.2) which is written for unitary matrices possesses a solution for finite N.

A continuum (operator) analog of Eq. (15.1) is

$$\boldsymbol{D}(x) = \prod_{\mu=1}^{d} e^{-i\boldsymbol{P}_\mu x_\mu} \tag{15.89}$$

and similarly for Eq. (14.4):

$$\varphi^{ij}(x) \overset{N\rightarrow\infty}{\longrightarrow} \boldsymbol{D}^\dagger(x)\tilde{\varphi}\boldsymbol{D}(x). \tag{15.90}$$

We can change the order of operators in the product on the RHS of Eq. (15.89) by introducing a more general path-dependent operator

$$\boldsymbol{D}(C_{x0}) = \boldsymbol{P}e^{-i\int_{C_{x0}} d\xi^\mu \boldsymbol{P}_\mu}, \tag{15.91}$$

where the integration contour C_{x0} connects the origin and the point x, but is arbitrary otherwise. Changing the shape of the contour results in an extra factor

$$\boldsymbol{P}\,\mathrm{e}^{-\mathrm{i}\oint \mathrm{d}\xi^\mu \boldsymbol{P}_\mu} \;=\; \mathrm{e}^{-\mathrm{i}B_{\mu\nu}\int \mathrm{d}\sigma^{\mu\nu}} \tag{15.92}$$

which is a c-number and cancels in the reduction formula (15.90). This is quite similar to the consideration in Problem 15.1 on p. 352.

In particular, we can always pass in Eq. (15.89) from the normal ordering of the operators to a symmetric ordering:

$$\boldsymbol{D}(x) \;=\; \mathrm{e}^{-\mathrm{i}\sum_{\mu=1}^d \boldsymbol{P}_\mu x_\mu}. \tag{15.93}$$

This is an operator analog of Eq. (15.60).

The action of the continuum twisted reduced model is given by the same formula (14.21) as for the continuum quenched reduced model except that $\boldsymbol{P}_\mu$ obey the commutation relation (15.87) rather than commuting as in the quenched reduced model. A "volume element" v is again given for the lattice regularization by Eq. (14.23). Just as in the case of the quenched reduced model, the very formulation of the continuum twisted reduced model implies a regularization.

What remains to be defined are two related issues: how to understand the trace in Eq. (14.21) and how to introduce a regularization directly within the continuum theory.

We begin with a two-dimensional case where $B_{\mu\nu} = B\varepsilon_{\mu\nu}$. The operators $\boldsymbol{P}_1$ and $\boldsymbol{P}_2$ are then similar to the position and momentum operators in one-dimensional quantum mechanics, with B playing the role of Planck's constant.

A Hilbert space is spanned either by $|p_1\rangle$ or $|p_2\rangle$ states which are the eigenstates of either $\boldsymbol{P}_1$ or $\boldsymbol{P}_2$:

$$\boldsymbol{P}_1|p_1\rangle \;=\; p_1|p_1\rangle, \qquad \boldsymbol{P}_2|p_2\rangle \;=\; p_2|p_2\rangle, \tag{15.94}$$

and are normalized to $\langle p'|p\rangle = \delta^{(1)}(p - p')$.

In either basis the trace of an operator $\boldsymbol{O}$ on the Hilbert space is defined via its (diagonal) matrix elements by

$$\mathrm{tr}_{\mathcal{H}}\,\boldsymbol{O} \;=\; \int \mathrm{d}p\,\langle p|\boldsymbol{O}|p\rangle. \tag{15.95}$$

The matrix element can be easily calculated, representing $\boldsymbol{O}$ by the use of the commutator (15.87) in a normal form, where all $\boldsymbol{P}_1$ are to the left of $\boldsymbol{P}_2$, and

$$\mathrm{e}^{-\mathrm{i}k_1\boldsymbol{P}_2/B}|p_1\rangle \;=\; |p_1 - k_1\rangle, \qquad \mathrm{e}^{\mathrm{i}k_2\boldsymbol{P}_1/B}|p_2\rangle \;=\; |p_2 - k_2\rangle. \tag{15.96}$$

There exists a simple operator representation of the $N = \infty$ limit of the basis elements J_m introduced in Sect. 15.2. Substituting the operator limit (15.86) of $\Gamma_1 = \mathcal{P}$ and $\Gamma_2 = \mathcal{Q}$ into Eq. (15.20), we obtain

$$J_m^{ij} \;\longrightarrow\; e^{-iam_1 \boldsymbol{P}_1} \, e^{-iam_2 \boldsymbol{P}_2} \, e^{-i\pi m_1 m_2 / L} \;\equiv\; \boldsymbol{J}_m. \qquad (15.97)$$

The order of operators on the RHS of Eq. (15.97) is normal. Applying the Baker–Campbell–Hausdorff formula

$$e^A \, e^B \, e^{-\frac{1}{2}[A,B]} \;=\; e^{A+B}, \qquad (15.98)$$

which is exact when the commutator $[A, B]$ is a c-number as in our case, it can be represented conveniently in a symmetric- or Weyl-ordered form

$$\boldsymbol{J}_m \;=\; e^{-ia(m_1 \boldsymbol{P}_1 + m_2 \boldsymbol{P}_2)}. \qquad (15.99)$$

The continuum operator counterparts of the formulas of Sect. 15.2 are obvious.

Introducing the continuum momentum variable $k_\mu = 2\pi\varepsilon_{\mu\nu} m_\nu / aL$ which is a $d = 2$ version of Eq. (15.50) and using the substitution (15.51), we have

$$\boldsymbol{f} \;=\; \int \prod_\mu \frac{dk_\mu}{2\pi} \boldsymbol{J}_k \, f(k) \qquad (15.100)$$

which is quite analogous to the Fourier transform of a function

$$f(x) \;=\; \int \prod_\mu \frac{dk_\mu}{2\pi} \, e^{ikx} \, f(k). \qquad (15.101)$$

Here

$$\boldsymbol{J}_k \;=\; e^{i(k_2 \boldsymbol{P}_1 - k_1 \boldsymbol{P}_2)/B} \;=\; e^{ik_2 \boldsymbol{P}_1/B} \, e^{-ik_1 \boldsymbol{P}_2/B} \, e^{-ik_1 k_2 / 2B} \qquad (15.102)$$

as follows from Eq. (15.97).

The coefficients $f(k)$ on the RHSs of Eqs. (15.100) and (15.101) are the same. Therefore, these equations relate operators in Hilbert space and functions to each other. This relation is often called the *Weyl transform*.*

Given Eqs. (15.100) and (15.101) and using Eqs. (15.94) and (15.96), we can alternatively write down the Weyl transform via the matrix element

$$f(k_1, k_2) \;=\; \frac{2\pi}{B} \int dp_1 \, e^{-ik_2 p_1/B} \, \langle p_1 + \tfrac{1}{2}k_1 \, | \, \boldsymbol{f} \, | \, p_1 - \tfrac{1}{2}k_1 \rangle. \qquad (15.103)$$

An extension to d dimensions is straightforward. Say, k and x in Eqs. (15.100) and (15.101) were to simply become d-dimensional vectors. Similarly, the integration as well as the matrix element are

---

* More rigorous mathematical definitions can be found in the book [Won98].

taken in Eq. (15.103) with respect to half of the momentum variables: $p_1, p_3, \ldots, p_{d-1}$.

The Weyl transform can, of course, be formulated without any reference to the discrete formulas of Sect. 15.2. We simply followed the spirit of Weyl's original book [Wey31].

However, an advantage of such an approach which starts from a lattice discretization is that it provides an ultraviolet cutoff, making the continuum twisted reduced model well-defined. The values of momenta are bounded by $|k_\mu| \leq \pi/a$, which introduces the cutoff. Instead of the lattice regularization, we can use a Lorentz-invariant regularization of [GK83] directly for the continuum theory restricting $k^2 \leq \Lambda^2$ in the integral over k_μ in Eqs. (15.100) and (15.101). This will both regularize perturbation theory and bound operators on the Hilbert space.

The action of the continuum twisted reduced model regularized in such a way can be represented in the form

$$S_{\text{TRM}} = \frac{(2\pi)^{d/2}}{\text{Pf}\,(B_{\mu\nu})} \text{tr}_{\mathcal{H}} \left\{ -\frac{1}{2}\,[P_\mu, \tilde{\varphi}]^2 + \tilde{V}\,(\tilde{\varphi}) \right\}, \qquad (15.104)$$

where we have substituted

$$vN = \frac{(2\pi)^{d/2}}{\text{Pf}\,(B_{\mu\nu})} \qquad (15.105)$$

and $\text{Pf}\,(B_{\mu\nu}) = \sqrt{\det_{\mu\nu} B_{\mu\nu}}$. This substitution is justified by the definition (15.88) of $B_{\mu\nu}$ and v is again a volume element given by Eq. (14.23) for the lattice regularization.

We have already met the factor (15.105) for $d = 2$ in Eq. (15.103). It appears whenever the trace over the Hilbert space is substituted by the integral over space as

$$\frac{(2\pi)^{d/2}}{\text{Pf}\,(B_{\mu\nu})} \text{tr}_{\mathcal{H}}\,\mathcal{L} = \int \text{d}^d x\,\mathcal{L}(x), \qquad (15.106)$$

where $\mathcal{L}(x)$ is the Weyl transform of $\mathcal{L}$. This formula is a counterpart of Eq. (15.53)

The proof of how the continuum twisted reduced model reproduces planar graphs is quite similar to that of Sect. 15.2 on the lattice. The integral over space is reproduced according to Eq. (15.106). Nonplanar graphs are again suppressed as $\theta_{\mu\nu} = B_{\mu\nu}^{-1} \to \infty$.

Remark on the number of states in Hilbert space

For the matrix approximation, the Hilbert space is spanned by N states. A question arises as to what is the number of states in the Hilbert space regularized in a Lorenz-invariant way.

This can be easily understood from an analogy with the semiclassical limit of quantum mechanics when B, which plays the role of Planck's constant, is small. The volume occupied by the N states in a phase space is given semiclassically by the Bohr–Sommerfeld formula. It can be written in our notation as

$$\prod_\mu \frac{\Delta p_\mu}{2\pi} = N \frac{\mathrm{Pf}\,(B_{\mu\nu})}{(2\pi)^{d/2}}. \tag{15.107}$$

Dividing by N, we conclude that the factor on the RHS of Eq. (15.105) is related semiclassically to the inverse volume of a cell in the phase space.

Given a regularization which determines the LHS of Eq. (15.107) via the cutoff Λ, we can solve Eq. (15.107) for N which gives the number of states in the regularized Hilbert space.

Of course, all of these formulas become exact for the lattice regularization when $\mathrm{Pf}\,(B_{\mu\nu}) \sim 1/N \to 0$.

15.5 Continuum version of TEK

The continuum version of the twisted Eguchi–Kawai model can be constructed [GK83] from the lattice counterpart of Sect. 15.3 by substituting

$$U_\mu = \mathrm{e}^{\mathrm{i}aA_\mu}, \qquad \Gamma_\mu = \mathrm{e}^{-\mathrm{i}aP_\mu}, \tag{15.108}$$

when the lattice spacing $a \to 0$ and $N \to \infty$. Here A_μ and P_μ are $N \times N$ Hermitian matrices which become Hermitian operators when $N \to \infty$ as is described in the previous section. We shall imply, but not explicitly use the operator notation.

To derive the action of the continuum twisted Eguchi–Kawai model, we first obtain from Eqs. (15.108) and (15.87)

$$U_\nu^\dagger U_\mu^\dagger U_\nu U_\mu = \mathrm{e}^{a^2[A_\mu,A_\nu]}, \qquad Z_{\mu\nu} = \mathrm{e}^{\mathrm{i}a^2 B_{\mu\nu}} \tag{15.109}$$

to order a^2. Finally, we arrive at the following action of the continuum twisted Eguchi–Kawai model:

$$S_{\mathrm{TEK}}[A] = -\frac{v}{4g^2}\,\mathrm{tr}\,\left([A_\mu, A_\nu] + \mathrm{i}B_{\mu\nu}\right)^2, \tag{15.110}$$

where v is again a "volume element" given for the lattice regularization by Eq. (14.23). Just as in the case of the quenched Eguchi–Kawai model, the very formulation of the continuum twisted Eguchi–Kawai model implies a regularization.

It is worth mentioning here a subtlety associated with the fact that A_μ are Hermitian operators (infinite-dimensional matrices). The point is that

$$\text{tr}\,[A_\mu, A_\nu] \neq 0 \qquad (15.111)$$

in this case so that $B_{\mu\nu}$ cannot be omitted. This is a well-known property of operators obeying the Heisenberg commutation relation (15.87) as has already been pointed out.

Nevertheless, the presence of the $B_{\mu\nu}$ does not affect the classical equation of motion for the continuum twisted Eguchi–Kawai model which coincides with Eq. (14.71) since $B_{\mu\nu}$ is a c-number.

Owing to the presence of $B_{\mu\nu}$ in the action (15.110), the vacuum configuration of the continuum twisted Eguchi–Kawai model is given by

$$A_\mu^{\text{cl}} = -P_\mu \qquad (15.112)$$

modulo a gauge transformation $A_\mu^{\text{cl}} \to \Omega A_\mu^{\text{cl}} \Omega^\dagger$. The minimum of the action is reached when P_μ obey Eq. (15.87) rather than being diagonal matrices.

The continuum limit of Eqs. (15.68) and (15.69) determines the averages of Wilson loops in the continuum twisted Eguchi–Kawai model:

$$W_{\text{TEK}}(C_{yx}) = \left\langle \frac{1}{N} \text{tr}\, D^\dagger(C_{yx}) \frac{1}{N} \text{tr}\, \boldsymbol{P} \, e^{\text{i} \int_{C_{yx}} \text{d}\xi^\mu A_\mu} \right\rangle_{\text{TEK}}, \qquad (15.113)$$

where $D(C_{yx})$ is defined in Eq. (15.91). They are nontrivial since A_μ do not commute.

The trace of $D^\dagger(C_{yx})$ on the RHS of Eq. (15.113) vanishes for open loops. This provides the vanishing of the averages of open Wilson loops as is prescribed by the R^d symmetry (14.61) of the action (15.110).

For closed loops this factor does not vanish and represents the flux of the $B_{\mu\nu}$-field through a surface bounded by the contour C. It is needed to provide the equivalence with planar graphs of d-dimensional Yang–Mills theory, since the classical extrema of the continuum twisted Eguchi–Kawai model are given by Eq. (15.112) and perturbation theory is constructed by expanding around this classical solution. The equivalence can be demonstrated perturbatively using the theorem stated at the end of Sect. 15.2.

The proof of the equivalence between the large-N limit of d-dimensional Yang–Mills theory and the continuum Eguchi–Kawai model can be given using the continuum loop equation, for which the lattice regularization was considered in Problem 15.5 on p. 365. The loop equation for the continuum twisted Eguchi–Kawai model coincides with Eq. (14.65) for the continuum naive Eguchi–Kawai model. This is because the loop operator on the LHS of Eq. (14.65) is of first order (obeys the Leibnitz rule of the type of Eq. (12.96)). For this reason, the first trace in Eq. (15.113) produces

$$[P_\mu, [P_\mu, P_\nu]] \;=\; 0 \tag{15.114}$$

which vanishes since the commutator of P_μ with P_ν is a c-number. The manipulations with the result of acting with the loop operator on the second trace in Eq. (15.113) is exactly the same as for the naive Eguchi–Kawai model with an unbroken R^d symmetry, which are described in Sect. 14.4. Also the treatment of the averages of open Wilson loops according to Eqs. (14.68) and (14.69) remains the same. This shows, in particular, that the "volume factor" v for the twisted Eguchi–Kawai model is the same as for the quenched Eguchi–Kawai model if integrals over momentum are regularized in the same way.

Problem 15.7 Calculate $\mathrm{tr}_{\mathcal{H}} D^\dagger(C_{yx})$ for a straight line connecting x and y.

Solution Using the formulas of Sect. 15.4, we obtain in $d = 2$

$$\mathrm{tr}_{\mathcal{H}}\, e^{i(y_1-x_1)\boldsymbol{P}_1+i(y_2-x_2)\boldsymbol{P}_2}$$
$$= \int dp_1\, \langle p_1 |\, e^{i(y_1-x_1)\boldsymbol{P}_1}\, e^{i(y_2-x_2)\boldsymbol{P}_2}\, e^{-i(x_1-y_1)(x_2-y_2)B/2}\, | p_1 \rangle$$
$$= \frac{2\pi}{B}\delta^{(1)}(y_1 - x_1)\,\delta^{(1)}(y_2 - x_2). \tag{15.115}$$

An extension to d dimensions is straightforward:

$$\mathrm{tr}_{\mathcal{H}}\, e^{i(y-x)\boldsymbol{P}} \;=\; \frac{(2\pi)^{d/2}}{\mathrm{Pf}\,(B_{\mu\nu})}\delta^{(d)}(x - y). \tag{15.116}$$

This demonstrates how the averages of open Wilson loops vanish in the continuum twisted Eguchi–Kawai model.

Remark on TEK with fundamental matter

As has already been discussed in Sect. 11.5, matter in the fundamental representation of the gauge group $SU(N)$ can survive the large-N limit of Yang–Mills theory only in the Veneziano limit when the number N_{f} of flavors is proportional to the number N of colors.

Such a limit with $N_{\mathrm{f}} = n_{\mathrm{f}}N$ can be described [Das83] for an integral n_{f} by the following generalization of the twisted Eguchi–Kawai model.

We begin for simplicity with a scalar field on the lattice, whose free action for Hermitian matrices is given by the first line in Eq. (15.31). An interaction with the gauge field is introduced by gauging the first of the two matrix indices of the general complex matrix $\tilde{\varphi}^{ij}$, i.e. by replacing the second Γ_μ by U_μ, which is essentially an exponential of the covariant derivative as has already been pointed out.

The generalized action is given as

$$S = S_{\text{TEK}} + N \operatorname{tr} \left[M \tilde{\varphi}^\dagger \tilde{\varphi} - \sum_\mu \left(\Gamma_\mu \tilde{\varphi}^\dagger U_\mu^\dagger \tilde{\varphi} + \Gamma_\mu^\dagger \tilde{\varphi}^\dagger U_\mu \tilde{\varphi} \right) \right], \qquad (15.117)$$

where S_{TEK} is the action (15.64) describing self-interactions of the gauge field. Repeating the analysis of Sect. 15.2, we see that this model reproduces planar graphs of the d-dimensional Yang–Mills theory with $N_{\text{f}} = N$ species of scalars in the fundamental representation.

We can easily associate an extra index running from 1 to n_{f} to the matrix $\tilde{\varphi}$ in order to have a theory with $N_{\text{f}} = n_{\text{f}}N$ flavors.

A similar generalization of the twisted Eguchi–Kawai model can be made by incorporating fermions which belong to the fundamental representation, thereby describing the Veneziano limit of QCD. Introducing Grassmann-valued matrices $\tilde{\psi}$ and $\bar{\tilde{\psi}}$, we write down the action as

$$S = S_{\text{TEK}} + N \operatorname{tr} \left[M \bar{\tilde{\psi}} \tilde{\psi} - \sum_\mu \left(\Gamma_\mu \bar{\tilde{\psi}} P_\mu^- U_\mu^\dagger \tilde{\psi} + \Gamma_\mu^\dagger \bar{\tilde{\psi}} P_\mu^+ U_\mu \tilde{\psi} \right) \right],$$

$$(15.118)$$

where $P_\mu^\pm$ are the projectors for lattice fermions that are defined in Chapter 8.

The continuum counterparts of Eqs. (15.117) and (15.118) can be easily written down by noting that the interaction with the gauge field can be incorporated by the substitution

$$[P_\mu, \tilde{\varphi}] \quad \rightarrow \quad -A_\mu \tilde{\varphi} - \tilde{\varphi} P_\mu \qquad (15.119)$$

in the free actions (cf. Eq. (15.104) for Hermitian scalars), since A_μ is associated with the covariant derivative in the fundamental representation.

Finally for the action of the continuum twisted Eguchi–Kawai model with fundamental matter we find

$$S = S_{\text{TEK}} + vN \operatorname{tr} \left[m^2 \tilde{\varphi}^\dagger \tilde{\varphi} + \sum_\mu |A_\mu \tilde{\varphi} + \tilde{\varphi} P_\mu|^2 \right] \qquad (15.120)$$

for scalars and

$$S = S_{\text{TEK}} + vN \operatorname{tr} \left[m \bar{\tilde{\psi}} \tilde{\psi} - i \sum_\mu \bar{\tilde{\psi}} \gamma_\mu \left(A_\mu \tilde{\psi} + \tilde{\psi} P_\mu \right) \tilde{\psi} \right] \qquad (15.121)$$

for fermions. Here S_{TEK} is given by Eq. (15.110).

When formulated in terms of operators on the Hilbert space, vN in Eqs. (15.120), (15.121) and (15.110) is to be substituted according to Eq. (15.105).

16

Noncommutative gauge theories

We have seen in the previous chapter that the twisted reduced models reproduce planar graphs of the d-dimensional quantum field theories as $N \to \infty$. However, the twisted reduced models make sense order by order in $1/N^2$. For the continuum twisted reduced models, the topological expansion goes in the parameter $\det(B_{\mu\nu})$.

At finite $B_{\mu\nu}$, the twisted reduced models are mapped [CDS98, AII00] into quantum field theories on noncommutative space characterized by a (dimensional) parameter of noncommutativity $\theta_{\mu\nu} = B_{\mu\nu}^{-1}$. The noncommutative gauge field is no longer matrix-valued as in Yang–Mills theory but noncommutativity of matrices in the reduced models is transformed into noncommutativity of coordinates in the noncommutative gauge theory. The planar limit of ordinary Yang–Mills theory is reproduced at large noncommutativity parameter $\theta_{\mu\nu} \to \infty$, while ordinary quantum electrodynamics is reproduced as $\theta_{\mu\nu} \to 0$.

Noncommutative gauge theories possess a number of remarkable properties. The noncommutative extension of Maxwell's theory is interacting and asymptotically free. The group of noncommutative gauge symmetry is very large and incorporates space-time symmetries, in particular, translation, Lorentz transformation, parity reflection. This restricts a set of observables in noncommutative gauge theory which are built out of both closed and open Wilson loops. At rational values of a (dimensionless) noncommutativity parameter, noncommutative gauge theories on a torus are equivalent to ordinary Yang–Mills theories on a smaller torus with twisted boundary conditions representing the 't Hooft flux.

We begin this chapter by mapping the twisted reduced models into noncommutative theories. Then we discuss some properties of noncommutative scalar and gauge theories including their lattice regularization.

16.1 The noncommutative space

As we have seen in the previous chapter, the twisted reduced models make sense order by order of the topological expansion in the parameter $\det (B_{\mu\nu})$. We start this section by showing how the twisted reduced models are mapped into noncommutative quantum field theories. We simply repeat the consideration of Sect. 15.2 using the continuum operator notation of Sect. 15.4.

Substituting the expansion (15.100) into the action (15.104) of the continuum twisted reduced model and using the orthogonality condition,

$$\frac{(2\pi)^{d/2}}{\mathrm{Pf}\,(B_{\mu\nu})} \, \mathrm{tr}_{\mathcal{H}} \, \boldsymbol{J}_k \boldsymbol{J}_q^\dagger \;=\; (2\pi)^d \delta^{(d)}(k-q), \tag{16.1}$$

we obtain for the kinetic term

$$\begin{aligned}
S^{(2)} &= \frac{1}{2} \int \frac{\mathrm{d}^d k}{(2\pi)^d} \, (k^2 + m^2) \, \varphi(-k)\varphi(k) \\
&= \frac{1}{2} \int \mathrm{d}^d x \Big\{ [\partial_\mu \varphi(x)]^2 + m^2 \varphi^2(x) \Big\}.
\end{aligned} \tag{16.2}$$

Here $\varphi(x)$ is given by Eq. (15.101), i.e. it is related to the operator $\tilde{\varphi}$ by the Weyl transformation. The RHS of Eq. (16.2) is simply the free action for a scalar field in d dimensions.

Let us now repeat the calculation for the cubic self-interaction. Using Eq. (16.1), we find

$$\frac{(2\pi)^{d/2}}{\mathrm{Pf}\,(B_{\mu\nu})} \, \mathrm{tr}_{\mathcal{H}} \, \tilde{\varphi}^3 \;=\; \int \frac{\mathrm{d}^d p}{(2\pi)^d} \int \frac{\mathrm{d}^d q}{(2\pi)^d} \varphi(-p-q)\varphi(p)\varphi(q) \, \mathrm{e}^{-\mathrm{i} p\theta q/2}, \tag{16.3}$$

where $\theta_{\mu\nu} = B_{\mu\nu}^{-1}$. This is a continuum analog of Eq. (15.35).

The RHS of Eq. (16.3) involves the phase factor $\mathrm{e}^{-\mathrm{i} p\theta q/2}$ representing noncommutativity of the generators $\boldsymbol{J}_k$. Relabeling the operators by introducing

$$\boldsymbol{x}_\mu \;=\; -\,\theta_{\mu\nu} \boldsymbol{P}_\nu, \tag{16.4}$$

which obey

$$[\boldsymbol{x}_\mu, \boldsymbol{x}_\nu] \;=\; \mathrm{i}\,\theta_{\mu\nu} \boldsymbol{1}, \tag{16.5}$$

we obtain

$$\boldsymbol{J}_k \;=\; \mathrm{e}^{\mathrm{i} k \boldsymbol{x}} \tag{16.6}$$

and

$$\boldsymbol{J}_k \boldsymbol{J}_q \;=\; \boldsymbol{J}_{k+q} \, \mathrm{e}^{-\mathrm{i} k\theta q/2} \tag{16.7}$$

according to the Baker–Campbell–Hausdorff formula (15.98).

In order to represent the multiplication rule (16.7) by the Fourier-basis functions e^{ikx}, we introduce a *noncommutative* product of functions:

$$f_1(x) \star f_2(x) \overset{\text{def}}{=} f_1(x) \exp\left(\frac{i}{2} \overleftarrow{\partial}_\mu \theta_{\mu\nu} \overrightarrow{\partial}_\nu\right) f_2(x). \tag{16.8}$$

Here $\overleftarrow{\partial}_\mu$ acts on $f_1(x)$ and ∂_ν acts on $f_2(x)$. It is noncommutative but associative, i.e.

$$\left[f_1(x) \star f_2(x)\right] \star f_3(x) = f_1(x) \star \left[f_2(x) \star f_3(x)\right], \tag{16.9}$$

similarly to the product of matrices (operators).

The product (16.8) is called the *star product* or the *Moyal product*. It becomes the ordinary product when $\theta_{\mu\nu} \to 0$ since

$$f_1(x) \star f_2(x) = f_1(x)f_2(x) + \frac{i}{2}\theta_{\mu\nu}\left[\partial_\mu f_1(x)\right]\left[\partial_\nu f_2(x)\right] + \mathcal{O}(\theta^2) \tag{16.10}$$

to the linear order in θ.

Given the star product (16.8), the function $f(x)$ can be viewed as a coordinate-space representation of the operator $\boldsymbol{f}$ to which it is related by the Weyl transform. Whenever we have a product of two operators, its coordinate-space representation is given by the star product of the two functions associated with the operators by the Weyl transform.

In particular, the function x is the coordinate-space representation of the operator $\boldsymbol{x}$ and the commutation relation (16.5) has the coordinate-space representation

$$x_\mu \star x_\nu - x_\nu \star x_\mu = i\theta_{\mu\nu}. \tag{16.11}$$

Equation (16.11) holds as a result of the definition (16.8) with $f_1 = x_\mu$ and $f_2 = x_\nu$.

Similarly, we have

$$e^{ikx} \star e^{iqx} = e^{i(k+q)x}e^{-ik\theta q/2} \tag{16.12}$$

reproducing the coordinate-space representation of Eq. (16.7).

With the aid of the star product, we can represent Eq. (16.3) in the coordinate space as

$$\frac{(2\pi)^{d/2}}{\text{Pf}\,(B_{\mu\nu})}\,\text{tr}_{\mathcal{H}}\,\tilde{\varphi}^3 = \int d^d x\, \varphi(x) \star \varphi(x) \star \varphi(x) \tag{16.13}$$

and similarly for higher interaction terms

$$\frac{(2\pi)^{d/2}}{\text{Pf}\,(B_{\mu\nu})}\,\text{tr}_{\mathcal{H}}\,\tilde{\varphi}^n = \int d^d x\, \overbrace{\varphi(x) \star \cdots \star \varphi(x)}^{n} \tag{16.14}$$

so that

$$\frac{(2\pi)^{d/2}}{\mathrm{Pf}\,(B_{\mu\nu})}\,\mathrm{tr}_{\mathcal{H}}\,\widetilde{V}(\tilde{\varphi}) \;=\; \int \mathrm{d}^d x\, V(\star\varphi(x)). \qquad (16.15)$$

The prescription for writing down the action of the noncommutative theory that comes from the mapping of the operator action (15.104) of the continuum twisted reduced model is obvious. We simply replace products of operators by the star products of their Weyl transforms and substitute the trace over the Hilbert space by the integral over coordinate space according to Eq. (15.106). In fact, the star product (16.8) is defined precisely in the way needed for this prescription to be valid!

One could ask why there is a usual product rather than the star product in the kinetic term (16.2)? The point is that it does not matter what product we write for the integral of a product of two functions: the ordinary product or the star product. It is easy to show that

$$\int \mathrm{d}^d x\, f_1(x) \star f_2(x) \;=\; \int \mathrm{d}^d x\, f_1(x) f_2(x) \;=\; \int \mathrm{d}^d x\, f_2(x) \star f_1(x)$$

$$(16.16)$$

for functions decreasing with their derivatives at infinity as a consequence of the definition (16.8). This is a counterpart of the cyclic symmetry of the trace.[*]

Finally we obtain the following action:

$$S[\varphi] \;=\; \int \mathrm{d}^d x \left[\frac{1}{2} \partial_\mu \varphi(x) \star \partial_\mu \varphi(x) + V(\star\varphi(x)) \right]. \qquad (16.17)$$

The parameter of noncommutativity $\theta_{\mu\nu}$ enters the action via the star product (16.8).

The action (16.17) is associated with a *noncommutative scalar theory*. In the limit of $\theta_{\mu\nu} \to 0$, it reproduces the ordinary theory of a single scalar field. In the opposite limit of $\theta_{\mu\nu} \to \infty$, only planar graphs survive and the noncommutative scalar theory is equivalent to the theory of a matrix-valued scalar field with the action (14.20) at $N = \infty$. This can be easily shown directly using the theorem of Sect. 15.2 which was considered for noncommutative quantum field theories in [Fil96], where the phase factor associated with a generic nonplanar diagram was calculated.

Problem 16.1 Prove associativity of the star product (16.8).

Solution Equation (16.9) can easily be proven by expanding in θ. To the quadratic order in θ, we are to verify that

$$\theta_{\mu\nu}\theta_{\lambda\rho}\,(\partial_\mu f_1)\,(\partial_\lambda\partial_\nu f_2)\,(\partial_\rho f_3) \;=\; \theta_{\mu\nu}\theta_{\lambda\rho}\,(\partial_\mu f_1)\,(\partial_\nu\partial_\lambda f_2)\,(\partial_\rho f_3) \qquad (16.18)$$

which is true since ∂_μ commute. It is similar to the next orders.

---

[*] To avoid confusion, let us mention that Eq. (16.16) is *not* valid when f_1 and f_2 are not decreasing at infinity, say, for $f_1 = x_\mu$ and $f_2 = x_\nu$. The trace of a commutator is then reproduced as a surface term.

If ∂_μ were noncommutative derivatives (say, like covariant derivatives in an external electromagnetic field), the star product would not be, generally speaking, associative.

Remark on the Weyl transformation

The Weyl transformation from operators to functions and vice versa can be written conveniently with the aid of the operator-valued function

$$\boldsymbol{\Delta}(x) = \int \frac{d^d k}{(2\pi)^d} e^{-ikx} \boldsymbol{J}_k = \int \frac{d^d k}{(2\pi)^d} e^{ik(\boldsymbol{x}-x)} \tag{16.19}$$

which is an operator counterpart of Eq. (15.40).

We then represent the Weyl transformation by

$$\boldsymbol{f} = \int d^d x \, \boldsymbol{\Delta}(x) f(x) \tag{16.20}$$

and

$$f(x) = \frac{(2\pi)^{d/2}}{\mathrm{Pf}\,(B_{\mu\nu})} \mathrm{tr}_{\mathcal{H}}\Big[\boldsymbol{f}\,\boldsymbol{\Delta}(x)\Big]. \tag{16.21}$$

Remember that

$$\frac{1}{\mathrm{Pf}\,(B_{\mu\nu})} = \mathrm{Pf}\,(\theta_{\mu\nu}). \tag{16.22}$$

Note that $\boldsymbol{\Delta}(x)$ becomes an ordinary delta-function as $\theta \to 0$ when x_μ commute.

Problem 16.2 Derive Eq. (16.8) by calculating the Weyl transform of the product of two operators.

Solution The star product can be defined via the Weyl transform of the product of two operators:

$$f_1(x) \star f_2(x) \stackrel{\mathrm{def}}{=} \frac{(2\pi)^{d/2}}{\mathrm{Pf}\,(B_{\mu\nu})} \mathrm{tr}_{\mathcal{H}}\Big[\boldsymbol{f}_1 \boldsymbol{f}_2 \boldsymbol{\Delta}(x)\Big]. \tag{16.23}$$

Inserting Eq. (16.20) and using Eqs. (16.7) and (16.1), we obtain

$$
\begin{aligned}
f_1(x) \star f_2(x) &= \int d^d y \, d^d z \, f_1(y) f_2(z) \frac{(2\pi)^{d/2}}{\mathrm{Pf}\,(B_{\mu\nu})} \mathrm{tr}_{\mathcal{H}}\Big[\boldsymbol{\Delta}(x)\boldsymbol{\Delta}(y)\boldsymbol{\Delta}(z)\Big] \\
&= \int d^d y \, d^d z \, f_1(y) f_2(z) \int \frac{d^d k}{(2\pi)^d} \frac{d^d q}{(2\pi)^d} e^{i(x-y)k + i(x-z)q - ik\theta q/2} \\
&= \int d^d y \, d^d z \, f_1(y) f_2(z) \frac{1}{\pi^d \,|\det \theta|} e^{-2i(x-y)\theta^{-1}(x-z)}. \tag{16.24}
\end{aligned}
$$

This is just the representation of the operator in Eq. (16.8) via a kernel.

The associativity of the star product is clear from the integral representation (16.24) as a consequence of the associativity of the matrix (operator) multiplication.

Remark on the Moyal bracket

The term linear in θ in Eq. (16.10) for the star product is in $d = 2$ just the Poisson bracket. It is a "semiclassical" limit of the Moyal bracket of the two functions:

$$\{f, g\}_{\mathrm{MB}} \overset{\mathrm{def}}{=} -\mathrm{i}\,(f \star g - g \star f). \tag{16.25}$$

The Moyal bracket represents the Weyl transform of the commutator of two operators. It allows one to construct quantum mechanics without operators using instead functions on noncommutative phase space. It is known as the Weyl–Wigner–Moyal approach [Wey27, Wig32, Moy49] to quantum mechanics. Generically, the Moyal bracket appears in various physical problems whenever the large-N limit of a matrix commutator is represented by functions. It was associated [FZ89] with the commutator (15.55) and discussed in early works [Li96, Far97] on the star product in matrix theory.

Remark on nonlocality of the star product

A nonlocal structure of the star product is obvious from the integral representation (16.24), while part of the integration region is suppressed by oscillations of the kernel. If f_1 and f_2 has support on a small region of size ϵ, their star product $f_1(x) \star f_2(x)$ is nonvanishing over a larger region of size $|\theta|/\epsilon$. In particular, we find

$$\delta^{(d)}(x) \star \delta^{(d)}(x) = \frac{1}{\pi^d\,|\det\theta|} \tag{16.26}$$

for $\epsilon \to 0$.

Remark on the double scaling limit

It has been recognized recently that the continuum noncommutative quantum field theories can be obtained as a large-N limit of the twisted reduced models.

In the previous chapter we considered the limit of the twisted reduced model when $N \to \infty$ at fixed a. Then $\theta \to \infty$ according to Eq. (15.38) and this limit is associated [EN83, GO83b] with the 't Hooft limit of lattice matrix theory, where only planar diagrams survive.

Alternatively, one can approach the continuum limit of the twisted reduced models keeping θ fixed as $N \to \infty$, which requires $a \sim 1/\sqrt{L} \sim N^{-1/d}$. The period $\ell = aL \sim \sqrt{L} \sim N^{1/d} \to \infty$ in this limit so noncommutative theories on $\mathbb{R}^d$ are reproduced [AII00]. This limit is of the same type as the double scaling limit which is considered in Sect. 13.5 for the matrix models.

In Sect. 16.6 we shall describe how the original construction of [CDS98] for a torus can also be reproduced from the twisted reduced models.

16.2 The $U_\theta(1)$ gauge theory

A noncommutative gauge theory can be constructed from the continuum version of the twisted Eguchi–Kawai model described in Sect. 15.5. We should only remember that the operator $\boldsymbol{A}_\mu$ represents the reduction of the covariant derivative, as has already been pointed out, so we first substitute $\boldsymbol{A}_\mu = -\boldsymbol{P}_\mu + \tilde{\boldsymbol{A}}_\mu$ and then identify the noncommutative gauge field $\mathcal{A}_\mu(x)$ with the Weyl transform (W.t.) of the operator $\tilde{\boldsymbol{A}}_\mu$.

Proceeding in this way, we have

$$
[\boldsymbol{A}_\mu, \boldsymbol{A}_\nu] + \mathrm{i}B_{\mu\nu}\mathbf{1} = -\left[\boldsymbol{P}_\mu, \tilde{\boldsymbol{A}}_\nu\right] + \left[\boldsymbol{P}_\nu, \tilde{\boldsymbol{A}}_\mu\right] + \left[\tilde{\boldsymbol{A}}_\mu, \tilde{\boldsymbol{A}}_\nu\right]
$$
$$
\xrightarrow{\text{W.t.}} \mathrm{i}\mathcal{F}_{\mu\nu}(x), \tag{16.27}
$$

where $\mathcal{F}$ denotes the noncommutative field strength

$$
\mathcal{F}_{\mu\nu} = \partial_\mu \mathcal{A}_\nu - \partial_\nu \mathcal{A}_\mu - \mathrm{i}\left(\mathcal{A}_\mu \star \mathcal{A}_\nu - \mathcal{A}_\nu \star \mathcal{A}_\mu\right). \tag{16.28}
$$

It appeared as the Weyl transform of the LHS of Eq. (16.27).

Using Eqs. (16.27) and (15.106), we rewrite the action (15.110) as

$$
S[\mathcal{A}] = \frac{1}{4\lambda} \int \mathrm{d}^d x \, \mathcal{F}^2, \tag{16.29}
$$

where $\lambda = g^2 N$ coincides with the 't Hooft coupling of the twisted Eguchi–Kawai model. This action determines the noncommutative $U_\theta(1)$ gauge theory [CR87].

The action (16.29) involves cubic and quartic interactions of $\mathcal{A}_\mu$. This is quite the same as Yang–Mills theory! For this reason the noncommutative gauge theory is often called the noncommutative Yang–Mills theory (NCYM).

The action (16.29) is invariant under the gauge transformation

$$
\mathcal{A}_\mu \xrightarrow{\text{g.t.}} \Omega \star \mathcal{A}_\mu \star \Omega^* + \mathrm{i}\Omega \star \partial_\mu \Omega^* \tag{16.30}
$$

which is related to the Weyl transform of the gauge transformation (14.56) for the twisted Eguchi–Kawai model, where $\Omega(x)$ is the Weyl transform of $\tilde{\Omega}$. Correspondingly, the complex conjugate function $\Omega^*(x)$ is the Weyl transform of $\tilde{\Omega}^\dagger$. The transformation (16.30) is often termed the *star gauge* transformation.

Note that

$$(f_1 \star f_2)^* = f_1^* \exp\left(-\frac{i}{2}\overleftarrow{\partial}_\mu \theta_{\mu\nu} \overrightarrow{\partial}_\nu\right) f_2^* = f_2^* \star f_1^* \tag{16.31}$$

owing to the definition (16.8) of the star product. This represents the rule for Hermitian conjugation of the product of two operators.

The function $\Omega(x)$ in Eq. (16.30) is *star unitary*, i.e. it obeys

$$\Omega \star \Omega^* = 1 = \Omega^* \star \Omega. \tag{16.32}$$

This is, of course, just the Weyl transform of the unitarity condition for the operator $\widetilde{\Omega}$.

A star unitary function can be constructed via the *star exponential*

$$\Omega(x) = e_\star^{i\alpha(x)}, \qquad \Omega^*(x) = e_\star^{-i\alpha(x)}, \tag{16.33}$$

where

$$e_\star^{i\alpha(x)} \stackrel{\text{def}}{=} \sum_{n=0}^{\infty} \frac{i^n}{n!} \overbrace{\alpha(x) \star \cdots \star \alpha(x)}^{n} \tag{16.34}$$

is defined via the Taylor expansion with the ordinary product substituted by the star product and α is real. This is simply the Weyl transform of the exponential of i times a Hermitian operator $\boldsymbol{\alpha}$.

Problem 16.3 Prove that the action (16.29) is invariant under the star gauge transformation (16.30).

Solution The noncommutative field strength (16.28) is changed under the star gauge transformation (16.30) as

$$\mathcal{F}_{\mu\nu} \xrightarrow{\text{g.t.}} \Omega \star \mathcal{F}_{\mu\nu} \star \Omega^*. \tag{16.35}$$

Correspondingly, we have

$$\int \mathrm{d}^d x \, \mathcal{F}^2 \xrightarrow{\text{g.t.}} \int \mathrm{d}^d x \, \Omega \star \mathcal{F}^2 \star \Omega^* = \int \mathrm{d}^d x \, \Omega^* \star \Omega \star \mathcal{F}^2 = \int \mathrm{d}^d x \, \mathcal{F}^2 \tag{16.36}$$

as a result of Eqs. (16.16) and (16.32).

Note that only the integral of $\mathcal{F}_{\mu\nu}^2$ over space is star gauge invariant rather than $\mathcal{F}_{\mu\nu}^2$ itself.

The group of the star gauge transformations (16.30) of the noncommutative gauge theory is much larger than the group of the gauge transformations (5.4) of ordinary Yang–Mills theory and contains some of the space-time symmetries.

Let us illustrate this statement by the simplest example of a star gauge transformation given by the star unitary function

$$\Omega(x) = e^{-i\eta\theta^{-1}x}. \tag{16.37}$$

We have

$$\Omega(x) \star \varphi(x) \star \Omega^*(x) \;=\; \varphi(x+\eta) \qquad (16.38)$$

which means that the star gauge transformation with the function (16.37) results in a translation by the d-vector η_μ.

We have considered here the star gauge transformation of a field $\varphi(x)$ which is uniformly transformed. This could be a scalar field (in the adjoint representation), the field strength $\mathcal{F}_{\mu\nu}(x)$ or the covariant derivative $i\partial_\mu + \mathcal{A}_\mu(x)$.

A similar formula can be written down for a Lorentz rotation of a scalar field in a noncommutative plane, say, the $(1,2)$-plane. It is generated by the star gauge transformation with the star unitary function

$$\Omega(x) \;=\; \sqrt{1+\alpha^2\theta^2}\, e^{i\alpha(x_1^2+x_2^2)}, \qquad (16.39)$$

where $\theta = \theta_{12}$. Then we obtain

$$\Omega(x_1, x_2) \star \varphi(x_1, x_2, \ldots) \star \Omega^*(x_1, x_2) \;=\; \varphi(x_1', x_2', \ldots) \qquad (16.40)$$

with

$$\left.\begin{aligned} x_1' &= \cos\gamma\, x_1 + \sin\gamma\, x_2\,, \\ x_2' &= -\sin\gamma\, x_1 + \cos\gamma\, x_2\,, \end{aligned}\right\} \qquad (16.41)$$

which is a rotation in the $(1,2)$-plane through the angle $\gamma = 2\arctan\alpha\theta$.

Finally, the parity reflection is represented by the star gauge transformation with the star unitary function

$$\Omega(x) \;=\; \pi^{d/2}\, \mathrm{Pf}\,(\theta_{\mu\nu})\, \delta^{(d)}(x) \qquad (16.42)$$

(cf. Eq. (16.26)). It acts as

$$\Omega(x) \star \varphi(x) \star \Omega^*(x) \;=\; \varphi(-x)\,. \qquad (16.43)$$

We shall see later in Sect. 16.5 how these properties of the star gauge transformation restrict observables in noncommutative gauge theory.

Problem 16.4 Prove Eqs. (16.40) and (16.43) for the Ωs given, respectively, by Eqs. (16.39) and (16.42).

Solution It is convenient to use the integral representation (16.24) of the star product. For the star product of three functions, we have

$$f_1(x) \star f_2(x) \star f_3(x)$$
$$= \frac{1}{\pi^d\,|\det\theta|} \int d^d\xi\, d^d\eta\, e^{-2i\xi\theta^{-1}\eta} f_1(x+\xi) f_2(x+\xi+\eta) f_3(x+\eta)\,. \qquad (16.44)$$

Substituting $f_1 = f_3 = \Omega$ given by Eq. (16.42) into Eq. (16.44), we obtain Eq. (16.43). Choosing there $\varphi(x) = 1$, we prove that $\Omega(x)$ given by Eq. (16.42) is star unitary (cf. Eq. (16.26)).

Analogously, it is easy to derive Eq. (16.40) substituting for $\varphi(x)$ a Fourier decomposition (15.101) and performing the Gaussian integral over ξ and η in Eq. (16.44) for $f_1 = \Omega$, $f_3 = \Omega^*$ with Ω given by Eq. (16.39). Again, it is easy to show that this $\Omega(x)$ is star unitary choosing $\varphi(x) = 1$.

Remark on the Lorentz invariance

The presence of a tensor $\theta_{\mu\nu}$ in Eq. (16.5) superficially breaks the Lorentz invariance for $d > 2$. We can always represent $\theta_{\mu\nu}$ in a canonical skew-diagonal form

$$
\theta_{\mu\nu} \;=\; \begin{pmatrix} 0 & -\theta_1 & & & \\ +\theta_1 & 0 & & & \\ & & \ddots & & \\ & & & 0 & -\theta_{d/2} \\ & & & +\theta_{d/2} & 0 \end{pmatrix}
\tag{16.45}
$$

by a unitary transformation. In noncommutative gauge theory, this unitary matrix can be gauged away by a star gauge transformation of the type (16.39). Therefore, the only dependence on $\theta_{\mu\nu}$ is via $\theta_1, \ldots, \theta_{d/2}$ and the Lorentz invariance is preserved.

Remark on the $U_\theta(n)$ gauge theory

An extension of the results of this section to the group $U_\theta(n)$ is obvious. The noncommutative gauge field becomes an $n \times n$ matrix-valued field $\mathcal{A}_\mu^{ij}(x)$. The field strength is again given by Eq. (16.28) since the ordering in matrix multiplication is consistent with the ordering in the star product. The action of the noncommutative $U_\theta(n)$ gauge theory is

$$
S[\mathcal{A}] \;=\; \frac{1}{4g^2} \int d^d x \, \mathrm{tr}_{(n)} \, \mathcal{F}^2,
\tag{16.46}
$$

where $\mathrm{tr}_{(n)}$ denotes the $n \times n$ matrix trace.

The noncommutative $U_\theta(n)$ gauge theory can be obtained from the twisted Eguchi–Kawai model by choosing a more general twist with $n_i = nL^{d/2-1}$, which is described at the end of Problem 15.3 on p. 354.

16.3 One-loop renormalization

One of the main original motivations for studying quantum field theory on noncommutative spaces was the expectation that noncommutativity provides an ultraviolet regularization. We shall see in this section that this is not quite the case, while ultraviolet properties of noncommutative theories are somewhat better than those of their ordinary counterparts.

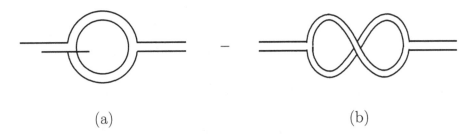

(a)   (b)

Fig. 16.1.   One-loop correction to the gauge-field propagator in the noncommutative $U_\theta(1)$ gauge theory. Diagram (a) is planar and has logarithmic ultraviolet
divergence. Diagram (b) is nonplanar and converges for $\theta \neq 0$. The diagrams involve momentum integrals shown in Eqs. (16.47) and (16.48). The contribution
of diagram (b) is suppressed as $\theta \to \infty$ according to Eq. (16.50).

We start from the noncommutative $U_\theta(1)$ gauge theory described in
the previous section, the Feynman diagrams of which have the form of
ribbon graphs for a $U(N)$ Yang–Mills theory. We shall use for them the
double-line notation similar to that of Sect. 11.1.

In order to construct the perturbation theory, we should first treat the
gauge symmetry properly by adding the gauge-fixing and ghost terms to
the action (16.29). They can be obtained easily once again by the Weyl
transformation from those in the twisted Eguchi–Kawai model which are
simply the reduction of the standard gauge-fixing and ghost term to zero
dimensions. In the one-loop calculation of Sect. 14.5, they are given by
Eq. (14.75) to quadratic order.

The one-loop corrections to the propagator are depicted in Fig. 16.1.
The diagrams are the same as for a $U(N)$ Yang–Mills theory rather than
$SU(N)$. The diagram in Fig. 16.1a is planar and that in Fig. 16.1b is
nonplanar. The latter is not usually considered in the ordinary $SU(N)$
Yang–Mills theory, since it is associated with propagation of diagonal
elements $\langle \mathcal{A}_\mu^{ii}(x)\mathcal{A}_\nu^{jj}(y)\rangle$, while the former describes an off-diagonal propagator $\langle \mathcal{A}_\mu^{ij}(x)\mathcal{A}_\nu^{ji}(y)\rangle$ with $i \neq j$.

The relative sign of the two diagrams in Fig. 16.1 is minus since the
relative sign of the two terms in the commutator of $\mathcal{A}_\mu$ and $\mathcal{A}_\nu$ is minus.

For this reason, the two diagrams cancel each other in the $\theta \to 0$
limit associated with ordinary commutative Maxwell theory where there
is no interaction between photons. They do not cancel however in the
noncommutative case where the contributions of planar and nonplanar
diagrams are different.

There is nothing special about the planar diagram in Fig. 16.1a, the
contribution of which is the same as in ordinary Yang–Mills theory. Accounting for ghosts, we obtain in the Feynman gauge for the self-energy

correction

$$\text{Fig. 16.1a} = \frac{20}{3}\lambda \int^{\Lambda} \frac{d^4k}{(2\pi)^4} \frac{1}{k^2(p-k)^2} \approx \frac{5}{12\pi^2}\lambda \ln\frac{\Lambda^2}{p^2} \qquad (16.47)$$

in $d = 4$ with logarithmic accuracy.

Similarly, we obtain for the nonplanar diagram in Fig. 16.1b:

$$\text{Fig. 16.1b} = -\frac{20}{3}\lambda \int^{\Lambda} \frac{d^4k}{(2\pi)^4} \frac{e^{ip\theta k}}{k^2(p-k)^2} \approx -\frac{5}{12\pi^2}\lambda \ln\frac{\Lambda_{\text{eff}}^2}{p^2},$$

$$(16.48)$$

where

$$\Lambda_{\text{eff}}^{-2} = |\theta p|^2 + \Lambda^{-2} \qquad (16.49)$$

and we have assumed that $|p||\theta p| \ll 1$ for the logarithmic domain of integration to exist.

In the opposite limit of $|p||\theta p| \gg 1$, the integrand in Eq. (16.48) oscillates rapidly and the integral vanishes as

$$\text{Fig. 16.1b} = -\frac{20}{3}\lambda \int^{\Lambda} \frac{d^4k}{(2\pi)^4} \frac{e^{ip\theta k}}{k^2(p-k)^2} \sim \frac{1}{p^{2d}\det(\theta_{\mu\nu})}. \qquad (16.50)$$

This last formula is in accord with the general consideration of Sect. 15.2 (cf. Eq. (15.39)).

At arbitrary finite θ, the integral in Eq. (16.48) is convergent at $|k| \sim 1/|\theta p|$ so we can disregard the Λ-dependence of Λ_{eff}. Consequently, only the planar graph in Fig. 16.1a has an ultraviolet logarithmic divergence.

A very important consequence of these results is that only planar graphs contribute to the Gell-Mann–Low function of the noncommutative $U_\theta(1)$ gauge theory. For this reason it coincides [VG99, CDP00, MS99, She99, KW00] with that for ordinary Yang–Mills theory in the 't Hooft limit.

Another intriguing property of noncommutative theories is a mixing of ultraviolet and infrared scales [MRS99]. It has already been seen from Eq. (16.48) for which the RHS becomes singular at $\Lambda = \infty$ as $|\theta p| \to 0$, which plays the role of an effective ultraviolet cutoff in the coordinate space. Therefore turning on θ replaces the ultraviolet divergence by a singular infrared behavior. In other words, the infinite-cutoff limit $\Lambda \to \infty$ does not commute with the low-momentum limit $p \to 0$.

The ordinary $U(1)$ theory, where the coupling is not renormalized, is recovered at very small momenta $p \lesssim 1/\Lambda|\theta|$, of the order of the inverse momentum cutoff for finite θ, which are associated with very large distances $\sim \Lambda|\theta|$. This result is quite surprising since naively we would expect from Eq. (16.5) that the ordinary theory would be recovered at distances of the order of $\sqrt{|\theta|}$. We shall return to this aspect of the UV/IR mixing in the next section when discussing noncommutative quantum electrodynamics.

Remark on the UV/IR mixing

If we introduce an infrared cutoff by putting a noncommutative theory in a box of size ℓ, the minimal value of momentum is $p_{\min} = 2\pi/\ell$. It is related to the ultraviolet cutoff $p_{\max} = \Lambda$ by

$$p_{\min}|\theta|p_{\max} = 2\pi \tag{16.51}$$

because the position operator x_μ and the momentum operator P_ν are related by Eq. (16.4).

For the lattice regularization which is described in Sect. 15.2, we have $p_{\min} = 2\pi/aL$, $|\theta| = a^2L/\pi$, $p_{\max} = \pi/a$ and Eq. (16.51) is obviously satisfied.

16.4 Noncommutative quantum electrodynamics

A noncommutative extension of quantum electrodynamics (NCQED) can be constructed by the Weyl transformation of the continuum twisted Eguchi–Kawai model with fermions in the fundamental representation, the action of which is given by Eq. (15.121).

The action of noncommutative quantum electrodynamics is

$$S_{\text{NCQED}} = \int d^d x \left[\frac{1}{4\lambda} \mathcal{F}^2 + \bar{\psi}\gamma_\mu(\partial_\mu - i\mathcal{A}_\mu\star)\psi + m\bar{\psi}\psi \right], \tag{16.52}$$

where ψ and $\bar{\psi}$ are in the fundamental representation of the star gauge group, contrary to the adjoint scalar field φ described previously in this chapter. The fields ψ and $\bar{\psi}$ are associated with "noncommutative" electrons and positrons.

Under the star gauge transformation when the gauge field is changed according to Eq. (16.30), they are transformed by

$$\psi \xrightarrow{\text{g.t.}} \Omega \star \psi, \qquad \bar{\psi} \xrightarrow{\text{g.t.}} \bar{\psi} \star \Omega^*. \tag{16.53}$$

The action (16.52) is invariant under the star gauge transformation (16.30) and (16.53).

The ordinary quantum electrodynamics with $e^2 = \lambda$ is obviously reproduced as $\theta \to 0$.

Feynman graphs of noncommutative quantum electrodynamics recall those described in Sect. 11.4 for a $U(N)$ Yang–Mills theory with quarks. It is most important that the vertex for emitting the gauge field by the fundamental matter is oriented owing to the presence of the noncommutative product. The gauge field can be emitted only to one side of the fermionic line but not to the other.

Table 16.1.   Limits of noncommutative $U_\theta(1)$ gauge theory at various distances.

Distances	Theories
$r \ll \sqrt{\theta}$	Veneziano limit of QCD
$\sqrt{\theta} \lesssim r \ll \theta\Lambda$	Noncommutative $U_\theta(1)$ gauge theory
$\theta\Lambda \lesssim r$	Quantum electrodynamics

For this reason only the diagram in Fig. 11.14a on p. 230 describes the one-loop correction to the gauge-field propagator coming from fermions. This diagram is planar and there are no nonplanar diagrams with fermionic loops to this order.

This allows us to immediately conclude that the one-loop Gell-Mann–Low function of noncommutative quantum electrodynamics coincides with that (11.83) of multicolor QCD in the Veneziano limit. Given $n_{\rm f}$ species of the fermions, the one-loop Gell-Mann–Low function of noncommutative quantum electrodynamics is [Hay00]

$$\mathcal{B}(\lambda) \;=\; \frac{\lambda^2}{12\pi^2}\left(-11 + 2n_{\rm f}\right). \qquad (16.54)$$

This formula shows that noncommutative quantum electrodynamics is asymptotically free at small distances for $n_{\rm f} \leq 5$, in contrast to ordinary QED.

Singular infrared behavior in noncommutative quantum electrodynamics is the same as in the pure noncommutative $U_\theta(1)$ gauge theory since there are no nonplanar diagrams with a fermionic loop. The usual Gell-Mann–Low function of QED is therefore reproduced at very large distances $r \gtrsim \theta\Lambda$.

To say it once again, the $\theta \to 0$ limit is not interchangeable with the $\Lambda \to \infty$ limit. Ordinary QED is reproduced for all distances when $\theta \to 0$ at fixed Λ.

In the opposite limit of $\theta \to \infty$, noncommutative quantum electrodynamics is equivalent to multicolor QCD in the Veneziano limit when the number of flavors in QCD is $N_{\rm f} = n_{\rm f} N$. This is because only the same planar diagrams survive in both cases. We have already pointed out this property in the Remark in Sect. 15.5 when describing the twisted Eguchi–Kawai model with matter in the fundamental representation. It is utilized to obtain the one-loop Gell-Mann–Low function (16.54).

These results are summarized in Table 16.1.

Remark on large but finite Λ

The value of the cutoff Λ in quantum electrodynamics (or φ^4-theory) cannot be infinite because of its "triviality" dictated by the positive sign of the Gell-Mann–Low function. The renormalized charge would vanish as $\Lambda \to \infty$. Therefore, the theory cannot be fundamental, but rather is an effective theory applicable down to the distances $\sim 1/\Lambda$ where new degrees of freedom become essential. The property of renormalizability tells us that nothing depends on this scale so the effective theory is self-consistent at large distances.

A standard way to cure the "triviality" of quantum electrodynamics is to embed it (or, strictly speaking, the $SU(2) \otimes U(1)$ electroweak theory) into an asymptotically free theory with a compact gauge group at a grand unified scale.

Remark on phenomenology in NCQED

The current experimental bound on the value of θ in our world is $\theta < (10 \text{ TeV})^{-2}$. Phenomenological consequences of noncommutative quantum electrodynamics are discussed, in particular, in the recent papers [BGH01, CHK01, HPR00, Mat01, MPR00].

16.5 Wilson loops and observables

Observables in noncommutative gauge theory are to be invariant under the star gauge transformation. As has been mentioned already in Sect. 16.2, the star gauge invariance strongly restricts the allowed set of observables.

Just as in ordinary Yang–Mills theory, observables can be expressed via the Wilson loops. The standard way to derive proper formulas is to integrate over fundamental matter fields by performing the Gaussian path integral. This strategy can be repeated for noncommutative gauge theory. We describe in this section what kinds of Wilson loops then emerge.

We first define a path-dependent phase factor associated with parallel transport from the point x to the point y in an external gauge field $\mathcal{A}_\mu(x)$. The analogy with Yang-Mills theory prompts one to define

$$\mathcal{U}(C_{yx}) \stackrel{\text{def}}{=} \prod_{\xi \,:\, x+\xi \in C_{yx}} \star \Big[1 + \mathrm{i}\, \mathrm{d}\xi^\mu \mathcal{A}_\mu(x + \xi)\Big], \qquad (16.55)$$

where the product on the RHS is the star product with respect to x, $\xi(0) = 0$, $\xi(1) \equiv \eta = y - x$ is a d-vector pointing from the initial point x of the contour to its final point y, and the ordering is along the contour C_{yx}.

Under the star gauge transformation (16.30), $\mathcal{U}(C_{yx})$ is changed as

$$\mathcal{U}(C_{yx}) \xrightarrow{\text{g.t.}} \Omega(y) \star \mathcal{U}(C_{yx}) \star \Omega^*(x) \tag{16.56}$$

quite similarly to the phase factor in Yang–Mills theory. The property (16.56) shows that $\mathcal{U}(C_{yx})$ is indeed a parallel transporter.

This analogy with Yang–Mills theory can be made precise by returning to the continuum twisted Eguchi–Kawai model and noting that $\mathcal{U}(C_{yx})$ is the Weyl transform of the product

$$\boldsymbol{D}^\dagger(C_{\eta 0}) \, \boldsymbol{U}(C_{\eta 0}) \xrightarrow{\text{W.t.}} \mathcal{U}(C_{yx}), \tag{16.57}$$

where $\boldsymbol{D}(C_{\eta 0})$ is given by Eq. (15.91) and we have denoted

$$\boldsymbol{U}(C_{\eta 0}) = \boldsymbol{P} \, \mathrm{e}^{\mathrm{i}\int_{C_{yx}} \mathrm{d}\xi^\mu \boldsymbol{A}_\mu} \tag{16.58}$$

to emphasize that it does not depend on the position of the initial point x of the contour since $\boldsymbol{A}_\mu$ is constant.

Similarly, Eq. (16.56) is related to the Weyl transform of the operator formula

$$\boldsymbol{U}(C_{\eta 0}) \xrightarrow{\text{g.t.}} \Omega \boldsymbol{U}(C_{\eta 0}) \, \Omega^\dagger \tag{16.59}$$

which resulted from the unitary transformation

$$\boldsymbol{A}_\mu \xrightarrow{\text{g.t.}} \Omega \boldsymbol{A}_\mu \Omega^\dagger. \tag{16.60}$$

We demonstrate this by explicit formulas for the lattice regularization in Problem 16.8.

Multiplying by $\exp\left(\mathrm{i}\eta\theta^{-1}\boldsymbol{x}\right)$ from the left, Eq. (16.57) can be represented equivalently as

$$Z(C_{\eta 0}) \, \boldsymbol{U}(C_{\eta 0}) \xrightarrow{\text{W.t.}} \mathrm{e}^{\mathrm{i}\eta\theta^{-1}x} \star \mathcal{U}(C_{yx}), \tag{16.61}$$

where a c-number phase factor

$$Z(C_{\eta 0}) = \mathrm{e}^{\mathrm{i}\eta\theta^{-1}\boldsymbol{x}} \boldsymbol{D}^\dagger(C_{\eta 0}) \tag{16.62}$$

resulted from the difference between the path ordering of operators in $\boldsymbol{D}^\dagger(C_{\eta 0})$, given by Eqs. (15.91) and (16.4), and the symmetric ordering in $\exp\left(\mathrm{i}\eta\theta^{-1}\boldsymbol{x}\right)$. For a straight line, the difference disappears and we have $Z(C_{\eta 0}) = 1$.

In Yang–Mills theory, the trace of a phase factor for a closed loop is gauge invariant. Since the trace over the Hilbert space is substituted by the integral according to Eq. (15.106), we define

$$\mathcal{W}_{\text{clos}}(C) = \int \mathrm{d}^d x \, \mathcal{U}(C_{xx}) \tag{16.63}$$

which is star gauge invariant as can be easily seen using Eq. (16.16). This determines closed Wilson loops in noncommutative gauge theory.

The role of the integration over the initial point x of the contour C_{xx} is to parallel transport the contour over the space. Since an average over quantum fluctuations of the field $\mathcal{A}_\mu$ is invariant under translation, the (normalized) average of the closed Wilson loop is given simply by

$$W_{\mathrm{NC}}(C) \;=\; \frac{1}{V}\langle \mathcal{W}_{\mathrm{clos}}(C)\rangle \;=\; \langle \mathcal{U}(C_{xx})\rangle. \qquad (16.64)$$

It recovers the average of the closed Wilson loop in Maxwell's theory as $\theta \to 0$.

Quite surprisingly there exists yet another kind of star gauge-invariant object in noncommutative gauge theory – open Wilson loops. They are given by [IIK00]

$$\mathcal{W}_{\mathrm{open}}(C_{\eta 0}) \;=\; \int \mathrm{d}^d x\, \mathcal{U}(C_{yx})\, \mathrm{e}^{\mathrm{i}\eta_\mu \theta^{-1}_{\mu\nu} x_\nu}, \qquad (16.65)$$

where the integration over x translates the contour as a whole so $\eta = y - x$ does not change under the translation. The closed Wilson loop (16.63) corresponds to $y = x$ (or $\eta = 0$) in Eq. (16.65).

Problem 16.5 Show that the open Wilson loop (16.65) is star gauge invariant.

Solution The star gauge invariance of the open Wilson loop in noncommutative gauge theory can be shown as follows:

$$\int \mathrm{d}^d x\, \mathcal{U}(C_{yx})\, \mathrm{e}^{\mathrm{i}\eta \theta^{-1} x}$$

$$\xrightarrow{\text{g.t.}} \int \mathrm{d}^d x\, \Omega(x+\eta) \star \mathcal{U}(C_{yx}) \star \Omega^*(x)\, \mathrm{e}^{\mathrm{i}\eta \theta^{-1} x}$$

$$\overset{(16.16)}{=} \int \mathrm{d}^d x\, \mathcal{U}(C_{yx})\, \Omega^*(x) \star \mathrm{e}^{\mathrm{i}\eta \theta^{-1} x} \star \Omega(x+\eta)$$

$$\overset{(16.38)}{=} \int \mathrm{d}^d x\, \mathcal{U}(C_{yx})\, \mathrm{e}^{\mathrm{i}\eta \theta^{-1} x} \star \Omega^*(x+\eta) \star \Omega(x+\eta)$$

$$\overset{(16.32)}{=} \int \mathrm{d}^d x\, \mathcal{U}(C_{yx})\, \mathrm{e}^{\mathrm{i}\eta \theta^{-1} x}. \qquad (16.66)$$

The noncommutative Wilson loop is simply related to the integral over space of the Weyl transform (16.61) with $\boldsymbol{U}(C_{\eta 0})$ given by the operator expression (16.58) for the (open or closed) contour $C_{\eta 0}$. Since the trace over the Hilbert space and the integral are related by Eq. (15.106), the open Wilson loop (16.65) is simply proportional to the trace of $\boldsymbol{U}(C_{\eta 0})$ for an open contour:

$$\mathcal{W}_{\mathrm{open}}(C_{\eta 0}) \;=\; Z(C_{\eta 0})\,(2\pi)^{d/2}\,\mathrm{Pf}\,\theta\, \mathrm{tr}_{\mathcal{H}}\,\boldsymbol{U}(C_{\eta 0}), \qquad (16.67)$$

where the c-number phase factor $Z(C_{\eta 0})$ is given by Eq. (16.62).

The RHS of Eq. (16.67) is obviously invariant under the unitary transformation (16.60) both for closed and open contours. It is of the same type as the expression under the average in Eq. (15.113), while the first trace is replaced by $Z(C_{\eta 0})$. These are the same for the closed Wilson loops.

As was shown in [AMN00a], the closed Wilson loops (16.63) naturally appear in the sum-over-path representation of the matter correlator $\langle \bar{\psi}(x) \star \psi(x) \rangle_\psi$, while the open Wilson loops (16.65) do so for the correlator $\langle \bar{\psi}(x+\eta) \star \exp\left(i\eta\theta^{-1}x\right) \star \psi(x) \rangle_\psi$. Both correlators are invariant under the star gauge transformation (16.53). The second one is invariant owing to Eq. (16.38).

Noncommutative extensions of local operators, such as $\mathrm{tr}\,\mathcal{F}^2$, are constructed using the open Wilson loops in [GHI00].

Remark on the definition of open Wilson loops

The RHS of Eq. (16.67) looks slightly different from the expression under the sign of averaging in Eq. (15.113) since the first trace is replaced by $Z(C_{\eta 0})$, which only coincide for the closed Wilson loops. This difference between the two factors becomes inessential for the averages of open Wilson loops since they vanish in the noncommutative gauge theory owing to translational invariance

$$
\begin{aligned}
\left\langle \mathcal{W}_{\text{open}}(C_{yx}) \right\rangle &= W_{\text{NC}}(C) \int d^d x \, e^{i\eta\theta^{-1}x} \\
&= (2\pi)^d \det\theta \, \delta^{(d)}\,(x-y)\, W_{\text{NC}}(C)\,.
\end{aligned} \tag{16.68}
$$

Note that the vanishing of the open Wilson loops in the twisted Eguchi–Kawai model was guaranteed by the first trace in Eq. (15.113).

This difference of the two definitions of the open Wilson loops is a result of historical reasons. Once again they are essentially the same in the large-N limit where the averages factorize.

Problem 16.6 Obtain an explicit expression for the noncommutative phase factor expanding in $\mathcal{A}_\mu$.

Solution The calculation is similar to that in Problem 5.2 on p. 89. Expanding in $\mathcal{A}_\mu$, we obtain finally

$$
\begin{aligned}
\mathcal{U}(C_{yx}) &= \sum_{k=0}^{\infty} i^k \int_0^\eta d\xi_1^{\mu_1} \cdots \int_0^\eta d\xi_{k-1}^{\mu_{k-1}} \int_0^\eta d\xi_k^{\mu_k} \, \theta(k, k-1, \ldots, 1) \\
&\quad \times \mathcal{A}_{\mu_k}(x+\xi_k) \star \mathcal{A}_{\mu_{k-1}}(x+\xi_{k-1}) \star \cdots \star \mathcal{A}_{\mu_1}(x+\xi_1)\,,
\end{aligned} \tag{16.69}
$$

where the star product is with respect to x and the theta function orders the points ξ_i along the contour. This formula is simply the Weyl transform of an operator version of Eq. (5.27).

Problem 16.7 Derive the loop equation in noncommutative $U_\theta(1)$ gauge theory as $\theta \to \infty$.

Solution The derivation is similar to that in Yang–Mills theory. Applying the operator $\partial_\mu \delta/\delta\sigma_{\mu\nu}$ to $\mathcal{U}(C_{xx})$ at the point $z \in C_{xx}$, we have

$$\partial_\mu \frac{\delta}{\delta\sigma_{\mu\nu}(z)}\mathcal{U}(C_{xx})$$
$$= \mathcal{U}(C_{xz}) \star (i\partial_\mu\mathcal{F}_{\mu\nu} + \mathcal{A}_\mu \star \mathcal{F}_{\mu\nu} - \mathcal{F}_{\mu\nu} \star \mathcal{A}_\mu)(z) \star \mathcal{U}(C_{zx}).$$
(16.70)

This calculation is purely geometrical and results in the insertion of the noncommutative Maxwell equation at the point z. Replacing it by $-i\delta/\delta\mathcal{A}_\nu(z)$ in the average, we obtain [AMN99]

$$\partial_\mu \frac{\delta}{\delta\sigma_{\mu\nu}(z)} W_{\mathrm{NC}}(C) = \lambda \int_0^\eta d\xi^\nu \left\langle \mathcal{U}(C_{xz}) \star \delta^{(d)}(x + \xi - z) \star \mathcal{U}(C_{zx}) \right\rangle.$$
(16.71)

Using translational invariance of the average and the identity

$$\int d^d x\, f_1(x) \star \delta(x + \xi - z) \star f_2(x)\Big|_{z=x}$$
$$= \frac{1}{(2\pi)^d |\det\theta_{\mu\nu}|} \int d^d x\, f_1(x)\, e^{-i\xi\theta^{-1}x} \int d^d y\, f_2(y)\, e^{i\xi\theta^{-1}y}$$
(16.72)

which can be easily derived from Eq. (16.44), Eq. (16.71) finally takes the form [AD01]

$$\partial_\mu \frac{\delta}{\delta\sigma_{\mu\nu}(x)} \langle W_{\mathrm{clos}}(C) \rangle = \frac{\lambda}{(2\pi)^d |\det\theta_{\mu\nu}|} \oint_C dz^\nu \langle W_{\mathrm{open}}(C_{xz}) W_{\mathrm{open}}(C_{zx}) \rangle.$$
(16.73)

Note that Eq. (16.73) relates the average of closed Wilson loops to the correlator of two open Wilson loops. The latter has a factorized part and a connected part:

$$\langle W_{\mathrm{open}}(C_{xz}) W_{\mathrm{open}}(C_{zx}) \rangle$$
$$= \langle W_{\mathrm{open}}(C_{xz}) \rangle \langle W_{\mathrm{open}}(C_{zx}) \rangle + \langle W_{\mathrm{open}}(C_{xz}) W_{\mathrm{open}}(C_{zx}) \rangle_{\mathrm{conn}}.$$
(16.74)

The resulting equation for the factorized parts is the Weyl transform of the loop equation in the continuum twisted Eguchi–Kawai model owing to Eq. (16.67) relating the Wilson loops in both cases. Remember, that the volume $V = N(2\pi)^{d/2}\,\mathrm{Pf}\,\theta$ provides the correct normalization.

Each average in the factorized part is proportional to a delta-function as a result of Eq. (16.68), which should be treated by introducing a regularization as is discussed in Sect. 14.4. Since the connected correlator is suppressed at large θ as $1/\det\theta$, we arrive at Eq. (12.59) as $\theta \to \infty$.

16.6 Compactification to tori

To describe compactification of noncommutative theories to tori, we start again from the twisted reduced models. We consider a lattice regularization of noncommutative gauge theories in order to make the results of this and the next sections rigorous.

A compactification of reduced models to a d-torus $\mathbb{T}^d$ can be described [CDS98] by imposing the quotient condition

$$A_\mu + 2\pi R_\mu \delta_{\mu\nu} = \Omega_\nu A_\mu \Omega_\nu^\dagger \qquad (16.75)$$

on A_μ. Here Ω_ν are unitary transition matrices like those in Eq. (15.76).

In a moment we shall see that the twisted Eguchi–Kawai model with imposed quotient condition (16.75) and a certain choice of Ω_ν describes, at $N = \infty$, the noncommutative $U_\theta(1)$ gauge theory on a torus. This explains the terminology used in this section.

Taking the trace of Eq. (16.75), we see that a solution only exists for infinite matrices (= Hermitian operators).

Motivated by the discretization (15.2) of the Heisenberg commutation relation (15.87) at finite N by the unitary matrices, we exponentiate A_μ according to Eq. (15.108) with a dimensional parameter a to obtain

$$e^{2\pi i a \delta_{\mu\nu} R_\mu} U_\mu = \Omega_\nu U_\mu \Omega_\nu^\dagger. \qquad (16.76)$$

This U_μ is unitary and Eq. (16.76) is an $N \times N$ matrix discretization of Eq. (16.75) which has solutions (described below) for finite N.

Taking the trace of Eq. (16.76), we conclude that U_μ should be traceless, which is the case for the twist eaters Γ_μ. Taking the determinant of Eq. (16.76), we conclude that $aR_\mu N$ should be integral. The consistency of Eq. (16.76) also requires

$$\Omega_\mu \Omega_\nu = Z_{\mu\nu} \, \Omega_\nu \Omega_\mu \qquad (16.77)$$

with $Z_{\mu\nu} \in Z(N)$. The quotient condition (16.76) is compatible with the gauge symmetry (14.39) if Ω commutes with the transition matrices Ω_ν.

Let us choose

$$\Omega_\mu = \prod_\nu \Gamma_\nu^{m\varepsilon_{\mu\nu}}, \qquad (16.78)$$

where m is an integer and

$$\varepsilon_{\mu\nu} = \begin{pmatrix} 0 & +1 & & & \\ -1 & 0 & & & \\ & & \ddots & & \\ & & & 0 & +1 \\ & & & -1 & 0 \end{pmatrix}. \qquad (16.79)$$

These Ω_μ obviously obey Eq. (16.77).

Then a particular solution to Eq. (16.76) with $aR_\mu = m/L$ is given by $U_\mu = \Gamma_\mu$, while a general solution is

$$U_\mu = \Gamma_\mu \tilde{U}_\mu, \tag{16.80}$$

with $\tilde{U}_\mu$ obeying

$$\tilde{U}_\mu = \Omega_\nu \tilde{U}_\mu \Omega_\nu^\dagger. \tag{16.81}$$

We are interested in a very special solution [AMN99] to Eq. (16.81) at finite N when m is a divisor of L so that $n = L/m$ is an integer. Then a solution to Eq. (16.81) can be written as

$$\tilde{U}_\mu^{ij} = \frac{1}{m^d} \sum_{k \in \mathbb{Z}_m^d} (J_k^n)^{ij} U_\mu(k), \tag{16.82}$$

where J_k are defined in Eq. (15.22). Here k_μ runs from 1 to m since $\Gamma_\mu^L = 1$. This $\tilde{U}_\mu$ obviously commutes with Ω_ν given by Eq. (16.78).

Given the c-number coefficients $U_\mu(k)$ which describe the dynamical degrees of freedom, we can use a Fourier transformation to obtain the field

$$\mathcal{U}_\mu(x) = \frac{1}{m^d} \sum_{k \in \mathbb{Z}_m^d} e^{2\pi i\, x \varepsilon k/am} U_\mu(k) \tag{16.83}$$

which is periodic on an m^d lattice (or equivalently on a discrete torus $\mathbb{T}_m^d$). The spatial extent of the lattice is therefore $\ell = am$.

The field $\mathcal{U}_\mu(x)$ describes the same degrees of freedom as the (constraint) $N \times N$ matrix U_μ^{ij}, while the unitarity condition $U_\mu U_\mu^\dagger = 1$ is rewritten as

$$\mathcal{U}_\mu(x) \star \mathcal{U}_\mu^*(x) = 1 \tag{16.84}$$

similarly to Eq. (16.32) in the continuum.

The lattice star product in Eq. (16.84) is given by

$$f_1(x) \star f_2(x) = \frac{1}{m^d} \sum_{y,z} e^{-2iy_\mu \theta_{\mu\nu}^{-1} z_\nu} f_1(x+y) f_2(x+z) \tag{16.85}$$

with

$$\theta_{\mu\nu} = -\frac{a^2 mn}{\pi} \varepsilon_{\mu\nu} = -\frac{\ell^2}{\pi} \frac{n}{m} \varepsilon_{\mu\nu}. \tag{16.86}$$

This expression for $\theta_{\mu\nu}$ is of the same type as Eq. (15.38) for the given simplest twist with*

$$n_{\mu\nu} = 2L^{d/2-1}\varepsilon_{\mu\nu}. \tag{16.87}$$

These formulas follow from comparing the expansions (16.82) with (16.83) and using Eq. (15.26). As $a \to 0$, Eq. (16.85) recovers the integral representation (16.24) of the star-product (16.8) in the continuum.

The twisted Eguchi–Kawai model (15.65) (in general, with the quotient condition (16.76)) can be rewritten identically as a noncommutative $U_\theta(1)$ lattice gauge theory. Given the relations (16.82) and (16.83) between matrices and fields, we rewrite the action (15.64) of the twisted Eguchi–Kawai model as

$$\frac{N}{g^2}S_{\text{TEK}} = \frac{1}{2\lambda} \sum_{x \in \mathbb{T}_m^d} \sum_{\mu \neq \nu} \left(1 - \mathcal{U}_\nu^*(x) \star \mathcal{U}_\mu^*(x + a\hat\nu) \star \mathcal{U}_\nu(x + a\hat\mu) \star \mathcal{U}_\mu(x)\right),$$

$$\tag{16.88}$$

where the coupling constant $\lambda = g^2 N$.

Analogously, the (constraint) measure $\mathrm{d}U_\mu$ turns into the Haar measure

$$\prod_{x,\mu} \mathrm{d}\mathcal{U}_\mu(x) = \prod_{k,\mu} \mathrm{d}U_\mu(k). \tag{16.89}$$

In fact, both (constraint) $\mathrm{d}U_\mu$ and $\prod_{x,\mu} \mathrm{d}\mathcal{U}_\mu(x)$ are simply given by the RHS of Eq. (16.89) since the degrees of freedom are the same.

The action (16.88) is invariant under the lattice star gauge transformations

$$\mathcal{U}_\mu(x) \xrightarrow{\text{g.t.}} \Omega(x + a\hat\mu) \star \mathcal{U}_\mu(x) \star \Omega^*(x), \tag{16.90}$$

where $\Omega(x)$ is star unitary.

The usual twisted Eguchi–Kawai model (without the quotient condition) is associated with $n = 1$. Then $\Omega_\mu = 1$ (remember that $\Gamma_\mu^L = 1$) and Eq. (16.76) becomes trivial. The large-N limit of the usual twisted reduced models can be associated [AII00] with noncommutative theories on $\mathbb{R}^d$ as has already been discussed in the Remark on p. 382.

For $n > 1$ (that is the twisted Eguchi–Kawai model with the quotient condition), the noncommutativity parameter (16.86) can be kept finite as $N \to \infty$, even for a finite ℓ if the dimensionless noncommutativity parameter

$$\Theta_{\mu\nu} \overset{\text{def}}{=} \frac{2\pi}{\ell^2}\theta_{\mu\nu} \tag{16.91}$$

---

* The discrepancy in a factor of 4 is because we have changed the definition of the lattice momentum: $p_\mu = 2\pi\varepsilon_{\mu\nu}k_\nu/aL = \pi n_{\mu\nu}k_\nu/aN$, which is more natural for even $n_{\mu\nu}$ and odd L (see the footnote on p. 356).

is kept finite (and becomes an irrational number as $N \to \infty$). This means that the resulting continuum noncommutative theory lives on a torus [CDS98]. The case of finite N corresponds [AMN99] to the noncommutative lattice gauge theory (16.88) which is a lattice regularization of this continuum theory. The noncommutative theory on $\mathbb{R}^d$ is reached as $\ell \to \infty$.

Remark on finite Heisenberg–Weyl group

The noncommutative lattice gauge theory can be constructed [BM99, AMN00b] on the basis of finite-dimensional representations of the Heisenberg–Weyl group without the use of the matrix approximation. Equation (16.86) relating θ and the lattice size then emerges as a consistency condition.

Remark on Wilson loops on the lattice

A lattice contour C consisting of J links is given by the set of unit vectors $\hat{\mu}_j$ associated with the direction of each link j $(j = 1, \ldots, J)$ forming the contour (cf. Eq. (6.40)). The parallel transporter from the point x to the point $y = x + \eta$ $(\eta = a \sum_j \hat{\mu}_j)$ along C_{yx} is given by

$$\mathcal{U}(C_{yx}) = \mathcal{U}_{\mu_J}\left(x + a \sum_{j=1}^{J-1} \hat{\mu}_j\right) \star \cdots \star \mathcal{U}_{\mu_2}(x + a\hat{\mu}_1) \star \mathcal{U}_{\mu_1}(x). \quad (16.92)$$

It is star gauge covariant,

$$\mathcal{U}(C_{yx}) \xrightarrow{\text{g.t.}} \Omega(x + \eta) \star \mathcal{U}(C) \star \Omega^*(x), \quad (16.93)$$

under the star gauge transformation (16.90) of the link variable.

The lattice analog of the open Wilson loop (16.65) is

$$\mathcal{W}_{\text{open}}(C_{\eta 0}) = \sum_x e^{i\eta_\mu \theta_{\mu\nu}^{-1} x_\nu} \star \mathcal{U}(C_{yx}), \quad (16.94)$$

where $\eta_\mu = anj_\mu$ with integer-valued j_μ (modulo possible windings). It is star gauge invariant.

The continuum limit of Eq. (16.94) determines star gauge-invariant Wilson loops in noncommutative gauge theory. The open loops (16.65), which exist in addition to closed Wilson loops, can have an arbitrary value of η on $\mathbb{R}^d$ [IIK00]. On $\mathbb{T}^d$ the open Wilson loops are star gauge invariant only for discrete values of η measured in the units of $2\pi\theta/\ell$ [AMN99].

We shall see in the next section that the open Wilson loops in noncommutative $U_\theta(1)$ gauge theory for integral $m/n = \tilde{p}$ are Morita equivalent to the Polyakov loops in a $U(\tilde{p}^{d/2})$ Yang–Mills theory on a smaller torus of period $\ell/\tilde{p}$ with twisted boundary conditions.

Problem 16.8 Find a map between the Wilson loops in the twisted Eguchi–Kawai model and the noncommutative lattice gauge theory.

Solution The map between the Wilson loops in the twisted Eguchi–Kawai model and the noncommutative $U_\theta(1)$ gauge theory on the lattice can be written down explicitly using the relation (15.47) and (15.48) between matrices and fields. We consider for simplicity the twisted Eguchi–Kawai model without the quotient condition.

We mention first that $\Delta^{ij}(x)$ defined in Eq. (15.40) obeys the property

$$\Delta(y) \;=\; D(C_{yx})\Delta(x)D^\dagger(C_{yx})\,, \tag{16.95}$$

where $D(C_{yx})$ is given by Eq. (15.7) but the RHS is path-independent as is shown in Problem 15.1 on p. 352.

It also satisfies

$$\Delta^{ij}(x) \star \Delta^{kl}(x) \;=\; \frac{1}{N}\delta^{il}\Delta^{kj}(x) \tag{16.96}$$

as a consequence of the formula

$$\sum_{m\in\mathbb{Z}_L^d} J_m^{ij} J_{n-m}^{kl}\, e^{\pi i \sum_{\mu,\nu} m_\mu n_{\mu\nu} n_\nu/N} \;=\; N\delta^{il} J_n^{kj} \tag{16.97}$$

which recovers the completeness condition (15.25) for $n = 0$.

Given Eqs. (16.95) and (16.96), we have

$$N\,\mathrm{tr}\left[A\Delta(x+\eta)\right] \star N\,\mathrm{tr}\left[B\Delta(x)\right] \;=\; N\,\mathrm{tr}\left[D^\dagger(C_{yx})AD(C_{yx})B\Delta(x)\right], \tag{16.98}$$

where $y = x + \eta$. Equation (16.98) is simply a matrix analog of the extension of Eq. (16.23) for the case when y does not coincide with x.

Noting that

$$\mathcal{U}_\mu(x) \;=\; N\,\mathrm{tr}\left[\tilde{U}_\mu\Delta(x)\right] \tag{16.99}$$

and applying Eq. (16.98) several times, we obtain

$$\mathcal{U}(C_{yx}) \;=\; N\,\mathrm{tr}\left[D^\dagger(C_{yx})\,U(C_{yx})\,\Delta(x)\right] \tag{16.100}$$

with $U(C_{yx}) = \prod_{C_{yx}} U_{\mu_i}$. This is a lattice analog of Eq. (16.57).

Under the gauge transformation (14.39) the RHS of Eq. (16.100) transforms as

$$N\,\mathrm{tr}\left[D^\dagger(C_{yx})\,U(C_{yx})\,\Delta(x)\right] \;\overset{\text{g.t.}}{\longrightarrow}\; N\,\mathrm{tr}\left[D^\dagger(C_{yx})\,\Omega\,U(C_{yx})\,\Omega^\dagger\Delta(x)\right]$$

$$=\; N\,\mathrm{tr}\left[D^\dagger(C_{yx})\,\Omega\,D(C_{yx})\,D^\dagger(C_{yx})\,U(C_{yx})\,\Omega^\dagger\Delta(x)\right]$$

$$\overset{(16.98)}{\longrightarrow}\; \Omega(y) \star \mathcal{U}(C_{yx}) \star \Omega^*(x) \tag{16.101}$$

which is a lattice version of Eq. (16.56).

For a closed contour, $D(C_{xx})$ in Eq. (16.100) is a c-number and summing over x by $\sum_x \Delta(x) = 1$, we arrive at

$$\sum_x \mathcal{U}(C_{xx}) = \operatorname{tr} D^\dagger(C) \operatorname{tr} U(C). \tag{16.102}$$

This reproduces Eq. (15.68) for the Wilson loops in the twisted Eguchi–Kawai model after averaging and dividing by the volume factor of $L^d = N^2$ which appears on the LHS owing to Eq. (16.64).

For an open contour, we obtain analogously

$$\sum_x \mathcal{U}(C_{yx}) \, e^{i\eta\theta^{-1}x} = Z(C_{\eta 0}) \, N \operatorname{tr} U(C_{\eta 0}), \tag{16.103}$$

where $Z(C_{\eta 0}) = J_{\eta/a} D^\dagger(C_{\eta 0})$ is a lattice analog of the c-number phase factor (16.62) (cf. Eq. (15.60)). Equation (16.103) is a lattice analog of Eq. (16.67).

16.7 Morita equivalence

The continuum noncommutative gauge theory with rational values of the dimensionless noncommutativity parameter Θ defined in Eq. (16.91) has an interesting property known as Morita equivalence [Sch98].[*] We shall describe it for the lattice regularization associated with the simplest twist (16.87), assuming that the ratio $m/n = \tilde{p}$ is an integer.

Then the noncommutative $U_\theta(1)$ gauge theory on an m^d periodic lattice is equivalent to ordinary $U(p)$ Yang–Mills theory with $p = \tilde{p}^{d/2}$ on a smaller $n^d = (m/\tilde{p})^d$ lattice with twisted boundary conditions and the coupling $g^2 = \lambda/p$ (where λ is the coupling of the $U_\theta(1)$ gauge theory).

In the previous section we have discussed the equivalence of the twisted Eguchi–Kawai model (with the quotient condition in general) with $N = (mn)^{d/2}$ and the noncommutative $U_\theta(1)$ gauge theory on $\mathbb{T}_m^d$. Both theories have the same m^d degrees of freedom, which are described either by the (constraint) $N \times N$ matrix (16.82) or the lattice field (16.83).

In the matrix language, the noncommutativity emerges since

$$J_k^n J_q^n = J_{k+q}^n \, e^{2\pi i k\varepsilon q n/m} \tag{16.104}$$

as it follows from the general Eq. (15.26) for the given simplest twist.

In the noncommutative language, this noncommutativity resides in the star product

$$e^{2\pi i k\varepsilon x/\ell} \star e^{2\pi i q\varepsilon x/\ell} = e^{2\pi i (k+q)\varepsilon x/\ell} \, e^{2\pi i k\varepsilon q n/m} \tag{16.105}$$

as follows from the definition (16.85).

---

[*] It is often defined in a broader sense relating the values of $\Theta_{\mu\nu}$ in two equivalent noncommutative theories (see the review [KS00] and references therein).

When $m = \tilde{p}n$, a third equivalent model exists where the same dynamical degrees of freedom are described by a $p \times p$ matrix-valued field $\tilde{U}_\mu^{ab}(\tilde{x})$ on an n^d lattice, the sites of which are denoted by $\tilde{x}$.

Let us introduce $p \times p$ twist eaters $\tilde{\Gamma}_\nu$ obeying the Weyl–'t Hooft commutation relation

$$\tilde{\Gamma}_\mu \tilde{\Gamma}_\nu = \tilde{Z}_{\mu\nu} \tilde{\Gamma}_\nu \tilde{\Gamma}_\mu, \qquad \tilde{Z}_{\mu\nu} = e^{4\pi i \varepsilon_{\mu\nu}/\tilde{p}} \qquad (16.106)$$

and $\tilde{p}$ is also assumed to be odd. The integers $\tilde{p}$ and p play, respectively, the same role as L and N above.

A solution to Eq. (16.81) can then be represented as

$$\tilde{U}_\mu^{ab}(\tilde{x}) = \frac{1}{m^d} \sum_{k \in \mathbb{Z}_m^d} \tilde{J}_k^{ab} e^{2\pi i \tilde{x}\varepsilon k/a\tilde{p}n} U_\mu(k). \qquad (16.107)$$

Here we have introduced a basis on $gl(p; \mathbb{C})$ given by the formulas similar to Eq. (15.22):

$$\tilde{J}_k = \prod_\mu \tilde{\Gamma}_\mu^{k_\mu} e^{-2\pi i \sum_{\mu<\nu} k_\mu \varepsilon_{\mu\nu} k_\nu/\tilde{p}}, \qquad (16.108)$$

where $k \in \mathbb{Z}_{\tilde{p}}^d$. They obey, in particular,

$$\tilde{J}_k \tilde{J}_q = \tilde{J}_{k+q} e^{2\pi i k\varepsilon q/\tilde{p}} \qquad (\text{mod } \tilde{p}). \qquad (16.109)$$

The action of the third model is just the ordinary Wilson lattice action

$$S = \frac{1}{2} \sum_{\tilde{x}\in\tilde{\mathbb{T}}_n^d} \sum_{\mu\neq\nu} \left(1 - \frac{1}{p} \operatorname{tr}_{(p)} \tilde{U}_\nu^\dagger(\tilde{x}) \tilde{U}_\mu^\dagger(\tilde{x}+a\hat{\nu}) \tilde{U}_\nu(\tilde{x}+a\hat{\mu}) \tilde{U}_\mu(\tilde{x})\right),$$

$$(16.110)$$

while the coupling constant $g^2 = \lambda/p$. The field $\tilde{U}_\mu(\tilde{x})$ is quasi-periodic on $\tilde{\mathbb{T}}_n^d$ and obeys the twisted boundary conditions

$$\tilde{U}_\mu(\tilde{x} + an\hat{\nu}) = \tilde{\Gamma}_\nu^\dagger \tilde{U}_\mu(\tilde{x}) \tilde{\Gamma}_\nu \qquad (16.111)$$

since

$$\tilde{\Gamma}_\mu \tilde{J}_k \tilde{\Gamma}_\mu^\dagger = \tilde{J}_k e^{4\pi i \varepsilon_{\mu\nu} k_\nu/\tilde{p}}. \qquad (16.112)$$

It is of the type of Eq. (15.78) with $\Omega_\nu = \tilde{\Gamma}_\nu^\dagger$. Therefore, $\tilde{Z}_{\mu\nu} \in Z(p)$ in Eq. (16.106) represents the non-Abelian 't Hooft flux.

The number of degrees of freedom of the third model is $n^d p^2 = m^d$ for $p = \tilde{p}^{d/2}$ and coincides with those in the other two equivalent models.

For $n = 1$ when $\tilde{p} = m$ and $p = N$, the third model lives on a unit hypercube with twisted boundary conditions and coincides with the twisted

Eguchi–Kawai model as is shown in Problem 15.6 on p. 366. Therefore, the derivation of noncommutative gauge theories from the twisted
Eguchi–Kawai model is the simplest example of Morita equivalence.

In the continuum limit ($N \to \infty$) when the twisted Eguchi–Kawai
model is formulated via operators, the noncommutative $U_\theta(1)$ gauge theory lives on $\mathbb{T}^d$ with period ℓ and is Morita equivalent at the rational
value of $\Theta_{\mu\nu}$ to the ordinary $U(p)$ gauge theory on a smaller torus $\tilde{\mathbb{T}}^d$
with twisted boundary conditions and period $\tilde{\ell} = \ell/\tilde{p}$. Its coupling constant is given by $g^2 = \lambda/p$. This twisted torus is precisely the one which
first appeared in Eq. (16.75) since $\tilde{\ell}_\mu = 1/R_\mu$.

The lattice regularization makes these results rigorous. An arbitrary
rational value of $\Theta_{\mu\nu}$ can be obtained for the most general twist described
in Problem 15.3 on p. 354. And vice versa, a continuum noncommutative
theory with an arbitrary irrational value of the $\Theta_{\mu\nu}$ can be obtained starting from the ordinary Yang–Mills theory on a twisted torus as $p \to \infty$ or,
equivalently, the (constraint) twisted Eguchi–Kawai model as $N \to \infty$.

Remark on constraint TEK

The results of this section show that ordinary Yang–Mills theory on a
twisted torus (i.e. with the 't Hooft flux) can be represented as the twisted
Eguchi–Kawai model with constraint matrices. This fact was not known
in the 1980s.

Remark on fundamental matter

The results of this and the previous sections can be extended [AMN00a]
to the presence of matter. Let $\varphi(x)$ be a scalar matter field in the fundamental representation of $U_\theta(1)$. The matter part of the action is

$$S_{\text{mat}} = -\sum_{x,\mu} \varphi^*(x + a\hat{\mu}) \star \mathcal{U}_\mu(x) \star \varphi(x) + M^2 \sum_x \varphi^*(x)\varphi(x)$$

$$(16.113)$$

and is invariant under the star gauge transformation

$$\varphi(x) \xrightarrow{\text{g.t.}} \Omega(x) \star \varphi(x), \qquad \varphi^*(x) \xrightarrow{\text{g.t.}} \varphi^*(x) \star \Omega^*(x) \qquad (16.114)$$

simultaneously with that (16.90) for $\mathcal{U}_\mu(x)$. Equation (16.114) is similar
to Eq. (16.53) in the continuum.

At a rational value of Θ, the action (16.113) on a torus is Morita equivalent to

$$S_{\text{mat}} = -\sum_{\tilde{x},\mu} \text{tr}_{(p)}\, \phi^\dagger(\tilde{x} + a\hat{\mu})\tilde{U}_\mu(\tilde{x})\phi(\tilde{x}) + M^2 \sum_{\tilde{x}} \text{tr}_{(p)}\, \phi^\dagger(\tilde{x})\phi(\tilde{x}),$$

$$(16.115)$$

where $\phi^{ab}(\tilde{x})$ is a $p \times p$ matrix-valued field on a n^d lattice which obeys the twisted boundary conditions

$$\phi(\tilde{x} + \tilde{\ell}\hat{\nu}) \;=\; \tilde{\Gamma}_\nu^\dagger \phi(\tilde{x}) \tilde{\Gamma}_\nu \tag{16.116}$$

similar to Eq. (16.111) for the gauge field.

The index a of ϕ^{ab} plays the role of color, while b plays the role of flavor (labeling species). The color symmetry is local, while the flavor symmetry is global. In particular, the model (16.115) reduces for $n = 1$ to the twisted Eguchi–Kawai model with fundamental matter [Das83], the action of which is given by Eq. (15.117).

The continuum limit of the above formulas is obvious. The continuum $U_\theta(1)$ gauge theory with fundamental matter (noncommutative QED) is reproduced as $N \to \infty$. For $\theta \to \infty$ it is equivalent to large-N QCD on $\mathbb{R}^d$ in the Veneziano limit when the number of flavors of fundamental matter is proportional to the number of colors so the matter survives in the large-N limit. This makes the results of Sect. 16.4 rigorous since they are now obtained with regularization.

Remark on classical solutions

Noncommutative theories admit a whole zoo of classical solutions: instantons [NS98], solitons [GMS00], monopoles [GN00]. Some of them, such as instantons in the noncommutative $U_\theta(1)$ gauge theory or solitons in a four-dimensional noncommutative scalar theory, are new in the sense that they do not exist in ordinary cases. Some of them, such as monopoles in the noncommutative $U_\theta(1)$ gauge theory, are counterparts of the known solutions, which are usually associated with an infinite action. Now they become essential since turning on θ regularizes tension of the Dirac string. More on these classical solutions can be found in the reviews [Nek00, Har01].

It is intriguing whether or not these classical solutions in noncommutative gauge theories can give us a new insight into the problems of large-N QCD.

Bibliography to Part 4

Reference guide

There are no books on the reduced models. The existing reviews include those by Migdal [Mig83] and Das [Das87]. I would recommend to read the well-written original papers [GK82, GO83b] as well as the others cited in the text. A modern survey of the twist-eating solutions is given by González-Arroyo [Gon98].

The reduced models were discovered by Eguchi and Kawai [EK82]. The quenched Eguchi–Kawai model was introduced by Bhanot, Heller and Neuberger [BHN82] and elaborated in [Par82, GK82, DW82, Mig82]. The twisted Eguchi–Kawai model is constructed by González-Arroyo and Okawa [GO83a, GO83b] for the Yang–Mills theory and by Eguchi and Nakayama [EN83] for scalar fields.

The literature on noncommutative theories is vast. A mathematical background is described in the books by Connes [Con94], Landi [Lan97], and Madore [Mad99a]. The Weyl transformation is presented in the books by Weyl [Wey31] and Wong [Won98]. The properties of the star product are considered in [BFF78]. Its relation to the large-N limit is discussed in the review [Ran92] and the original papers cited therein. Some recent reviews on noncommutative theories and their applications are [Mad99b, KS00, DN01], which contain a comprehensive list of references. The classical solutions in noncommutative theories are discussed in the lectures by Nekrasov [Nek00] and Harvey [Har01].

The recent interest in noncommutative gauge theories came from string theory [SW99]. Their relation to the twisted reduced models was pointed out in [AII00]. Its extension to the original toroidal construction of [CDS98] was given in [AMN99]. The Wilson loops in noncommutative gauge theories are constructed in [IIK00]. The lattice regularization of noncommutative gauge theories is described in [AMN00b]. The issue of the UV/IR mixing in noncommutative quantum field theories is discussed in the paper [MRS99]. Subtleties with renormalization of noncommutative field theories are discussed in [CR00].

Some other papers on noncommutative quantum field theories are cited in the text.

References

[AD01]   ABOU-SEID M. AND DORN H. 'Dynamics of Wilson observables in noncommutative gauge theory'. *Phys. Lett.* **B504** (2001) 165.

[AII00]   AOKI H., ISHIBASHI N., ISO S., KAWAI H., KITAZAWA Y., AND TADA T. 'Noncommutative Yang–Mills in IIB matrix model'. *Nucl. Phys.* **B565** (2000) 176.

[AIK00]   AOKI H., ISO S., KAWAI H., KITAZAWA Y., AND TADA T. 'Space-time structures from IIB matrix model'. *Prog. Theor. Phys.* **99** (1998) 713.

[AMN99]   AMBJØRN J., MAKEENKO Y.M., NISHIMURA J., AND SZABO R.J. 'Finite-N matrix models of noncommutative gauge theory'. *JHEP* **9911** (1999) 029.

[AMN00a] AMBJØRN J., MAKEENKO Y.M., NISHIMURA J., AND SZABO R.J. 'Nonperturbative dynamics of noncommutative gauge theory'. *Phys. Lett.* **B480** (2000) 399.

[AMN00b] AMBJØRN J., MAKEENKO Y.M., NISHIMURA J., AND SZABO R.J. 'Lattice gauge fields and discrete noncommutative Yang–Mills theory'. *JHEP* **0005** (2000) 023.

[Bar90]   BARS I. 'Strings from reduced large-N gauge theory via area-preserving diffeomorphisms'. *Phys. Lett.* **B245** (1990) 35.

[BFF78]   BAYEN F., FLATO M., FRONSDAL C., LICHNEROWICZ A., AND STERMHEIMER D. 'Deformation theory and quantization. 1. Deformations of symplectic structures'. *Ann. Phys.* **111** (1978) 61.

[BFS97]   BANKS T., FISCHLER W., SHENKER S.H., AND SUSSKIND L. 'M-theory as a matrix model: a conjecture'. *Phys. Rev.* **D55** (1997) 5112.

[BG86]   VAN BAAL P. AND VAN GEEMEN B. 'A simple construction of twist eating solutions'. *J. Math. Phys.* **27** (1986) 455.

[BGH01]   BAEK S., GHOSH D.K., HE X.-G., AND HWANG W.Y.P. 'Signatures of noncommutative QED at photon colliders'. *Phys. Rev.* **D64** (2001) 056001.

[BHN82]   BHANOT G., HELLER U.M., AND NEUBERGER H. 'The quenched Eguchi–Kawai model'. *Phys. Lett.* **B113** (1982) 47.

[BM99]   BARS I. AND MINIC D. 'Noncommutative geometry on a discrete periodic lattice and gauge theory'. *Phys. Rev.* **D62** (2000) 105018.

[CDP00]   CHAICHIAN M., DEMICHEV A., AND PRESNAJDER P. 'Quantum field theory on noncommutative space-time and the persistence of ultraviolet divergences'. *Nucl. Phys.* **B567** (2000) 360.

[CDS98]   CONNES A., DOUGLAS M.R., AND SCHWARZ A. 'Noncommutative geometry and matrix theory: compactification on tori'. *JHEP* **9802** (1998) 003.

[CHK01]   CARROLL S.M., HARVEY J.A., KOSTELECKY V.A., LANE C.D., AND OKAMOTO T. 'Noncommutative field theory and Lorentz violation'. *Phys. Rev. Lett.* **87** (2001) 141601.

[Con94]   CONNES A. *Noncommutative geometry* (Academic Press, New York, 1994).

[CR87]   CONNES A. AND RIEFFEL M. 'Yang–Mills for noncommutative two-tori'. *Contemp. Math.* **62** (1987) 237.

[CR00]   CHEPELEV I. AND ROIBAN R. 'Renormalization of quantum field theories on noncommutative $\mathbb{R}^d$. 1. Scalars'. *JHEP* **05** (2000) 037.

[Cre84]   CREWTHER R.J. 'Local $SL(2;\mathbb{R})$ invariance applied to the master field problem'. *Phys. Lett.* **B134** (1984) 207.

[Das83]   DAS S.R. 'Quark fields in twisted reduced large-N QCD'. *Phys. Lett.* **B132** (1983) 155.

[Das87]   DAS S.R. 'Some aspects of large-N theories'. *Rev. Mod. Phys.* **59** (1987) 235.

[DN01]   DOUGLAS M.R. AND NEKRASOV N.A. 'Noncommutative field theory'. *Rev. Mod. Phys.* **73** (2002) 977.

[DW82]   DAS S. AND WADIA S. 'Translation invariance and a reduced model for summing planar diagrams in QCD'. *Phys. Lett.* **B117** (1982) 228.

[EK82]   EGUCHI T. AND KAWAI H. 'Reduction of dynamical degrees of freedom in the large-N gauge theory'. *Phys. Rev. Lett.* **48** (1982) 1063.

[EN83]   EGUCHI T. AND NAKAYAMA R. 'Simplification of quenching procedure for large-N spin models'. *Phys. Lett.* **B122** (1983) 59.

[Far97]   FAIRLIE D.B. 'Moyal brackets in M-theory'. *Mod. Phys. Lett.* **A13** (1998) 263.

[FFZ89]   FAIRLIE D.B., FLETCHER P., AND ZACHOS C.K. 'Trigonometric structure constants for new infinite algebras'. *Phys. Lett.* **B218** (1989) 203.

[Fil96]   FILK T. 'Divergences in a field theory on quantum space'. *Phys. Lett.* **B376** (1996) 53.

[FIT89]   FLORATOS E.G., ILIOPOULOS J., AND TIKTOPOULOS G. 'A note on $SU(\infty)$ classical Yang–Mills theories'. *Phys. Lett.* **B217** (1989) 285.

[FZ89]   FARILIE D.B. AND ZACHOS C.K. 'Infinite dimensional algebras, sine brackets and $SU(\infty)$'. *Phys. Lett.* **B224** (1989) 101.

[GHI00]   GROSS D.J., HASHIMOTO A., AND ITZHAKI N. 'Observables of non-commutative gauge theories'. *Adv. Theor. Math. Phys.* **4** (2000) 893.

[GK82]   GROSS D.J. AND KITAZAWA Y. 'A quenched momentum prescription for large-N theories'. *Nucl. Phys.* **B206** (1982) 440.

[GK83]   GONZALEZ-ARROYO A. AND KORTHALS ALTES C.P. 'Reduced model for large-N continuum field theories'. *Phys. Lett.* **B131** (1983) 396.

[GMS00]   GOPAKUMAR R., MINWALLA S., AND STROMINGER A. 'Noncommutative solitons'. *JHEP* **0005** (2000) 020.

[GN00]   GROSS G.J. AND NEKRASOV N.A. 'Monopoles and strings in noncommutative gauge theory'. *JHEP* **0007** (2000) 034.

[GO83a]   GONZÁLEZ-ARROYO A. AND OKAWA M. 'A twisted model for large-N lattice gauge theory'. *Phys. Lett.* **B120** (1983) 174.

[GO83b]   GONZÁLEZ-ARROYO A. AND OKAWA M. 'The twisted Eguchi–Kawai model: a reduced model for large-N lattice gauge theory'. *Phys. Rev.* **D27** (1983) 2397.

[Gon98]   GONZÁLEZ-ARROYO A. 'Yang–Mills fields on the four-dimensional torus. Part I: Classical theory', in Proc. of Peñiscola 1997 Advanced School on nonperturbative quantum field physics (World Scientific, Singapore, 1998) p. 57.

[Har01]   HARVEY J.A. 'Komaba lectures on noncommutative solitons and D-branes'. `hep-th/0102076`.

[Hay00]   HAYAKAWA M. 'Perturbative analysis on infrared aspects of noncommutative QED on $\mathbb{R}^4$'. *Phys. Lett.* **B478** (2000) 394.

[Hoo78]   'T HOOFT G. 'On the phase transition towards permanent quark confinement'. *Nucl. Phys.* **B138** (1978) 1.

[Hoo79]   'T HOOFT G. 'A property of electric and magnetic flux in non-Abelian gauge theory'. *Nucl. Phys.* **B153** (1979) 141.

[Hoo81]   'T HOOFT G. 'Some twisted self-dual solutions for the Yang–Mills equations on a hypertorus'. *Commun. Math. Phys.* **81** (1981) 267.

[Hop89]   HOPPE J. 'Diffeomorphism groups, quantization and $SU(\infty)$'. *Int. J. Mod. Phys.* **A4** (1989) 5235.

[HPR00]   HEWETT J.L., PETRIELLO F.J., AND RIZZO T.G. 'Signals for noncommutative interactions at linear colliders'. *Phys. Rev.* **D64** (2001) 075012.

[IIK00]   ISHIBASHI N., ISO S., KAWAI H., AND KITAZAWA Y. 'Wilson loops in noncommutative Yang–Mills'. *Nucl. Phys.* **B573** (2000) 573.

[IKK97]   ISHIBASHI N., KAWAI H., KITAZAWA Y., AND TSUCHIYA A. 'A large-N reduced model as superstring'. *Nucl. Phys.* **B498** (1997) 467.

[KM82] KAZAKOV V.A. AND MIGDAL A.A. 'Weak-coupling phase of the Eguchi–Kawai model'. *Phys. Lett.* **B116** (1982) 423.

[KS00] KONECHNY A. AND SCHWARZ A. 'Introduction to M(atrix) theory and noncommutative geometry'. hep-th/0012145.

[KW00] KRAJEWSKI T. AND WULKENHAAR R. 'Perturbative quantum gauge fields on the noncommutative torus'. *Int. J. Mod. Phys.* **A15** (2000) 1011.

[Lan97] LANDI G. *An introduction to noncommutative spaces and their geometries* (Springer-Verlag, Berlin, 1997).

[Li96] LI M. 'Strings from IIB matrices'. *Nucl. Phys.* **B499** (1997) 149.

[LP86] LEBEDEV D.R. AND POLIKARPOV M.I. 'Extrema of the twisted Eguchi–Kawai action and the finite Heisenberg group'. *Nucl. Phys.* **B269** (1986) 285.

[LPR89] LEBEDEV D.R., POLIKARPOV M.I., AND ROSLY A.A 'Gauge fields on the continuum and lattice tori'. *Nucl. Phys.* **B325** (1989) 138.

[Mad99a] MADORE J. *An introduction to noncommutative geometry and its physical applications* (Cambridge Univ. Press, 1999).

[Mad99b] MADORE J. 'Noncommutative geometry for pedestrians'. gr-qc/9906059.

[Mat01] MATHEWS P. 'Compton scattering in noncommutative space-time at the NLC'. *Phys. Rev.* **D63** (2001) 075007.

[Mig82] MIGDAL A.A. 'Reduction of asymptotically free QCD at large N to the random matrix model'. *Phys. Lett.* **B116** (1982) 425.

[Mig83] MIGDAL A.A. 'Loop equations and $1/N$ expansion'. *Phys. Rep.* **102** (1983) 199.

[MK83] MKRTCHYAN R.L. AND KHOKHLACHEV S.B. 'Reduction of the $U(\infty)$ theory to a model of random matrices'. *JETP Lett.* **37** (1983) 160.

[Moy49] MOYAL J.E. 'Quantum mechanics as a statistical theory'. *Proc. Cambridge Philos. Soc.* **45** (1949) 99.

[MPR00] MOCIOIU I., POSPELOV M., AND ROIBAN R. 'Low-energy limits on the antisymmetric tensor field background on the brane and on the noncommutative scale'. *Phys. Lett.* **B489** (2000) 390.

[MRS99] MINWALLA S., VAN RAAMSDONK M., AND SEIBERG N. 'Noncommutative perturbative dynamics'. *JHEP* **0002** (2000) 020.

[MS99] MARTÍN C.P. AND SANCHEZ-RUIZ D. 'The one-loop UV divergent structure of $U(1)$ Yang–Mills theory on noncommutative $\mathbb{R}^4$'. *Phys. Rev. Lett.* **83** (1999) 476.

[Nek00] NEKRASOV N.A. 'Trieste lectures on solitons in noncommutative gauge theories'. hep-th/0011095.

[NS98]   NEKRASOV N. AND SCHWARZ A. 'Instantons on noncommutative $\mathbb{R}^4$ and $(2,0)$ superconformal field theory'. *Commun. Math. Phys.* **198** (1998) 689.

[Par82]   PARISI G. 'A simple expression for planar field theories'. *Phys. Lett.* **B112** (1982) 463.

[PS89]   POPE C.N. AND STELLE K.S. '$SU(\infty)$, $SU_+(\infty)$ and area-preserving algebras'. *Phys. Lett.* **B226** (1989) 257.

[Ran92]   RANKIN S.J. '$SU(\infty)$ and the large-N limit'. *Ann. Phys.* **218** (1992) 14.

[Sch98]   SCHWARZ A. 'Morita equivalence and duality'. *Nucl. Phys.* **B534** (1998) 720.

[She99]   SHEIKH-JABBARI M.M. 'One-loop renormalizability of supersymmetric Yang–Mills theories on noncommutative torus'. *JHEP* **9906** (1999) 015.

[SW99]   SEIBERG N. AND WITTEN E. 'String theory and noncommutative geometry'. *JHEP* **9909** (1999) 032.

[VG99]   VÁRILLY J.C. AND GRACIA-BONDÍA J.M. 'On the ultraviolet behavior of quantum fields over noncommutative manifolds'. *Int. J. Mod. Phys.* **A14** (1999) 1305.

[Wey27]   WEYL H. 'Quantenmechanik und Gruppentheorie'. *Z. Phys.* **46** (1927) 1.

[Wey31]   WEYL H. *The theory of groups and quantum mechanics* (Dover, New York, 1931); 2nd edn (1950).

[Wig32]   WIGNER E.P. 'Quantum corrections for thermodynamic equilibrium'. *Phys. Rev.* **40** (1932) 749.

[Won98]   WONG M.W. *Weyl transforms* (Springer-Verlag, Berlin, 1998).

Index

Printed in the United States
by Baker & Taylor Publisher Services